LIQUID CRYSTALS - RECENT ADVANCEMENTS IN FUNDAMENTAL AND DEVICE TECHNOLOGIES

Edited by **Pankaj Kumar Choudhury**

Liquid Crystals - Recent Advancements in Fundamental and Device Technologies
http://dx.doi.org/10.5772/68030
Edited by Pankaj Kumar Choudhury

Contributors

Boris I. Lembrikov, David Ianetz, Yossef Ben-Ezra, Emine Kemiklioglu, Mercedes Pérez Méndez, Carlos Gabriel Avendaño, Daniel Martínez, Ismael Molina, Yuji Yamashita, Ci-Ling Pan, Ru-Pin Pan, Chan-Shan Yang, Wei-Ta Wu, Chia-Jen Lin, Haimei Luo, Changjing Wang, Yinghua Ji, Wen Yuan, Martin Hernandez Contreras, Xia Xin, Zhaohua Song, Yanzhao Yang, Pankaj Kumar Choudhury, Lamees Abdulkareem Alqurainy, Kais A. M. Al Naimee, Jiun-Woei Huang

Published by InTech

Janeza Trdine 9, 51000 Rijeka, Croatia

Publishing Process Manager Iva Simcic
Technical Editor SPi Global
Cover InTech Design team

First published February, 2018
Printed in Croatia
Legal deposit, Croatia: National and University Library in Zagreb

Additional hard copies can be obtained from orders@intechopen.com

Liquid Crystals - Recent Advancements in Fundamental and Device Technologies, Edited by Pankaj Kumar Choudhury
p. cm.
Print ISBN 978-953-51-3871-6
Online ISBN 978-953-51-3872-3

For EU product safety concerns:
IN TECH d.o.o., Prolaz Marije Krucifikse Kozulić 3, 51000 Rijeka, Croatia,
info@intechopen.com or visit our website at intechopen.com.

Meet the editor

Pankaj K. Choudhury held academic/research positions in the Banaras Hindu University (Varanasi, India), Goa University (Goa, India), Laval University (Quebec, Canada), Gunma University (Kiryu, Japan), Multimedia University (Cyberjaya, Malaysia), and Telekom R&D (Malaysia). He is currently a professor at the Institute of Microengineering and Nanoelectronics, Universiti Kebangsaan Malaysia (Bangi, Malaysia). His research is focused on photonics. Prof. Choudhury has published over 210 research papers, 11 chapters in research-level books, and edited and coedited 3 books. He is an editorial board member of *Optik—The International Journal for Light and Electron Optics* (Elsevier, The Netherlands) and the editor in chief of the *Journal of Electromagnetic Waves and Applications* (Taylor & Francis, UK). Prof. Choudhury is also a senior member of IEEE, OSA, and SPIE.

Contents

Preface

This book encapsulates varieties of application-oriented phenomena of liquid crystals, contributed by authors from different countries. The spotlight shines on the aspects pivoted to the novel properties of this complex-structured medium that can be harnessed for real-life needs. In a sense, some of the chapters describe well-understood phenomena of liquid crystals, whereas the others highlight on how these could be conceived for the development of new devices. In every chapter, the authors review the recent developments in the area reported by the preeminent researchers and, also, touch upon their own contributions. As such, the book essentially provides readers a glimpse of the multitudes of liquid crystal research.

Liquid crystals are optically active in nature and exist in a few different forms of molecular structures, viz., nematic, smectic, and cholesteric. Chapter 1 of this book remains the introductory part, wherein the fundamentals behind liquid crystals and the involved research directives are briefly described. Within the context of the features of varieties of liquid crystals, the optical textures of cholesterol make the cholesteric kind of liquid crystals greatly interesting for the study of optical activity. The synthesis of a certain kind of cholesteric liquid crystal phase—the cationic cholesteric liquid crystals—is reported by Méndez in Chapter 2. This chapter shares the important result that such a form of liquid crystal polymer can act as non-viral vectors in gene therapy, transfecting DNA to the nucleus cell.

The director of cholesteric liquid crystal molecules exhibits periodic helical structure, the alteration of which due to dopants would modify the liquid crystal phase. In this stream, the blue phase of liquid crystals exhibits outstanding electro-optical properties, which makes it a promising introduction into display-related usage. Chapter 3 by Kemiklioglu discusses the stabilization and electro-optical properties of blue phases of liquid crystals emphasizing their potentials for applications in display technology and other photonic devices.

In the context of phase transition of liquid crystals, Contreras in Chapter 4 focuses on the discussions of thermotropic liquid crystals of nematic and smectic A types under the external magnetic field. Exploiting the linear stability theory, the author investigates the effects of thermal phase transition on nematics of finite thickness samples with the conceptual framework of Faraday wave propagation. Here, the author shares his quantitative understanding of the dynamics of surface phenomena in liquid crystals through the theoretical and experimental means.

Chapter 5 by Song et al. reviews the structure and property of lyotropic liquid crystal-based materials. In the ordered phase of liquid crystals, molecules tend to align along a common direction, thereby yielding orderly oriented macroscopic domains, which would provide a way to control the orientation of *additive* materials. Within the context, different kinds of additives, namely, carbon nanotubes, graphene, biomolecules, etc., may be used for investigation. This chapter describes the mechanical, electrical, and physicochemical properties of the hybrids composed of such lyotropics of amphiphilic molecules with different kinds of aforementioned additives. It is reported that the features of lyotropic liquid crystals are greatly improved upon the usage of additives—the phenomenon that can be prudent in nanotechnology, electrochemical, and biochemical arena.

Photonic crystals are known to exhibit band gaps. Liquid crystals also exhibit similar features due to periodicity. However, the physical and chemical features of these are highly dependent on the externally applied fields, and therefore, the band gaps may also be tailored on demand. Apart from this, varying thermal ambience also plays a great role in manipulating the behavior of liquid crystals. Technically speaking, thermal and electrical tuning of liquid crystals would alter the spectral characteristics. Pivoted to this concept, Avendaño et al. present multilayered slabs of nematic liquid crystals in twisted configuration to investigate the tuning property of photonic band gaps; the relevant descriptions are incorporated in Chapter 6. The temperature-dependent defect modes are also exploited in the study. The investigation yields electrical control of defect modes in nematic liquid crystal-based multilayer structures through the tuning of transmission and reflection bands. The authors claim the usefulness of study in the development of liquid crystal-based tunable optical filters and waveguides.

Chapter 7 by Lembrikov et al. discusses theoretically some of the interesting nonlinear phenomena in the case of liquid crystals of the smectic A kind. For this purpose, the authors implement cubic optical nonlinearity, which is determined by the normal displacement of the smectic layer —a particular kind of orientational mechanism of liquid crystal molecules. They claim the investigation to be useful in explaining certain nonlinear phenomena in liquid crystals.

The refractive index of liquid crystals depends on the temperature of ambiance—the feature that would greatly affect the birefringent property. This feature is highlighted by Al-Qurainy and Al-Naimee in Chapter 8, wherein they consider five different samples of liquid crystals in pure and mixed (of pure) forms. They share the idea to model the birefringence of liquid crystals and, also, put efforts to demonstrate the validity through experiments. Their study on the bistability of liquid crystals due to temperature reveals the extraordinary refractive index to have larger bistability than the ordinary one.

The use of liquid crystal-based optical fibers in chemical sensors is common. In Chapter 9 Luo et al. discuss micro-/nano-dimensional liquid crystal layer-based tunable optical fiber interferometers. Two different kinds of interferometers, namely, the optical fiber grating-based and locally bent microfiber taper-based structures, are described on the basis of theories involved, followed by the experimental investigations of the functional properties of devices. The results indicate potential applications of the suggested forms of interferometers as tunable *all-fiber* photonic devices, such as filters and *all-optical* switches.

Chapter 10 by Pan et al. is pivoted on the investigation of liquid crystal-based terahertz (THz) phase gratings and beam steerers theoretically and experimentally. The authors introduce such phase gratings as capable to function as THz polarizers and tunable THz beam splitters. However, the thickness of the used liquid crystal layers affects the response time of grating, which is reported to be relatively low, thereby causing unsuitability for *fast modulation*-related applications. Instead, the suitability of the device remains in the need of precise control requiring a fixed beam splitting ratio. The ways to overcome the issues are also proposed in the chapter. The developed liquid crystal phase grating-based electrically tunable THz beam steerer can steer broadband THz radiation by 8.5° (with respect to the incident beam). This can be achieved through electrical control, that is, by varying the driving voltages to yield the required design of phase gradient in the grating structure. Good agreement of the experimental results with the theoretical predictions puts the investigation on a level where further research has to be carried out to eliminate the lacunae.

The use of liquid crystals in flat panel electronic displays offers several advantages over the traditional ones wherein cathode-ray tubes (CRTs) are implemented. Though varieties of liquid crystal panels are available in the market, in Chapter 11, J. W. Huang makes an attempt to touch upon the design and fabrication of ultrashort throw projection systems for home cinema and display systems in automobiles. The author discusses the technique of liquid crystal on silicon for generating high-quality images. The emphasis is on the enhancement in viewing angle through reducing the throw ratio.

Another newly ventured application of liquid crystals would be in the pharmaceutical industry. Yuji Yamashita in Chapter 12 describes functional media comprised of unstable colloid dispersion systems using liquid crystals. The author specifically mentions the emulsification technology using liquid crystals constructed by self-assembly of several molecules and, also, the dispersion of liquid crystals, which would find prominent medicinal applications.

Finally, the book—*Liquid Crystals: Recent Advancements in Fundamental and Device Technologies*—overall highlights various research-related aspects that liquid crystal media have come across in recent years. Pioneering scientists from different countries contributed their research results; these are put in 11 different chapters. The editor hopes the book to be useful for novice as well as expert researchers—the former group of readers would remain abreast of recent research advancements in the relevant area, whereas the latter kind would be fueled to plan for new research ventures in the liquid crystal arena.

Prof. Pankaj Kumar Choudhury
Universiti Kebangsaan, Malaysia

Chapter 1

Introductory Chapter: Whither Liquid Crystals?

Pankaj Kumar Choudhury

Additional information is available at the end of the chapter

http://dx.doi.org/10.5772/intechopen.74413

1. Introduction

"Whither liquid crystals?"—a very general question that every researcher falling on the relevant research directives would ask about. A veteran researcher may also attempt to seek the answer to be abreast of the recent research developments in the area. Amazing physical and chemical properties of liquid crystals draw great attention of the research and development (R&D) community. These essentially make them indispensable for several technological applications, namely sensing [1, 2], communication systems [3], lasing actions [4, 5], flat panel displays [6], holography [7], and nanotechnology-enabled medicinal needs [8].

Though there are many different kinds of liquid crystals, nematic, smectic, and cholesteric exist as the three widely accepted phases (of these crystals) [9]. In short, the nematic version is characterized by molecules having no positional order, but aligned along the director with thread-like molecular formations. In the smectic phase, molecules have positional order only in one dimension, thereby having restricted movement within the planes. The cholesteric state is a kind of nematic phase wherein the molecular orientation undergoes helical rotation about the director. These may exist naturally or can be synthesized as well [10, 11]. Certain liquid-crystalline phases are abundant in living organisms, for example, proteins and cell membranes. Technologically developed liquid crystals are used for liquid crystal display (LCD) applications [6]. Apart from the LCD panels, there are host of other avenues where the synthesized versions of liquid crystals are used.

2. Liquid crystal properties

Liquid crystals exhibit chirality and possess very high electro-optic coefficient [12]. Chiral objects have the property to discriminate between the left-handed and right-handed electromagnetic fields [13]. These *optically active* mediums are classified into the categories of isotropic and structurally chiral ones. The isotropic chiral molecules can be formed by randomly dispersed, randomly oriented, electrically small, handed inclusions in an isotropic achiral host medium. On the other hand, the structurally chiral molecules, such as those of chiral nematic liquid crystals, are randomly positioned and exhibit helicoidal kind of orientation. One may exemplify biological structures of plants and animals, such as cholesterols, which represent chiral molecules. The director of cholesteric liquid crystal molecules exhibits periodic helical structure depending on the chirality of molecules, and may be altered due to external conditions—the feature that has great potential in technological applications [14]. For example, the changes in the helix (formed by the rotation of director) pitch due to chiral dopants would modify the phase of liquid crystals.

3. Application-oriented R&D

The unique properties of liquid crystals fueled scientists to invent new applications. Continuous research and development determine these mediums to gain increasingly important industrial and techno-scientific usages, and become vital in modern technological advancements. It is true that the research on liquid crystals tremendously bloomed after the invention of LCD panels [15]. Though the use of liquid crystals in flat panel electronic displays offers several advantages over the traditional ones, wherein cathode ray tubes (CRTs) are implemented [6], the LCDs have the drawback of having limited viewing angle, and also, higher manufacturing cost. However, these parameters have now become less significant with the advances in research, which becomes evident from the multitude of other applications of liquid crystals. This is primarily because external perturbations would introduce significant alterations in the macroscopic properties of liquid crystals [12, 13]. As an example, the chirality of liquid crystals allows these to acquire selective reflectance property, which can even be modified in the presence of external electric field. As such, these would be of great use in optical filters and imaging [16, 17] applications. Apart from these, the property of temperature dependence also makes liquid crystals to acquire the selectivity of reflection spectrum—the phenomenon that can be harnessed for devising temperature sensors [18].

Since the temperature plays a determining role to alter the refractive index values of liquid crystals, the birefringence property of these [19] allows splitting of light waves into the *slow* and *fast* components—the phenomenon which remains highly temperature dependent. A relatively higher temperature would induce a strong birefringence characteristic in certain form of liquid crystals, which would result due to higher temperature of the ambience. As such, a variation in temperature would introduce alterations in the phase difference between the incoming and outgoing light waves, thereby determining the polarization state of light.

As the properties of liquid crystals are affected by electric field, these mediums can be used to sense the field strength. Similarly, magnetic field also has effects on the properties of liquid crystals owing to the moving electric charges (magnetic dipoles are generated by the electrons moving around the nucleus in atoms). An externally applied magnetic field would make the liquid crystal molecules to align accordingly.

As stated before, the *optical activity* of liquid crystals opens up varieties of avenues. Under high external electrical fields, an optical material would exhibit nonlinear characteristics, that is, the refractive index of medium may not vary linearly with the field [20]. Liquid crystals, being optically anisotropic mediums, possess the birefringence property that remains of great potential in optics-based applications [21, 22]. The increase of birefringence happens owing to the nonlinear phenomenon that liquid crystals also exhibit. In fact, liquid crystals are characterized by extremely high optical nonlinearity. Some of the featured nonlinear phenomena would be self-phase modulation, four-wave mixing, stimulated Brillouin scattering, optical bistability, and so on. [23].

4. Liquid crystal-based fibers

Optical fibers with radially anisotropic liquid crystals have been greatly dealt with in the literature [24–26]. These have been much attractive owing to the fairly high optical anisotropic properties of liquid crystals—the feature that attracted the R&D community to introduce varieties of liquid crystal-based optical fibers in respect of geometry as well as material distributions. These include fibers of circular [25] and elliptical [26] cross-sections, and also, those with the loading of conducting helical structures, in order to achieve control over the dispersion characteristics [27–29]. It has been reported before that the radially anisotropic kind of nematic liquid crystal-loaded fibers become highly sensitive, and would be of potential for evanescent wave-based sensing applications [24, 25]. Indeed, the use of such fibers in chemical sensors would be one of the great avenues that liquid crystal mediums open up.

5. Liquid crystals for band gap features

Photonic crystals are known to exhibit band gaps, that is, the range of frequencies (or wavelengths) for which the propagation of waves remains forbidden [30–32]. Microstructured dielectric mediums can be engineered to exhibit such an excellent feature, which has been of great technological use in many optics-related applications [33]. Apart from dielectrics, liquid crystals may also be utilized due to the fact that the physical and/or chemical features of these are highly dependent on the externally applied fields. Varying thermal ambience also plays a great role in manipulating the behavior of liquid crystals, which essentially happens owing to the effects on the birefringence property. Technically speaking, thermal and electrical tuning of liquid crystals would alter the spectral characteristics. As such, photonic crystals, infiltrated with liquid crystal mediums, would exhibit tunable band gap features, which would be greatly interesting in optics-based needs.

6. Other miscellaneous applications and summary

Apart from the aforementioned applications of liquid crystals, there are many other varieties of usages that these materials offer [34]. Some of these include optical recording mediums, lasers, light modulators, and so on. Also, the area of biomedical applications is no more untouched on the exploitation of liquid crystals. For example, certain forms of liquid crystal polymers can act as nonviral vectors in gene therapy, transfecting DNA to the nucleus cell. Furthermore, functional mediums composed of specific colloid dispersion systems using liquid crystals would be greatly useful in pharmaceutical industry.

In summary, the book delineates several important advances occurring at the forefront of liquid crystal research. These are in terms of the development of fundamental theories as well as the exploitation of liquid crystals in inventing new devices. The subject matter of the book primarily focuses on the aspects of (i) varieties of liquid crystal polymer syntheses and their stability, (ii) physical and optical properties of complex liquid-crystalline states, and (iii) device applications of liquid crystals. The editor hopes that the topics included will be greatly useful for the R&D workers at universities and industries. The researchers use the book as springboard for their own thoughts in varieties of ways the different forms of liquid crystals can be exploited.

Author details

Pankaj Kumar Choudhury

Address all correspondence to: pankaj@ukm.edu.my

Universiti Kebangsaan Malaysia, Bangi, Selangor, Malaysia

References

[1] Khoo I-C. Liquid Crystals. Wiley: New York; 1994

[2] Choudhury PK. Liquid crystal optical fibers for sensing applications. Proceedings of SPIE. 2013;**8818**:88180E-1-88180E-10

[3] Wolinski TR, Lesiak P, Dabrowski R, Kedzierski J, Nowinowski E. Polarization mode dispersion in an elliptical liquid crystal core fiber. Molecular Crystals and Liquid Crystals. 2004;**421**:175-186

[4] Dolgaleva K, Wei SKH, Lukishova SG, Chen SH, Schwertz K, Boyd RW. Enhanced laser performance of cholesteric liquid crystals doped with oligofluorene dye. Journal of the Optical Society of America B: Optical Physics. 2009;**25**:1496-1504

[5] Coles H, Morris S. Liquid-crystal lasers. Nature Photonics. 2010;**4**:676-685

[6] O'Mara WC. Liquid Crystal Flat Panel Displays. New York: Van Nostrand Reinhold; 1993

[7] Berenberg VA, Venediktov VY. Liquid crystal valves as dynamic holographic correctors. In: Choudhury PK, editor. New Developments in Liquid Crystals and Applications. New York: Nova; 2013

[8] Lin Y-H. Liquid crystals for bio-medical applications. In: Lee C-C, editor. The Current Trends of Optics and Photonics. Vol. 129. Topics in Applied Physics. Dordrecht: Springer; 2014

[9] Choudhury PK, editor. New Developments in Liquid Crystals and Applications. New York: Nova; 2013

[10] Collings PJ. Liquid Crystals: Nature's Delicate Phase of Matter. Princeton: Princeton University Press; 1990

[11] de Gennes PG, Prost J. The Physics of Liquid Crystals. Clarendon Press: Oxford; 1993

[12] Wu S-T, Efron U. Optical properties of thin nematic liquid crystal cells. Applied Physics Letters. 1986;**48**:624-636

[13] Robbie K, Brett MJ, Lakhtakia A. Chiral sculptured thin films. Nature. 1996;**384**:616-617

[14] Moreno I. Liquid crystals for photonics. Optical Engineering. 2011;**50**:081201-081201

[15] Castellano JA. Liquid Gold: The Story of Liquid Crystal Displays and the Creation of an Industry. Singapore: World Scientific; 2005

[16] Gebhart SC, Stokes DL, Vo-Dinh T, Mahadevan-Jansen A. Instrumentation considerations in spectral imaging for tissue demarcation: comparing three methods of spectral resolution. Proceedings of SPIE. 2005;**5694**:41-52. DOI: 10.1117/12.611351

[17] Levenson RM, Lynch DT, Kobayashi H, Backer JM, Backer MV. Multiplexing with multispectral imaging: from mice to microscopy. ILAR Journal. 2008;**49**:78-88

[18] Plimpton RG. Pool Thermometer. U.S. Patent 4738549. 1988

[19] Madsen LA, Dingemans TJ, Nakata M, Samulski ET. Thermotropic biaxial nematic liquid crystals. Physical Review Letters. 2004;**92**:145505-1-145505-4

[20] Choudhury PK, Singh ON. Electromagnetic materials. In: Chang K, editor. Encyclopedia of RF and Microwave Engineering. New York: Wiley; 2005

[21] Green M, Madden SJ. Low loss nematic liquid crystal cored fiber waveguides. Applied Optics. 1989;**28**:5202-5203

[22] Lin H, Muhoray PP, Lee MA. Liquid crystalline cores for optical fibers. Molecular Crystals and Liquid Crystals. 1991;**204**:189-200

[23] Agrawal GP. Fiber Optic Communication Systems. Wiley: New York; 2012

[24] Choudhury PK, Soon WK. On the transmission by liquid crystal tapered optical fibers. Optik. 2011;**122**:1061-1068

[25] Choudhury PK. Evanescent field enhancement in liquid crystal optical fibers – A field characteristics based analysis. Advances in Condensed Matter Physics. 2013;**2013**: 504868-1-504868-9

[26] Moghaddas S, Choudhury PK, Ibrahim A-BMA. TE and TM mode power transmission through liquid crystal clad elliptical guides. Optik. 2017;**145**:113-120

[27] Ghasemi M, Choudhury PK. Propagation through complex structured liquid crystal optical fibers. Journal of Nanophotonics. 2014;**8**:083997-1-083997-13

[28] Ghasemi M, Choudhury PK. Conducting tape helix loaded radially anisotropic liquid crystal clad optical fiber. Journal of Nanophotonics. 2015;**9**:093592-1-0093592-15

[29] Ghasemi M, Choudhury PK. On the conducting sheath double-helix loaded liquid crystal optical fibers. Journal of Electromagnetic Waves and Applications. 2015;**29**:1580-1592

[30] Alkeskjold TT, Scolari L, Noordegraaf D, Laegsgaard J, Weirich J, Wei L, Tartarini G, Bassi P, Gauza S, Wu S-T, Bjarklev A. Biased liquid crystal infiltrated photonic bandgap fiber. Optical and Quantum Electronics. 2007;**39**:1009-1019

[31] Ibrahim A-BMA, Choudhury PK, Alias MSB. Analytical design of photonic band-gap fibers and their dispersion characteristics. Optik. 2005;**116**:169-174

[32] Ibrahim A-BMA, Choudhury PK, Alias MSB. On the analytical investigation of fields and power patterns in coaxial omniguiding Bragg fibers. Optik. 2006;**117**:33-39

[33] Choudhury PK. Electromagnetics of Micro- and Nanostructured Guides – Pathways to Nanophotonics. Malaysia: UKM Press; 2013

[34] Chandrasekhar S. Liquid Crystals. Cambridge University Press: Cambridge; 1992

Chapter 2

Synthetic Cationic Cholesteric Liquid Crystal Polymers

Mercedes Pérez Méndez

Additional information is available at the end of the chapter

http://dx.doi.org/10.5772/intechopen.70995

Abstract

We report the synthesis of six multifunctional cationic cholesteric liquid crystals polyesters functionalized with choline, amine, and amide groups to obtain new chemical formulations involving macromolecular features with new properties added to those of precursor chiral cholesteric polyesters. They are designed as PTOBDME-choline $[(C_{34}H_{36}O_8)_n–C_5H_{13}N]$; PTOBEE-choline $[(C_{26}H_{20}O_8)_n–C_5H_{13}N]$; PTOBDME-ammonium $[(C_{34}H_{36}O_8)_n–C_5H_{13}N]$; PTOBEE-ammonium $[(C_{26}H_{20}O_8)_n–C_5H_{13}N]$; PTOBUME-amide $(C_{33}H_{33}O_9N)_n$; and PTOBEE-amide $(C_{26}H_{19}O_9N)_n$. Structural characterization is performed by NMR. Thermal behavior is studied by thermogravimetry (TG) and differential scanning calorimetry (DSC), showing all the polymers endothermic transition from crystal phase to liquid crystal mesophase. Chirality is determined by optical rotatory dispersion (ORD). The cationic cholesteric liquid crystal polymers described here have proved to act as nonviral vectors in gene therapy, transfecting DNA to the nucleus cell.

Keywords: cholesteric LC, cationic polymers, chiral polyesters, synthesis, NMR, DSC

1. Introduction

Cholesteric liquid crystal polyesters have received much attention in the last few years for their interesting chemical, optical, mechanical, and biological properties. Due to their anisotropic formulation and amphiphilic nature, their molecules are able to self-associate and/or aggregate in blocks to form species with supramolecular ordered structure, which presents desirable material properties.

Two cholesteric liquid crystal polyesters, named PTOBDME and PTOBEE in **Figure 1**, were obtained by polycondensation reaction. Although only racemic materials were used in their synthesis, a cholesteric, chiral morphology, theoretically unexpected, was found. Evidence of this was obtained when a white solid, recrystallized, as the second fraction, from toluene mother

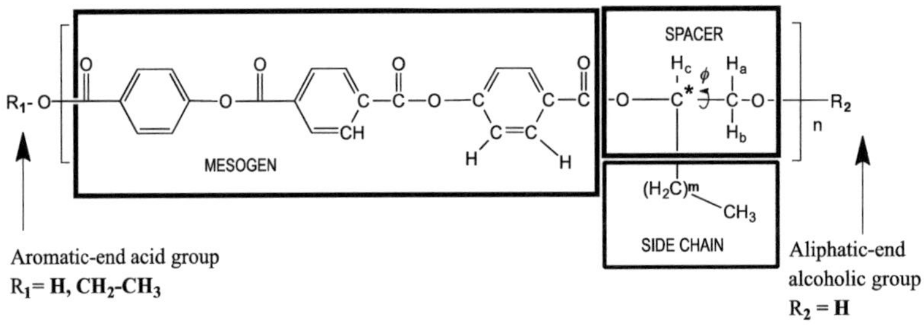

Figure 1. Monomeric unit of cholesteric liquid-crystalline PTOBEE (m = 1) and PTOBDME (m = 9). The three different zones of the monomer: *mesogen, spacer,* and *flexible side chain* are indicated. The asterisk indicates the chiral center. Torsion angle φ, along the bond containing the asymmetric carbon atom, is indicated. Aromatic-end acid and aliphatic-end alcoholic groups are also specified.

liquor after the filtration of the polymer, was identified as –PTOBDME, with $[\alpha]^{25}_{589} = -1.43$ (1.538g/100 ml, toluene) [1, 2] and –PTOBEE, with a value of $[\alpha]^{25}_{589} = -2.33$ (0.0056 mol/l, toluene) [3], respectively. The synthetic method [4], based on the previously reported by Bilibin [5], leads to obtain two or more fractions of different kinetic rates, with different enantiomeric excess. Not always, the enantiomer in excess is the same.

We are interested in the molecular design and chemical modifications of these multifunctional cholesteric liquid crystals to obtain new chemical formulations involving macromolecular features with new properties added. Our main interest being to introduce cationic charge, hence favoring the creation of hydrogen bonds, through intra and intermolecular interactions, giving secondary structures with long-range supramolecular order, and enabling to interact with molecules of interest, such as biological molecules (lipids, DNA, and oligonucleotides) and metal surfaces. The functional groups selected to be introduced at the end of the main chains were Choline [$-CH_2-CH_2-N-(CH_3)_3$] and ammonium [$-CH_2-CH_2-CH_2-NH-(CH_3)_2$] and amide groups ($-CONH_2$) at the end of the lateral hydrophobic chains.

The new synthetized cationic polymers reported here have proved to be able to interact with negatively charged DNA, forming polyplexes, which are able to condense and successfully transfect the new DNA into the nucleus cell, protecting it from damage during the transfection process, acting as nonviral vectors in Gene Therapy [6, 7]. Besides, they are sensitive to pH changes, acting as polycationic efficient transfection agents possessing substantial buffering capacity below physiological pH. These vectors have shown to deliver genes as well as oligonucleotides, both *in vitro* and *in vivo*, by protecting DNA from inactivation by blood components. Their efficiency relies on extensive endosome swelling and rupture that provides an escape mechanism for the polycation/DNA complexes [8].

2. Materials

The new cholesteric liquid crystal polymers so designed have been synthesized as follows: PTOBDME-choline [$(C_{34}H_{36}O_8)_n-C_5H_{13}N$]; PTOBEE-choline [$(C_{26}H_{20}O_8)_n-C_5H_{13}N$]; PTOBDME-ammonium [$(C_{34}H_{36}O_8)_n-C_5H_{13}N$]; PTOBEE-ammonium [$(C_{26}H_{20}O_8)_n-C_5H_{13}N$]; PTOBUME-amide [$(C_{33}H_{33}O_9N)_n$; and PTOBEE-amide $(C_{26}H_{19}O_9N)_n$.

2.1. Synthesis of cholesteric PTOBDME-choline [$(C_{34}H_{36}O_8)_n$–$C_5H_{13}N$]

Poly[oxy(1,2-dodecane)oxycarbonyl-1,4-phenylene-oxy-1,4-terephthaloyl-oxy-1,4-phenylene-carbonyl]-oxy-*N, N, N*-trimethylethan-1-ammonium (Choline) chloride, **II** in **Figure 2**, was obtained through polycondensation reaction between: 4,4′-(terephthaloyldioxydibenzoic

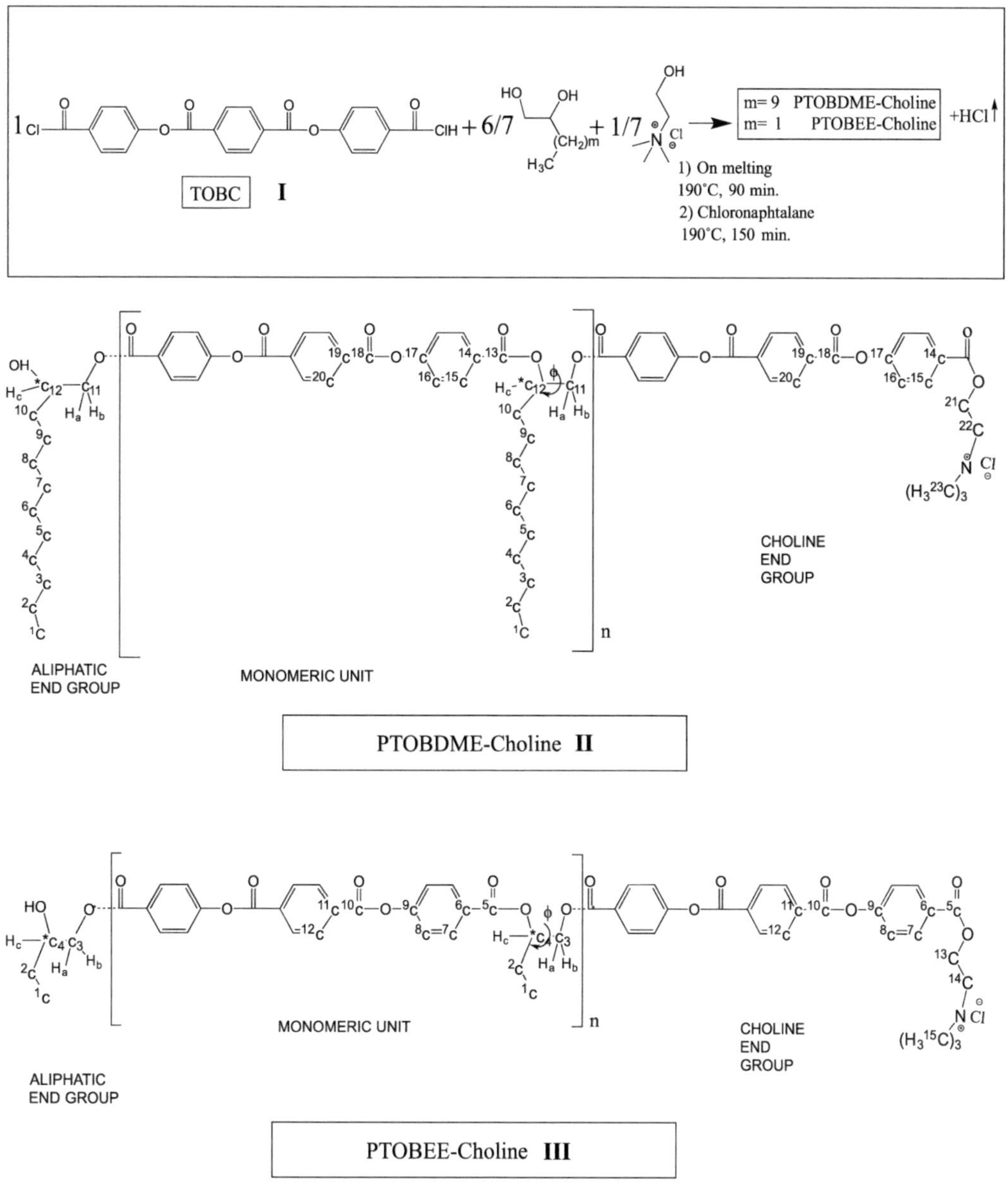

Figure 2. Synthetic process of cholesteric liquid-crystalline PTOBDME-choline (m = 9) (II) and PTOBEE-choline (m = 1) (III). Monomeric units are indicated, together with aliphatic end groups and choline aromatic end groups. The asterisks indicate the chiral centers ($^{12}C^*$) and ($^{4}C^*$), respectively. Torsion angles φ, along (^{12}C—^{11}C) and (^{4}C—^{3}C) bonds, respectively, are shown.

chloride) **TOBC**, **I** in **Figure 2**, the racemic mixture of DL-1,2-dodecanediol, and choline chloride. Notation similar to precursor cholesteric liquid crystal PTOBDME [1, 2] is used.

2.2. Synthesis of cholesteric PTOBEE-choline $[(C_{26}H_{20}O_8)_n\text{–}C_5H_{13}N]$

The structure of Poly[oxy(1,2-butane)oxycarbonyl-1,4-phenylene-oxy-1,4-terephthaloyl-oxy-1,4-phenylene-carbonyl]-oxy-*N,N,N*-trimethylethan-1-aminium (choline) chloride is shown in **III** of **Figure 2**. The polycondensation included DL-1,2-butanediol. Notation similar to precursor cholesteric liquid crystal PTOBEE [3, 4] is used.

2.2.1. *Preparation of PTOBDME-choline and PTOBEE-choline*

The dichloride, TOBC, was obtained by reaction between thionyl chloride and 4,4′-(terephthaloyldioxydibenzoic) acid (TOBA), previously synthesized from terephthaloyl chloride and 4-hydroxybenzoic acid [5].

The polycondensation reaction between TOBC and the racemic mixture, the corresponding glycol, takes place in presence of 1/7 equimolecular choline chloride. The preparation of these compounds was performed on melting due to the insolubility of choline chloride in the solvents used in the synthesis of PTOBDME or PTOBEE precursors, diphenyl oxide, or chloronaphthalene.

A mixture of 0.0054 mol of the glycol, either DL-1,2-dodecanediol or DL-1,2-butanediol, from Flucka Chemie GmBH (Buchs, Switzerland) and 0.000775 mol of choline chloride from Sigma-Aldrich Chemie GmBH (Steinheim, Germany) were placed into a flask of 50 ml contained in a bath with a high-temperature transfer agent, while a current of dry nitrogen from Praxair (Madrid, Spain) was used to purge the system at room temperature and then maintained in the rest of the reaction. The mixture was stirred and heated to 110°C to whole dissolution of the choline chloride into diol. The bath was cooled to 80°C, and 0.0062 mol of TOBC was added; this temperature was maintained for 15 minutes. The bath was heated up to 190°C, the mixture was melted, and emission of HCl was observed. After 60 minutes, 15 ml of chloronaphthalene from Sigma-Aldrich Chemie GmBH (Steinheim, Germany) was added. The reaction mix was maintained into the solvent stirring at 190°C for 150 minutes. Then, it was poured into 150 ml of toluene from Merck KGaA (Darmstadt, Germany), decanting PTOBDME-choline or PTOBEE-choline, respectively, which was filtered, washed with ethanol, and vacuum dried.

2.3. Synthesis of cholesteric PTOBDME-ammonium $[(C_{34}H_{36}O_8)_n\text{–}C_5H_{13}N]$

The structure of Poly[oxy (1,2-dodecane)-oxy-carbonyl-1,4-phenylene-oxy-1,4-terephthaloyl-oxy-1,4-phenylene-carbonyl]-oxy-3-dimethyl amine-1-propyl choride is shown in **II** of **Figure 3**.

2.4. Synthesis of Cholesteric PTOBEE-ammonium $[(C_{26}H_{20}O_8)_n\text{–}C_5H_{13}N]$

The structure of Poly[oxy(1,2-butane)oxycarbonyl-1,4-phenylene-oxy-1,4-terephthaloyl -oxy-1,4-phenylene-carbonyl]-oxy-3-dimethylamine-1-propyl choride is shown in **III** of **Figure 3**.

2.4.1. *Preparation of PTOBMDE-ammonium and PTOBEE-ammonium*

PTOBDME-ammonium chloride, **II** in **Figure 3**, and PTOBEE-ammonium chloride, **III** in **Figure 3**, were obtained through polycondensation reaction between 4 and 4′-(terephthaloyldioxydibenzoic chloride) **TOBC, I** in **Figure 3**, and the racemic mixture of **DL-1,2-dodecanediol** and **DL-1,2-butanediol,** respectively, and then reaction with 3-Dimethylamino-1-propanol. Notation of cholesteric liquid crystal PTOBDME and PTOBEE precursors is used. Next, a typical preparation of PTOBDME-ammonium chloride is shown.

Into a flask of 50 ml, TOBC (0.0079 mol) and 1,2-dodecanediol (0.0079 mol) from Flucka Chemie GmBH (Buchs, Switzerland) and diphenyl oxide (19.7 ml) of from Sigma-Aldrich Chemie GmBH (Steinheim, Germany) were mixed, while the system was purged with stream of dry nitrogen from Praxair (Madrid, Spain), for 30 min at room temperature. Then, while maintaining the gas current, the flask was transferred to a bath at 200°C for 2 hours; since the liberation of HCl is still observed, the temperature of the bath was descended to 160°C, the polycondensation was stopped and was not observed HCl formation. 3-Dimethylamino-1-propanol (0.2 ml, 0.00156 mol) was added to the reaction mix, and the liberation of HCl returned again. After 2 hours, the reaction finished. The result of the polycondensation reaction was poured into 200 ml of toluene from Merck KGaA (Darmstadt, Germany), decanting PTOBDME, which was filtered, washed with ethanol, and vacuum dried.

2.5. Synthesis of cholesteric PTOBUME-amide $[(C_{33}H_{33}O_9N)_n$

Poly[oxy(1,2-undecan-11-amidyl)-oxycarbonyl-1,4-phenylene-oxy-1,4-terephthaloyl-oxy-1,4-phenylene-carbonyl], **VII** in **Figure 4**, was obtained through polycondensation reaction between 4 and 4′-(terephthaloyldioxydibenzoic chloride) **TOBC** and the racemic mixture of **DL-10,11-dihydroxyundecanemide** (**V** in **Figure 4**) [11–15]. Similar notation has been used than with precursor cholesteric liquid crystal PTOBDME, **Figure 1**.

2.5.1. *Preparation of undec-10-enoyl chloride (**II** in **Figure 4**)*

To a stirred solution of 0.118 mol of undec-10-enoic acid in 100 ml of toluene, at 25°C, 0.078 mol of oxalyl chloride was added during 30 minutes. The solution was stirred for 30 minutes after emission of HCl gas had completed. The mixture reaction was concentrated to about half the initial volume by using a vacuum pump equipped with a sodium hydroxide trap. This solution was used directly to prepare undec-10-enamide (**III** in **Figure 4**).

2.5.2. *Preparation of undec-10-enamide*

A NH_3 gas stream was used to purge the stirred solution of undec-10-enoyl chloride, cooled in a bath of dry ice/acetone. The NH_3 stream was produced by boiling to reflux ammonia solution generated by reaction between 100 g of $ClHN_4$ solved into 300 ml of H_2O and 76 g of NaOH solved in 50 ml of water at 10°C. The reflux condenser and a NaOH trap were connected between the ammonia solution and the mixture reaction to prevent moisture. After 30 minutes of reaction, when a white solid had precipitated and HCl gas emission was not

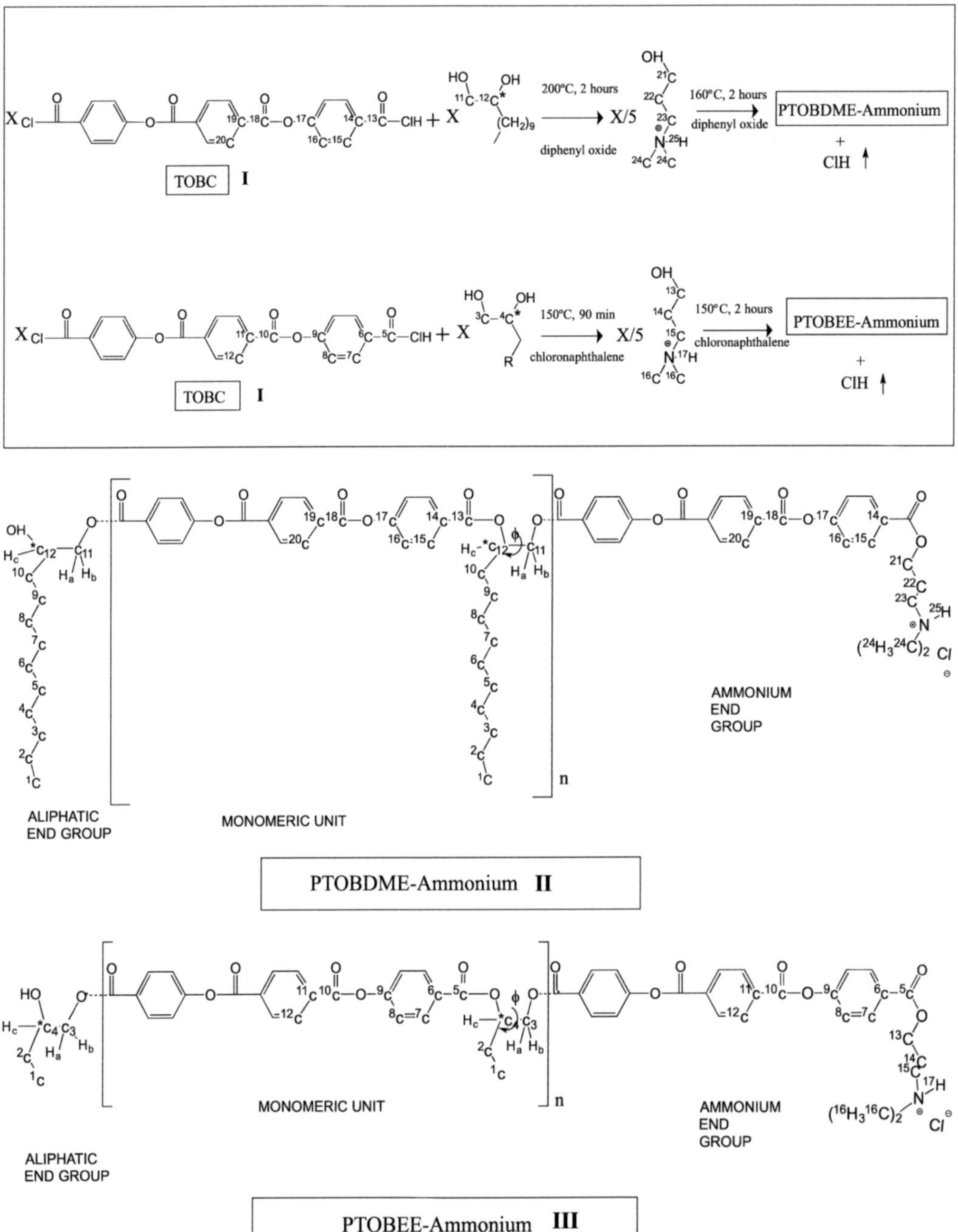

Figure 3. Synthetic process of cholesteric liquid-crystalline PTOBDME-ammonium (II), and PTOBEE-ammonium (III). Monomeric units are indicated, together with aliphatic end groups and ammonium aromatic end groups. The asterisks indicate the chiral centers ($^{4}C^{*}$) and ($^{12}C^{*}$), respectively. Torsion angles φ, along (^{4}C—^{3}C) and (^{12}C—^{11}C) bonds, are shown.

observed, the reaction flask was allowed to warm to room temperature. The mixture reaction was concentrated to a residue on a rotatory evaporator. The solid was partitioned between 10% aqueous sodium hydroxide and dichloromethane, and the aqueous phase was washed three times with additional dichloromethane. The combined dichloromethane extracts were washed with brine, dried with anhydrous sodium sulfate, and concentrated to a solid on a rotatory evaporator [12–15]. The solid was recrystallized in a mix chloroform/hexane (1:1) to give pure undec-10-enamide—yield (80%) and melting point 87°C (**III** in **Figure 4**).

^{1}H NMR ($CDCl_3$ 300 MHz, δ; (ppm)): δ 7.19, (dd, 2H) *(7.03)*, δ 5.80, (m, 1H) *(5.82)*, δ 5.32-5.22, (bs, 1H,) *(5.13)*, δ 4.96, (dd, 1H J = 7.2 Hz) *(4.88)*, δ 2.21, (t, 2H, J = 7.6 Hz) *(2.34)*; δ 2.02 *(2.13)*, (m, 2H), δ 1.62 (t, 2H, J = 7.4 Hz) *(1.53)*, δ 1.33-1.28, overlapped (10H) *(1.33, 1.30, 1.30, 1.30, 1.29)*. In tilted numbers are the calculated shifts.

^{13}C NMR ($CDCl_3$ 100 MHz, δ; (ppm)): 175.2 *(173.6)*, 139.6 *(139.1)*, 114.5 *(115.7)*, 36.3 *(38.7)*, 34.2 *(33.9)*, 29.4 *(29.7, 29.7, 29.6, 28.9, 28.6)* and 25.4 *(25.3)*. HRMS m/z calc. For $C_{11}H_{21}NONa$ + [M + Na] + 206.2; found 206.2.

2.5.3. Preparation of 10-11 epoxy undecanamide

To a stirred solution of undec-10-enamide (7 g;) in 108.4 ml of acetone, $NaHCO_3$ (26.4 g) was added, and then, 5.2 ml of water was added carefully. The resultant thick mixture was strongly stirred, while a solution of 40.6 g of oxone in 158 ml of water was added dropwise during 45 min. The reaction was monitored by thin layer chromatography (TLC) using a mix of ethyl acetate/hexane 2:1. After the reaction was complete, the acetone was removed by evaporation. The remaining solution was acidified with HCl 10% to pH 2 at 10°C and followed rapid extraction with 250 ml of dichloromethane. The aqueous phase was washed three times with additional dichloromethane. The combined organic phase was washed with brine, dried with anhydrous sodium sulfate, and concentrated to a white solid on a rotatory evaporator (**IV** in **Figure 4**).

^{1}H NMR ($CDCl_3$ 300 MHz, δ; (ppm)): δ 5.44, (bs, 2H), δ 2.90, (m, 1H), δ 2.74, (dd, 1H J = 4.6 Hz), δ 2.46, (dd, 1H J = 5.0 Hz), δ 2.21, (t, 2H, J = 7.6 Hz); δ 1.62 (t, 2H, J = 7.4 Hz), δ 1.51, (m, 2H), δ 1.44 (m, 2H), δ 1.33-1.28, (bs, 8H); ^{13}C NMR ($CDCl_3$ 100 MHz, δ; (ppm)): 173.8 (1C) (**CONH$_2$**), 137.6 (1C) (=**CH**—C), 115.7 (1C) (H_2**C**=), 38.7 (1C) (—H_2**C**—CONH$_2$), 29.7 (3C), 28.7 (2C), 25.3 (1C); HRMS m/z calc. For $C_{11}H_{21}NO_2Na$ + [M + Na]+; found.

2.5.4. Preparation of 10-11 of dihydroxyundecanamide

The previously obtained 10-11 epoxy undecanamide was stirred during 8 hours at 60°C in aqueous HCl 10%. The reaction was monitored by TLC using a mix of ethyl acetate/hexane 2:1. An oil, not miscible with water, was obtained. The mixture reaction was extracted with dichloromethane, and the aqueous phase was washed three times with additionally dichloromethane. The combined organic phase was washed with brine, dried with anhydrous sodium sulfate, and concentrated to a yellow oil on a rotatory evaporator. Yield (75%) (**V** in **Figure 4**).

^{1}H NMR ($CDCl_3$ 300 MHz, δ; (ppm)): δ 5.44, (bs, 2H), δ 3.78, (m, 1H), δ 3.60, (dd, 1H J = 11.0 Hz), δ 3.48, (dd, 1H J = 2.21, (t, 2H, J = 7.6 Hz); δ 1.62 (t, 2H, J = 7.4 Hz);), δ 1.51, (m, 2H), δ 1.44 (m, 2H), δ 1.33-1.28, (bs, 8H);

^{13}C NMR ($CDCl_3$ 100 MHz, δ; (ppm)): HRMS m/z calc. For $C_{11}H_{23}NO_3H_3O$+ [M + H_3O] + 236.2; found 236.2.

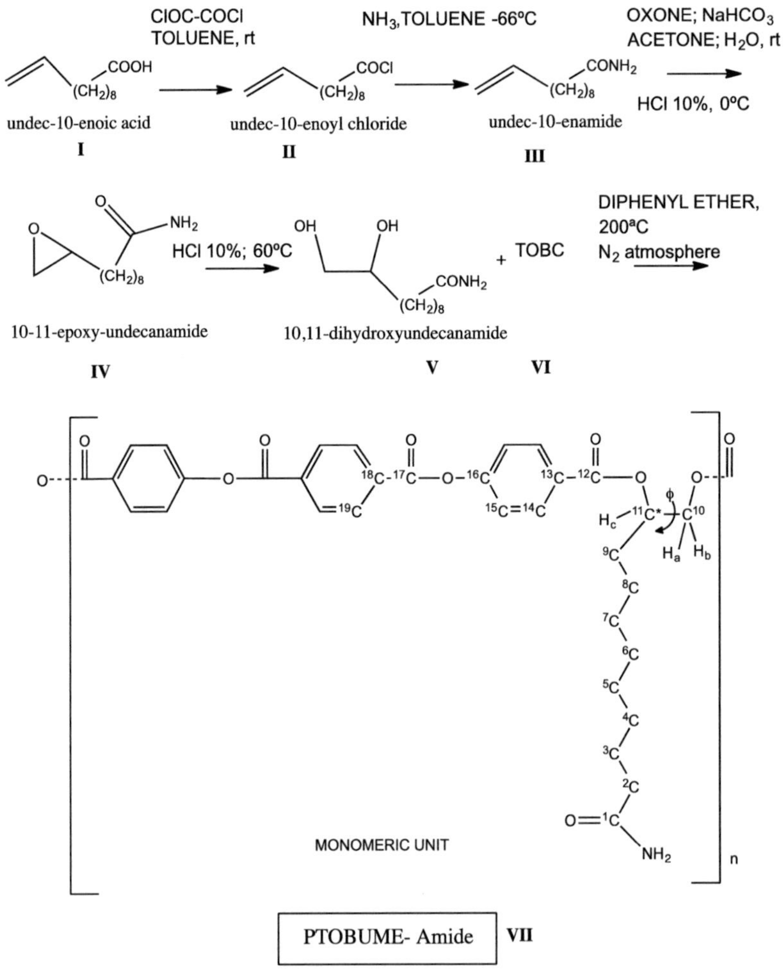

Figure 4. Synthetic method of cholesteric liquid-crystalline PTOBUME-amide. The asterisk indicates the chiral center ($^{11}C^*$). Torsion angle φ, along $^{10}C—^{11}C$ bond.

2.5.5. Preparation of TOBC

In the course of 20 minutes, 20 g TOBA were added to 350 ml thionyl chloride from Sigma-Aldrich Chemie GmBH (Steinheim, Germany), while stirring rapidly at room temperature (**VI** in **Figure 4**).

The solution was boiled with the reflux condenser. When the emission of HCl had finished and most of the sediment had dissolved, the hot solution was filtered and cooled down to 0°C for a day. The obtained product that separated out was filtered, vacuum dried, and recrystallized in chloroform, from SDS Votre Partenaire Chimie (Peypin, France).

Yield: 14 g (60%).

2.5.6. Preparation of PTOBUME-amide.

A mixture of TOBC (5.5 g; 0.012 mol), 10-11 of dihydroxyundecanamide (2.7 g; 0.012 mol) in 3 ml of diphenyl oxide from Sigma-Aldrich Chemie GmBH (Steinheim, Germany) was purged with dry nitrogen from Praxair (Madrid, Spain) for 25min at room temperature. Then, while maintaining the gas stream, the flask was transferred to a bath containing a high-temperature heat-transfer agent. The polycondensation was carried out for 360 minutes at 200°C. The reaction gets completed when emission of HCl had finished. The reaction mixture was poured into 300 ml of toluene from Merck KGaA (Darmstadt, Germany), decanting PTOBUME-amide. After 12 hours, it was filtered, washed with ethanol, and vacuum dried. After 3 weeks, a second fraction of polymer was precipitated of the toluene mother liquors, which was filtered, washed with ethanol, and vacuum dried.

Yield first fraction 3.0 g (38.5%); yield first and second fraction 0.1 g (40.0%).

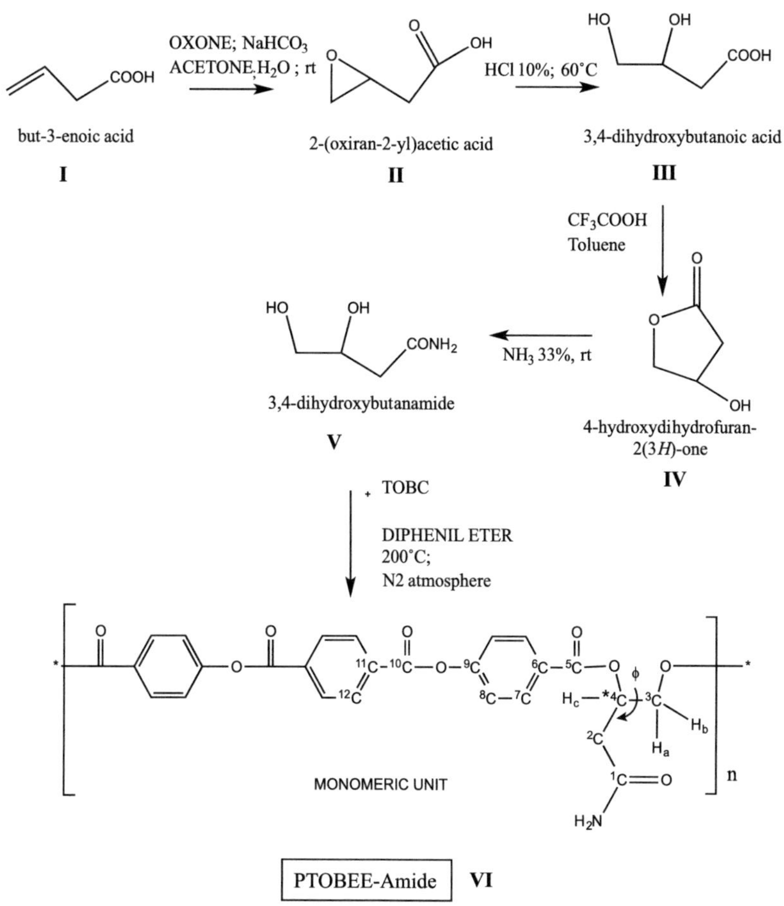

Figure 5. Synthetic method of cholesteric liquid-crystalline PTOBEE-amide.

2.6. Synthesis of Cholesteric PTOBEE-amide $(C_{26}H_{19}O_9N)_n$

Poly[oxy(1,2-butan-4-amidyl)-oxycarbonyl-1,4-phenylene-oxy-1,4-terephthaloyl-oxy-1,4-phenylene-carbonyl], **VI** in **Figure 5**, was obtained through poly-condensation reaction between 4 and 4′-(terephthaloyldioxydibenzoic chloride) **TOBC** and the racemic mixture of **DL-3,4-dihydroxybutanamide** (**IV** in **Figure 5**). The same notation has been used with precursor cholesteric liquid crystal PTOBEE, **Figure 1**.

2.6.1. Preparation of 3,4-dihydroxybutanoic acid

To a stirred mixture of 10 g of but-3-enoic acid solved in 130 ml acetone, 34 g. $NaHCO_3$ in 65 ml mili-Q water was added carefully. The resultant mixture was strongly stirred, while a solution of 51.1 g oxone in 200 ml of water was added dropwise during 120 min. The reaction was monitored by thin layer chromatography (TLC) using a mix of ethyl acetate/diethyl ether 1:1. After the reaction was complete, the acetone was removed by evaporation. The remaining solution was acidified with HCl 10% to pH 2 at 10°C and followed of rapid extraction with 250 ml of ethyl acetate. The aqueous phase was washed three times with additional ethyl acetate. The combined organic phase was washed with brine, dried with anhydrous sodium sulfate, and concentrated to a white solid on a rotatory evaporator (**III** in **Figure 5**).

2.6.2. Preparation of 4-hydroxydihydrofuran-2(3H)-one

To 3,4-dihydroxybutanoic acid into a flask equipped with a Dean Stark adapter filled with a toluene column finally connected to a refrigerant, 0.5 ml trifluoroacetic acid was added in 100 ml toluene heating to 110°C, mixing for 3 hours. The reaction product was removed with ethanol, washed in ethyl acetate, and dried. The reaction was monitored by thin layer chromatography (TLC) using a mix of ethyl acetate/diethyl ether 1:1 (**IV** in **Figure 5**).

2.6.3. Preparation of 3,4-dihydroxybutanamide

To the 4-hydroxydihydrofuran-2(3*H*)-one, 100 ml NH_3 33% was added stirring at 70°C with reflux for 12 h. The reaction product was removed with ethanol, filtered, and washed with water several times (**V** in **Figure 5**).

2.6.4. Preparation of PTOBEE-amide

In a three-neck round-bottom flask, 0.2 g 3,4-dihydroxybutanamide was added dropwise to 1 g TOBC solved in 100 ml 1,1,2,2-Tetrachloroethane. The reaction was stirred at 90°C for 20 hours. The reaction product was filtered, washed in 50 ml ethanol, 100 ml water, 200 ml $NaHCO_3$ (10%), 200 ml HCl (5%), 300 ml water, and 200 ml ethanol, and dried.

3. Characterization techniques

3.1. Conventional NMR techniques

The obtained polymers are characterized by ^{1}H-NMR, ^{13}C-NMR, COSY (Homonuclear Correlation Spectroscopy), TOCSY (Total Correlation Spectroscopy), NOESY (Nuclear Overhauser Effect

Spectroscopy, through-space correlation method), HSQC (Heteronuclear Single-Quantum Correlation spectroscopy), and HMBC (Heteronuclear Multiple Bond Correlation) for correlations between carbons and protons that are separated by two, three, and sometimes four bonds, in conjugated systems. Direct one-bond correlations being suppressed.

The experiments were performed in a Bruker 300 MHz NMR spectrometer and VARIAN 400 and 500 MHz spectrometers. The solvents used were DMSO-d_6 and $CDCl_3$, from Merck KGaA (Darmstadt, Germany), at 25°C. ^{1}H chemical shifts were referenced to the residual solvent signal at δ = 2.50 ppm (DMSO-d_6) relative to tetramethylsilane (TMS). All the spectra were processed and analyzed with MestReNova v.11.0.4 software [9]. Predicted ^{1}H and ^{13}C-NMR chemical shifts were calculated from the formula with ChemDraw Professional, v.15.1.0.144 [10].

3.2. Thermal behavior

Thermal stability was studied by Thermogravimetry on a Mettler TA4000-TG50 at heating rate of 10°C/min with nitrogen purge between 30 and 600°C. Thermal behavior was determined by differential scanning calorimetry (DSC) in a Mettler TA4000/DSC30/TC11 calorimeter, with series of heating/cooling cycles in a temperature range between 0 and 230°C.

3.3. The optical activity

The optical activity of the polymers was measured as optical rotatory dispersion (ORD) at 25°C in DMSO from Scharlau Chemie, in a Perkin Elmer 241 MC polarimeter with wavelengths: λ_{Na} = 589 nm, slit = 5 mm, integration time = 50 s; λ_{Hg} = 574 nm, slit = 14 mm, integration time = 50s; λ_{Hg} = 546 nm, slit = 30 mm, integration time = 50 s; λ_{Hg} = 435 nm, slit = 5 mm, integration time = 50s; λ_{Hg} = 365 nm, slit = 2.5 mm, integration time = 50 s.

4. Structural characterization by NMR

4.1. Structural characterization of PTOBDME-choline

The designation of the ^{1}H and ^{13}C-NMR chemical shifts, in DMSO-d_6, of the monomeric unit and the end groups of Polyester PTOBDME-choline, is given in **Table 1**. All the spectra have been analyzed and interpreted the help of MestReNova [9]. The predicted theoretical values, also in **Table 1**, have been calculated by ChemDraw [10]. Similar notations as those assigned with precursor cholesteric liquid crystal polyesters PTOBDME [1, 2] have been used.

Considering the monomer structure, three zones can be differentiated in the ^{1}H-NMR spectrum, corresponding to *the mesogen*, including aromatic protons between 11.0–7.00 ppm, *the spacer* where methylene and methine protons directly attached to oxygen atoms are observed, with signals between 6 and 3 ppm, and *the flexible side chain* formed by aliphatic protons between 2 and 0.8 ppm. The main feature of the proton spectrum is the presence of higher number of peaks than those expected for the monomeric unit. Hydrogen atoms H_a and H_b are bonded to 11C atom, allocated in α position with respect to the asymmetric carbon atom ^{12}C*. For that reason, they are diastereotopic and their 1H-NMR signals, usually indistinguishable, split in two easily differentiated. The same effect is observed for H_d and $H_{e'}$ bonded to ^{10}C, and for H_f and $H_{g'}$ both bonded to ^{9}C.

Set of signal of system (′) and (″)				Set of signal of system without apostrophe ( )				Calculated chemical shift	
Atom	**^{1}H(ppm)**	**Atom**	**^{13}C(ppm)**	**Atom**	**^{1}H(ppm)**	**Atom**	**^{13}C(ppm)**	**^{1}H(ppm)**	**^{13}C(ppm)**
$^{20'}$H	8.36	$^{20'}$C	130.4	^{20}H	8.36	^{20}C	130.4	8.04	130.2
		$^{19'}$C	133.3			^{19}C	133.3		135.4
		$^{18'}$C	163.5			^{18}C	163.4		165.2
		$^{17'}$C	153.9			^{17}C	154.4		155.6
$^{16'}$H	7.50	$^{16'}$C	122.1	^{16}H	7.55	^{16}C	122.2	7.26	121.5
$^{15'}$H	8.08	$^{15'}$C	130.8	^{15}H	8.15	^{15}C	131.0	8.13, 8.11	130.3
		$^{14'}$C	128.8			^{14}C	127.6		126.9
		$^{13'}$C	166.7			^{13}C	164.7		165.9
H_c'	5.45	$^{12'}$C	72.3	H_c	5.26	^{12}C	73.7	4.55	70.3
H_c''	4.39	$^{12''}$C	60.0					3.81	70.8
$H_a'H_b'$	4.63, 4.52	$^{11'}$C	65.7	$H_{a'}$ H_b	3.95, 3.89	^{11}C	46.4	4.80, 4.55	67.5
$H_a''H_b''$	4.52	$^{11''}$C	67.8					4.53, 4.28	70.8
H_d'	1.83	$^{10'}$C	30.0	H_d	1.77	^{10}C	31.2	1.71	30.7
H_d'',H_e''	1.92 1.78	$^{10''}$C						1.44	34.0
H_f', H_g'	1.45	$^{9'}$C	24.6	$H_{f'}$ H_g	1.35	^{9}C	24.4	1.29	23.3
H_f'', H_g''	1.53, 1.44	$^{9''}$C	25.6					1.29	23.1
$^{8'}$H	1.24	$^{8'}$C	28.9 m*	^{8}H	1.24	^{8}C	28.9 m*	1.29	29.6
$^{7'}$H	1.24	$^{7'}$C	28.9 m*	^{7}H	1.24	^{7}C	28.9 m*	1.29	29.6
$^{6'}$H	1.24	$^{6'}$C	28.9 m*	^{6}H	1.24	^{6}C	28.9 m*	1.26	29.6
$^{5'}$H	1.24	$^{5'}$C	28.9 m*	^{5}H	1.24	^{5}C	28.9 m*	1.26	29.6
$^{4'}$H	1.24	$^{4'}$C	28.6 m*	^{4}H	1.24	^{4}C	28.6 m*	1.26	29.3
$^{3'}$H	1.24	$^{3'}$C	31.3	^{3}H	1.24	^{3}C	31.3	1.26	31.8
$^{2'}$H	1.24	$^{2'}$C	22.1	^{2}H	1.24	^{2}C	22.1	1.26	22.7
$^{1'}$H	0.85	$^{1'}$C	14.0	^{1}H	0.85	^{1}C	14.0	0.86	14.1

The symbol (′) and without it ( ) distinguish the two independent system of the repeating unit, the symbol (″) is used to mark signals due to the aliphatic end group. *Signal of ^{4}C to ^{8}C at 28.8 ppm is a multiplet from 28.9 to 28.6.

Table 1. ^{1}H and ^{13}C-NMR chemical shifts (ppm) observed and calculated for the repeating unit and the aliphatic end group polyester PTOBDME-choline.

The presence of two independent ^{1}H-NMR sets of signals are observed in the spectrum, one marked with (′) and the other without it ( ). They are attributed to two conformers *gg* and *gt* of the spacer within the repeating unit respectively. The same effect has been reported for PTOBDME and PTOBEE, and accordingly, similar nomenclature is used to identify the signals. A third set of signals, marked with (″), is assigned to the aliphatic end group.

In the aromatic zone singlet at 8.36 ppm belongs to ^{20}H and doublets at 7.50 and 8.08 ppm are assigned to $^{16'}$H and $^{15'}$H, respectively, and doublets at 7.55 and 8.15 ppm to ^{16}H and ^{15}H; similar assignation was previously carried out in precursor PTOBDME [1]. In the spacer zone, multiplet at 5.45 ppm is interpreted due to H_c', and the double doublets at 4.63 and 4.52 ppm correspond to H_a' and H_b'. These peaks presented correlation signals in COSY and were related with other aliphatic signals $H_d'H_e'$ (1.83 ppm) and $H_f'H_g'$ (1.45 ppm) by TOCSY experiment. Multiplet at 5.26 ppm was assigned to H_c and double doublets at 3.95 and 3.89 ppm to H_a and H_b, and they showed COSY correlations and were related with signals at 1.77 ppm (H_d) and 1.35 ppm (H_f H_g) respectively by TOCSY experiment. The peaks assigned to H_a'', H_b'', and H_c'' due to aliphatic end group are overlapped with H_b' (4.52 ppm) and with ^{21}H (4.54 ppm) (in **Table 2**), and they were assigned through TOCSY correlations observed for signal at 4.52 ppm (not observed for 4.63 ppm, H_a'), with the multiplet at 4.39 ppm (H_c'') and confirmed by HSQC. By this method, carbon C$^{11'}$ (65.7 ppm) was correlated with signals at 4.63 (H_a') and 4.52 (H_b') and carbon $^{11''}$C (67.8 ppm) with 4.52 ppm (H_a'') and (H_b''). Signals at 1.83 and 1.45 ppm assigned to (H_d') and (H_f', H_g') and correlated by COSY with H_c'(5.45 ppm). Peaks at 1.77 ppm (H_d H_e) and 1.35 ppm (H_f H_g) are correlated by COSY with H_c (5.26 ppm). Signals at 1.92, 1.78 ppm are related with H_c'' (4.39 ppm) by COSY experiments, and they were assigned to H_d'' and H_e''. They are also related with Hf'' (1.53 ppm) and Hg''(1.44 ppm) by the same experiment.

Choline end group showed in **Table 2**, two set of signals probably due to conformational equilibrium: Multiplets assigned to ^{21}H (4.54 ppm), ^{22}H (3.33 ppm), and ^{23}H (2.74 ppm), correlated in COSY, and another set was multiplets $^{21'}$H (4.76 ppm), $^{22'}$H (3.85 ppm) and a singlet $^{23'}$H (3.21 ppm).

The HSQC experiment allowed the direct allocation of carbon atoms linked to hydrogens, confirming the assignation of the proton signals overlapped in the ^{1}HNMR experiment. The correlation of carbon atom $^{11'}$C (65.7 ppm) with H_a' (4.63 ppm) and H_b' (4.52 ppm); correlation between $^{11''}$C (67.8 ppm) and H_a'' and H_b''; and correlation between carbon ^{21}C at (62.6 ppm) and ^{21}H at (4.54 ppm), are observed in **Table 2**.

PTOBDME-choline						**PTOBEE-choline**					
Observed chemical shifts		**Calc. chemical shifts**				**Observed chemical shifts**		**Calc. chemical shifts**			
Atom ^{1}H(ppm)		**Atom ^{13}C(ppm)**				**Atom ^{1}H(ppm)**		**Atom ^{13}C(ppm)**			
	DMSO		**DMSO**	**^{1}H calc**	**^{13}C calc**		**DMSO**		**DMSO**	**^{1}H calc**	**^{13}C calc**
^{21}H,	4.54,	^{21}C	62.6	4.69	58.1	^{13}H	4.56	^{13}C	58.8	4.69	58.1
$^{21'}$H	4.76	$^{21'}$C	58.8			$^{13'}$H	4.75	$^{13'}$C	58.5		
^{22}H,	3.33	^{22}C	55.6	3.70	66.5	^{14}H	3.32	^{14}C	55.0	3.70	66.5
$^{22'}$H	3.85	$^{22'}$C	64.0			$^{14'}$H	3.84	$^{14'}$C			
^{23}H,	2.74	^{23}C	43.2	3.30	54.4	^{15}H	2.74	^{15}C	42.5	3.30	54.4
$^{23'}$H	3.21	$^{23'}$C	53.0			$^{15'}$H	3.22	$^{15'}$C	52.7		

Table 2. Observed ^{1}H and ^{13}C-NMR chemical shifts (ppm), in DMSO-d6, for the -N, N, N-trimethylethan- 1-ammonium (Choline) oxy benzoate hydrochloride end group, in Polyester PTOBDME-choline and Polyester PTOBEE-choline, and theoretical calculated values.

4.2. Structural characterization of PTOBEE-choline

The assignment of the ^{1}H and ^{13}C-NMR chemical shifts, in $CDCl_3$ and DMSO-d_6, of the monomeric unit and the end groups of PTOBEE-choline are given in **Table 3**, with the predicted values calculated by ChemDraw Professional [10]. Similar notations as those designated for precursor cholesteric liquid crystal PTOBEE [4] have been used.

In the ^{1}H-NMR experiment in CDC13, observed chemical shifts are ^{12}H singlet at (8.34 ppm), ^{8}H doublet at (7.34 ppm), ^{7}H doublet at (8.16 ppm), $^{8'}$H doublet at (7.36 ppm) and $^{7'}$H doublet at (8.18 ppm). Multiplets at 5.46 and at 5.25ppm are interpreted as H_c' and H_c, respectively. The double doublet at 4.60 ppm is assigned to H_a' and correlates in COSY with H_c' signal. An overlapped signal at 4.53 ppm is identified as H_b', with COSY and TOCSY cross signal

Set of signal of system (') and ('')						Set of signal of system without apostrophe ( )						Calculated chemical shifts	
Atom	^{1}H(ppm)		Atom	^{13}C(ppm)		Atom	^{1}H(ppm)		Atom	^{13}C(ppm)		Atom	Atom
	$CDCl_3$	DMSO		$CDCl_3$	DMSO		$CDCl_3$	DMSO	DMSO	$CDCl_3$	DMSO	^{1}H	^{13}C
$^{12'}$H	8.34	8.35	$^{12'}$C	130.4	130.1	^{12}H	8.34	8.35	^{12}C	130.4	130.0	8.04	130.2
			$^{11'}$C	133.8	133.2				^{11}C	132.2	131.6		135.4
			$^{10'}$C	163.6	163.3				^{10}C	163.6	163.3		165.2
			$^{9'}$C	154.5	154.2				^{9}C	154.5	153.8		155.6
$^{8'}$H	7.36	7.48, 7.50	$^{8'}$C	121.7	121.7	^{8}H	7.34	7.53, 7.51	^{8}C	121.7	122.0	7.26	121.5
$^{7'}$H	8.18	8.08, 8.06	$^{7'}$C	131.5	130.6	^{7}H	8.16	8.11, 8.09	^{7}C	131.5	130.6	8.13, 8.11	130.3
			$^{6'}$C	128.3	128.7				^{6}C	127.9	127.3		126.9
			$^{5'}$C	165.4	166.4				^{5}C	165.4	164.7		165.9
H_c'	5.46	5.38	$^{4'}$C	73.8	73.1	H_c	5.25	5.20	^{4}C	75.1	74.4	4.55	72.5
H_c''	4.15	4.36	$^{4''}$C		61.6*							3.81	73.0
H_a', H_b'	4.60, 4.53	4.63, 4.52	$^{3'}$C	65.6	64.8	$H_{a'}$ H_b	3.76, 3.74	3.94, 3.91	^{3}C	45.23	45.6	4.80, 4.55	67.2
H_a'', H_b''	4.53	4.52*	$^{3''}$C	67.8	67.1							4.53, 4.28	70.5
H_d'	1.90	1.86	$^{2'}$C	24.4	23.0	H_d	1.86	1.81	^{2}C	24.4	24.2	1.75	23.5
H_d'', H_e''	1.88, 1.13	1.96, 1.80	$^{2''}$C		*							1.48	26.8
$^{1'}$H	1.09	1.02	$^{1'}$C	9.8	9.2	^{1}H	1.03	0.95	^{1}C	9.6	9.8	0.96	7.8
$^{1''}$H	1.13	1.04*	$^{1''}$C		8.73*							0.96	7.6

*Overlapped signal

Table 3. Observed and calculated ^{1}H and ^{13}C-NMR chemical shifts (ppm) for polyester PTOBEE-choline in DMSO-d_6 and $CDCl_3$. Repeating unit and the aliphatic end group.

with H_a'. At 4.15, a weak multiplet is assigned to H_c'', it presented COSY correlation with signal 4.53 ppm, indicating the presence of H_a'' and H_b''. The two double doublets at 3.76 and 3.74 ppm were identified as H_a and H_b and presented the expected COSY correlation with H_c (5.25 ppm). The overlapped signal at 1.90 ppm is identified as H_d', with cross signal with H_c'. A very weak COSY cross signal between 4.15 and H_d''(1.88 ppm) is observed. Triplet at 1.09 ppm is due to $^{1'}H$, with TOCSY correlation with H_c', while triplet at (1.03 ppm) is ^{1}H, with TOCSY correlation with H_c. The weak triplet at 1.13 ppm corresponded with $^{1''}H$. As in PTOBDME-choline, the choline end group shows, in **Table 2**, two set of signals due to conformational equilibrium. Multiplets ^{13}H, ^{14}H are observed at 4.56 and 3.32 ppm, respectively, and ^{15}H at 2.74 ppm, correlated in COSY experiments, and another set was $^{13'}H$ and $^{14'}H$ multiplets at 4.75 and 3.84 ppm, respectively, singlet $^{15'}H$ at 3.22 ppm.

HSQC experiment was performed to determine the chemical shift of carbons bonded to the assigned hydrogen. The complex signal at 4.53 in proton presented several correlations with carbons. H_a'(4.60 ppm) and H_b' (4.53 ppm) showed correlation with carbon $^{3'}C$ (65.6 ppm). Another correlation with overlapped signal at 4.53 ppm $H_a''H_b''$ was observed with carbon $^{3''}C$ (67.8 ppm). Signals corresponding to ^{1}C, ^{2}C, and ^{4}C of aliphatic end group were not observed due to the low concentration. ^{13}C-NMR experiment allowed the assignation of the carbons not attached to hydrogens matching the calculated model.

4.3. Structural characterization of PTOBDME-ammonium

The structure of PTOBDME-ammonium, as depicted **II** in **Figure 3**, is confirmed by ^{1}H and ^{13}C-NMR, with the chemical shifts given in **Table 4**. In the aromatic zone of the ^{1}H-NMR spectrum of PTOBDME-ammonium, in DMSO-d_6, a singlet at 8.36 ppm belongs to ^{20}H and doublets at 7.50 and 8.07 ppm are assigned to $^{16'}H$ and $^{15'}H$, respectively, and doublets at 7.57 and 8.13 ppm to ^{16}H and ^{15}H. In the spacer zone where methylene and methines attached to oxygen are observed, a multiplet at 5.45 ppm, and the double doublets at 4.64 ppm, 4.48 ppm correspond to H_c', H_a', and H_b', respectively; these signals present correlation signals of COSY and are related with other aliphatic signals: H_d', H_e' (1.81 ppm) and H_f', H_g' (1.42 ppm) by TOCSY. Multiplet at 5.26 ppm is assigned to H_c and double doublets at 3.94 and 3.88 ppm, to H_a and H_b, they show COSY correlations and are related with signals H_d H_e (1.77 ppm) and H_f H_g (1.33 ppm), in the TOCSY experiment. In the set of signals due to aliphatic end group, H_a'', H_b'', and H_c'' are overlapped with H_b' (4.48 ppm) and with ^{21}H (4.38 ppm), according to TOCSY correlations observed for signal H_b' (4.48 ppm) and not observed for H_a' (4.63 ppm) and confirmed by HSQC. Aliphatic signals at 1.81 ppm (H_d', H_e') and 1.42 ppm (H_f', H_g') have TOCSY correlation with 5.45 ppm (H_c'), also COSY correlation. Signals at 1.77 ppm (H_d H_e) and 1.33 ppm (H_f H_g) show TOCSY correlation with H_c (5.26 ppm). The signal at 1.33 ppm cannot be observed in the ^{1}H sprectrum due to the overlapping with CH_2, but it was clearly observed in TOCSY 2D. Signals at 1.91 and 1.78 ppm, related with H_c''(4.38 ppm) by COSY and TOCSY, are assigned to H_d'' and H_e'' and are also related with H_f''(1.51 ppm) and H_g''(1.42 ppm) by the same experiment.

Ammonium end group shows ^{23}H, ^{22}H, ^{24}H multiplets at 3.28 ppm, 2.17 ppm, and 2.79 ppm, respectively . ^{23}H at 3.28 ppm was overlapped with signal of H_2O of the deuterated solvent,

Set of signal of system (′) and (″)				Set of signal of system without apostrophe ( )				Calculated chemical shift	
Atom	^{1}H (ppm	Atom	^{13}C(ppm)	Atom	^{1}H(ppm)	Atom	^{13}C(ppm)	^{1}H(ppm)	^{13}C(ppm)
$^{20'}$H	8.36	$^{20'}$C	130.4	^{20}H	8.36	^{20}C	130.4	8.04	130.2
		$^{19'}$C	133.3			^{19}C	133.3		135.5
		$^{18'}$C	163.5			^{18}C	163.4		165.2
		$^{17'}$C	153.8			^{17}C	154.3		155.7
$^{16'}$H	7.50	$^{16'}$C	122.1	^{16}H	7.57	^{16}C	122.4	7.26	121.5
$^{15'}$H	8.07	$^{15'}$C	131.0	^{15}H	8.13	^{15}C	131.0	8.13;8.11	130.3
		$^{14'}$C	128.9			^{14}C	127.6		127.0
		$^{13'}$C	166.6			^{13}C	164.8		166.0
H_c'	5.45	$^{12'}$C	72.6	H_c	5.26	^{12}C	73.4	4.55	70.4
H_c''	4.38	$^{12''}$C	60.4					3.81	70.9
H_a', H_b'	4.64, 4.48	$^{11'}$C	66.0	H_a, H_b	3.94, 3.88	^{11}C	46.4	4.80, 4.55	67.6
H_a'', H_b''	4.48	$^{11''}$C	67.8					4.53, 4.28	70.9
H_d', H_e'	1.81	$^{10'}$C	30.3	H_d, H_e	1.77	^{10}C	31.3	1.71	30.8
H_d'', H_e''	1.91 1.78	$^{10'}$C	33.6					1.44	34.1
H_f', H_g'	1.42	$^{9'}$C	24.6	H_f, H_g	1.33	^{9}C	24.3	1.29	23.4
H_f'', H_g''	1.51;1.42	$^{9''}$C	25.3					1.29	23.2
$^{8'}$H	1.22	$^{8'}$C	28.6 m*	^{8}H	1.22	^{8}C	28.6 m*	1.29	29.7
$^{7'}$H	1.22	$^{7'}$C	28.6 m*	^{7}H	1.22	^{7}C	28.6 m*	1.29	29.7
$^{6'}$H	1.22	$^{6'}$C	28.6 m*	^{6}H	1.22	^{6}C	28.6 m*	1.26	29.7
$^{5'}$H	1.22	$^{5'}$C	28.5 m*	^{5}H	1.22	^{5}C	28.5 m*	1.26	29.7
$^{4'}$H	1.22	$^{4'}$C	28.1 m*	^{4}H	1.22	^{4}C	28.1 m*	1.26	29.4
$^{3'}$H	1.22	$^{3'}$C	31.3	^{3}H	1.22	^{3}C	31.3	1.26	31.9
$^{2'}$H	1.22	$^{2'}$C	22.1	^{2}H	1.22	^{2}C	22.1	1.26	22.8
$^{1'}$H	0.84	$^{1'}$C	14.0	^{1}H	0.84	^{1}C	14.0	0.86	14.1

The symbol (′) and with no apostrophe ( ) distinguish the two independent systems of the repeating unit, the symbol (″) is used to mark the signals of the aliphatic end group.*Signal of ^{13}C at 28.8 ppm is a multiplet from 28.9 to 28.7.

Table 4. ^{1}H and ^{13}C-NMR chemical shifts (ppm) observed and calculated for the repeating unit of polyester PTOBDME-ammonium chloride and the aliphatic end group.

and it was assigned due to the COSY and TOCSY correlations with signals at 4.38 ppm (^{21}H) and at 2.17 ppm (^{22}H) and HSQC correlation with ^{23}C at 54.0 ppm. ^{21}H signal was overlapped at 4.38 ppm, and it was identified by COSY correlations with 2.17 ppm (^{22}H) and TOCSY correlations with 3.28 ppm (^{21}H). The polymer holds positive charge due to ammonium proton ^{25}H observed at 10.33 ppm.

The HSQC experiment confirmed the direct assignation of carbon atom $^{11'}$C (66.0 ppm) linked to protons H_a' (4.63 ppm) and H_b' (4.48 ppm). Signal H_b' exhibits correlation with $^{11''}$C (67.8 ppm), linked to H_a'' and H_b''. Two cross signal are observed for H_c'' (4.38 ppm), one with carbon atom at $^{12''}$C (60.4 ppm) and another with ^{21}H linked to carbon atom ^{21}C (62.3 ppm). The correlations of carbon atom $^{10''}$C (33.6 ppm) with H_d''(1.91 ppm) and H_e''(1.78 ppm), and carbon atom $^{9''}$C (25.6 ppm) with H_f''(1.51 ppm) and H_g''(1.42 ppm) confirmed the previous assignation.

4.4. Structural characterization of PTOBEE-ammonium

Table 6 shows the assignation of ^{1}H and ^{13}C-NMR chemical shifts (ppm) observed of polyester PTOBEE-ammonium chloride and calculated for the repeating unit and the aliphatic end group. In the ^{1}H-NMR experiment, in CDC_{13}, peaks observed at 8.34, 7.36, 7.34, 8.18, and 8.16 ppm are assigned to ^{12}H singlet, $^{8'}$H doublet, ^{8}H doublet and $^{7'}$H doublet, and ^{7}H doublet, respectively. Peak at 5.46 ppm is H_c' and 5.25 ppm is H_c. The double doublet at 4.60 ppm is interpreted as H_a' because of its shape and the COSY correlation with H_c'. An overlapped signal at 4.53 ppm is attributed to H_b', by COSY and TOCSY cross signal with H_a'. A weak multiplet at 4.15 ppm is assigned to H_c'', and this signal presented COSY correlation with H_b', indicating the presence of H_a'' and H_b''. The two double doublets at 3.76 and 3.74 ppm are identified as H_a and H_b and presented the expected COSY correlation with H_c. The overlapped signal at 1.90 ppm is H_d' correlated with H_c'. A very weak COSY cross signal is observed between H_c'' and 1.88 ppm (H_d''). Triplet signal at 1.09 ppm with TOCSY correlation with H_c' is interpreted as H_e', while triplet at 1.03 ppm with TOCSY correlation with H_c was assigned to H_e. The weak triplet at 1.13 ppm corresponds to H_e''.

Signals of proton ammonium end group in DMSO-d_6 are observed at (**Table 5**): ^{16}H singlet (2.81 ppm), ^{15}H multiplet (3.25 ppm), ^{14}H multiplet (2.18 ppm), and ^{13}H multiplet overlapped at (4.39 ppm) but presented COSY correlations with 2.18 ppm and TOCSY correlations with 3.25 ppm. The compound is positively charged, with the ammonium proton ^{17}H observed at 10.3 ppm in DMSO-d_6 and 13.2 ppm in $CDCl_3$.

HSQC experiment exhibits several correlations of the complex proton signal at 4.53 ppm, carbon atoms. Double doublet H_a'(4.60 ppm) H_b' (4.53) correlates with $^{3'}$C at (65.6 ppm). Another correlation is observed between proton H_a'' and $^{3''}$C (67.8 ppm). Correlation between the overlapped signal of proton ^{13}H (4.39 ppm), within the ammonium end group, and ^{13}C (61.9 ppm) in $CDCl_3$ is observed.

4.5. Structural characterization of PTOBUME-amide

The structures of undec-10-enamide, 10-11-epoxy-undecanamide, and 10,11-dihydroxyundecanamide (**III**, **IV** and **V** in **Figure 4**) were confirmed by ^{1}H-NMR, ^{13}C-NMR, registered in DMSO-d_6 at 25°C in a Bruker 300 MHz NMR spectrometer. Chemical shifts and Mass spectrometry results are given in Section 2.3.

The structure of PTOBUME-amide, **VII** in **Figure 4**, has also been confirmed by ^{1}H-NMR, ^{13}C-NMR, COSY and HSQC, obtained in VARIAN 400 and 500 MHz spectrometers, also at room

PTOBDME-Ammonium						PTOBEE-Ammonium							
Observed chemical shifts				Calc. chemical shifts		Observed chemical shifts						Calc. chemical shifts	
Atom	^{1}H (ppm)	Atom	^{13}C (ppm)	^{1}H (ppm)	^{13}C (ppm)	Atom	^{13}C (ppm)	^{1}H (ppm)	Atom	^{1}H (ppm)	^{13}C (ppm)	^{1}H (ppm)	^{13}C (ppm)
	DMSO		DMSO				DMSO	$CDCl_3$		DMSO	$CDCl_3$		
^{21}H	4.38	^{21}C	62.3	4.25	63.0	^{13}H	4.39	4.49	^{13}C	61.3	61.9	4.25	63.0
^{22}H	2.17	^{22}C	23.8	2.19	22.5	^{14}H	2.18	2.42	^{14}C	23.1	24.5	2.19	22.5
^{23}H	3.28	^{23}C	54.0	3.24	55.2	^{15}H	3.25	3.14	^{15}C	53.8	55.5	3.24	55.2
^{24}H	2.79	^{24}C	42.5	2.90	45.0	^{16}H	2.81	2.82	^{16}C	41.9	43.1	2.90	45.0
^{25}H	10.33			7.0		^{17}H	10.33	13.2				7.0	

Table 5. ^{1}H and ^{13}C-NMR chemical shifts (ppm) observed for the 3-dimethylamine-1-propyl benzoate hydrochloride end group in polyester PTOBDME-ammonium and polyester PTOBEE-ammonium, and calculated values.

Set of signal of system (′) and (″)						Set of signal of system without apostrophe ( )						Calculated chemical shift	
Atom	^{1}H(ppm)		Atom	^{13}C(ppm)		Atom	^{1}H(ppm)		Atom	^{13}C(ppm)		Atom	Atom
	$CDCl_3$	DMSO		$CDCl_3$	DMSO		$CDCl_3$		DMSO	$CDCl_3$	DMSO	^{1}H	^{13}C
$^{12\prime}H$	8.34	8.36	$^{12\prime}C$	130.4	129.9	^{12}H	8.34	8.36	^{12}C	130.4	129.9	8.04	130.2
			$^{11\prime}C$	133.8					^{11}C	132.2			135.4
			$^{10\prime}C$	163.6					^{10}C	163.6			165.2
			$^{9\prime}C$	154.5					^{9}C	154.5			155.6
$^{8\prime}H$	7.36	7.53	$^{8\prime}C$	121.7	121.7	^{8}H	7.34	7.53	^{8}C	121.9	121.7	7.26	121.5
$^{7\prime}H$	8.18	8.13	$^{7\prime}C$	131.5	130.4	^{7}H	8.16	8.11	^{7}C	131.5	130.4	8.13, 8.11	130.3
			$^{6\prime}C$	128.3					^{6}C	127.9			126.9
			$^{5\prime}C$	165.4					^{5}C	165.4			165.9
H_c'	5.46	5.39	$^{4\prime}C$	73.8	73.02	H_c	5.25	5.20	^{4}C	75.1	74.4	4.55	72.5
H_c''	4.15	4.40	$^{4\prime\prime}C$		*							3.81	73.0
H_a', H_b'	4.60, 4.53	4.64 4.53	$^{3\prime}C$	65.6	64.6	$H_{a'}$, H_b	3.76, 3.74	3.96, 3.92	^{3}C	45.23	45.6	4.80, 4.55	67.2
H_a'', H_b''	4.53	4.48*	$^{3\prime\prime}C$	67.8	*							4.53, 4.28	70.5
H_d'	1.90	1.86	$^{2\prime}C$	24.4	23.06	H_d	1.90	1.82	^{2}C	24.4	24.1	1.75	23.5
H_d'', H_e''	1.88, 1.13	1.90, 1.86*	$^{2\prime\prime}C$		*							1.48	26.8
$^{1\prime}H$	1.09	1.04	$^{1\prime}C$	9.8	8.73	H_e	1.03	0.96	^{1}C	9.6	8.38	0.96	7.8
$^{1\prime\prime}H$	1.13	1.06*	$^{1\prime\prime}C$		8.73*							0.96	7.6

*Overlapped signal.

Table 6. ^{1}H and ^{13}C-NMR chemical shifts (ppm) observed for polyester PTOBEE-ammonium chloride, both the repeating unit and the aliphatic end group, in DMSO-d6 and in $CDCl_3$, and calculated values.

temperature. The solvent used were DMSO-d_6 and $CDCl_3$ from Merck KGaA (Darmstadt, Germany). The spectra were processed and analyzed with the help of MestReNova 11.0.4 [9]. The chemical shifts are given in **Table 7**. Theoretical values predicted by ChemDraw Professional, v. 15.1.0.144. Tilted values are chemical shifts registered in $CDCl_3$.

4.6. Structural characterization of PTOBEE-amide

The structure of PTOBEE-amide, **VI** in **Figure 5**, has also been confirmed by ^{1}H-NMR, ^{13}C-NMR, COSY and HSQC, obtained in VARIAN 400 and 500 MHz spectrometers, at room temperature. The solvent used was DMSO-d_6 from Merck KGaA (Darmstadt, Germany). The experimental chemical shifts analyzed from the spectra are given in **Table 8**. Theoretical values predicted by ChemDraw Professional, v. 15.1.0.144.

System (′)				**System without apostrophe ( )**				**Theoretical chemical shifts**			
Atom	**^{1}H**	**Atom**	**^{13}C**	**Atom**	**^{1}H**	**Atom**	**^{13}C**	**Atom**	**^{1}H**	**Atom**	**^{13}C**
$^{19'}$H	8.36	$^{19'}$C		^{19}H	8.36	^{19}C	*130.4*	^{19}H	8.04	^{19}C	130.2
		$^{18'}$C				^{18}C	*133.7*			^{18}C	135.5
		$^{17'}$C				^{17}C	*163.7*			^{17}C	165.2
		$^{16'}$C				^{16}C	*154.7*			^{16}C	153.7
$^{15'}$H	8.07	$^{15'}$C		^{15}H	8.07	^{15}C	*121.6*	^{15}H	7.52	^{15}C	121.5
$^{14'}$H	7.50	$^{14'}$C		^{14}H	7.50	^{14}C	*131.3*	^{14}H	8.18	^{14}C	130.3
		$^{13'}$C				^{13}C	*128.2*			^{13}C	126.9
		$^{12'}$C				^{12}C	*165.7*			^{12}C	165.9
H_c'	5.76	$^{11'}$C		H_c	5.45	^{11}C		H_c	5.16	^{11}C	70.3
H_a', H_b'	4.95, 4.30	$^{10'}$C		H_a, H_b	4.23, 4.18	^{10}C		H_a,H_b	4.78, 4.53	^{10}C	66.0
H_d', H_e'	1.74	$^{9'}$C		H_d, H_e	1.55	^{9}C		H_d, H_e	1.67	^{9}C	30.7
H_f', H_g'	1.37	$^{8'}$C		H_f, H_g	1.22	^{8}C		H_f, H_g	1.25	^{8}C	23.3
$^{7'}$H	1.22	$^{7'}$C		^{7}H	1.22	^{7}C		^{7}H	1.25	^{7}C	29.6
$^{6'}$H	1.22	$^{6'}$C		^{6}H	1.22	^{6}C		^{6}H	1.26	^{6}C	29.6
$^{5'}$H	1.22	$^{5'}$C		^{5}H	1.22	^{5}C		^{5}H	1.30	^{5}C	28.9
$^{4'}$H	1.22	$^{4'}$C		^{4}H	1.22	^{4}C		^{4}H	1.30	^{4}C	28.6
$^{3'}$H	1.52	$^{3'}$C		^{3}H	1.52	^{3}C		^{3}H	1.53	^{3}C	25.3
$^{2'}$H	2.28	$^{2'}$C		^{2}H	2.28	^{2}C		^{2}H	2.34	^{2}C	38.7
NH_2'	7.0	$^{1'}$C		NH_2	7.0	^{1}C		NH_2	7.03	^{1}C	173.6
Experimental signals "end group"								Theoretical chemical shifts end group			
H_c''	4.18			$^{11''}$C				H_c''	4.57	$^{11''}$C	73.6

System (′)				System without apostrophe ( )				Theoretical chemical shifts			
Atom	^{1}H	Atom	^{13}C	Atom	^{1}H	Atom	^{13}C	Atom	^{1}H	Atom	^{13}C
H_a'', H_b''				$^{10''}C$				H_a'', H_b''	3.86, 3.80	$^{10''}C$	64.3
H_d'', H_e''	1.62			$^{9''}C$				H_d'', H_e''	1.67	$^{9''}C$	30.5
H_f'', H_g''	1.22			$^{8''}C$				H_f'', H_g''	1.25	$^{8''}C$	25.6
$^{7''}H$	1.22			$^{7''}C$				$^{7''}H$	1.25	$^{7''}C$	29.6
$^{6''}H$	1.22			$^{6''}C$				$^{6''}H$	1.26	$^{6''}C$	29.6
$^{5''}H$	1.22			$^{5''}C$				$^{5''}H$	1.30	$^{5''}C$	28.9
$^{4''}H$	1.22			$^{4''}C$				$^{4''}H$	1.30	$^{4''}C$	28.6
$^{3''}H$	1.52			$^{3''}C$				$^{3''}H$	1.53	$^{3''}C$	25.3
$^{2''}H$	2.28			$^{2''}C$				$^{2''}H$	2.34	$^{2''}C$	38.7
NH_2''	10.7			$^{1''}C$				NH_2''	7.03	$^{1''}C$	173.6
CH_2	4.36			CH_2		*61.2*		CH_2	4.29	CH_2	60.9
CH_3	1.35			CH_3		*14.3*		CH_3	1.30	CH_3	14.1

Table 7. ^{1}H and ^{13}C-NMR chemical shifts (ppm) observed and calculated for chiral Polyesteramide PTOBUME-amide.

System (′)				System without apostrophe ( )				Theoretical chemical shifts			
Atom	^{1}H	Atom	^{13}C	Atom	^{1}H	Atom	^{13}C	Atom	^{1}H	Atom	^{13}C
$^{12'}H$	8.32	$^{12'}C$	131.36	^{12}H	8.32	^{12}C	131.36	^{12}H	8.04	^{12}C	130.2
		$^{11'}C$				^{11}C				^{11}C	135.4
		$^{10'}C$				^{10}C				^{10}C	165.2
		$^{9'}C$				^{9}C				^{9}C	155.7
$^{8'}H$	8.08	$^{8'}C$	132.35	^{8}H	8.08	^{8}C	132.35	^{8}H	8.18	^{8}C	121.5
$^{7'}H$	7.45	$^{7'}C$	123.06	^{7}H	7.45	^{7}C	123.06	^{7}H	7.52	^{7}C	130.3
		$^{6'}C$				^{6}C				^{6}C	126.9
		$^{5'}C$				^{5}C				^{5}C	165.9
H_c'	5.67	$^{4'}C$		H_c	5.43	^{4}C		H_c	4.56	^{4}C	70.2
H_a', H_b'	4.63, 4.49	$^{3'}C$		H_a,H_b	4.32, 4.22	^{3}C		H_a,H_b	4.78, 4.53	^{3}C	66.5
H_d', H_e'	3.22, 3.10	$^{2'}C$		H_d, H_e	1.05	^{2}C		H_d, H_e	2.46	^{2}C	38.9
NH_2'	7.61	$^{1'}C$	17.47	NH_2	7.61	^{1}C		NH_2	7.03	^{1}C	173.6

Table 8. ^{1}H and ^{13}C-NMR chemical shifts (ppm) observed and calculated for chiral Polyesteramide PTOBEE-amide.

5. Thermal stability and differential scanning calorimetry (DSC)

The presence of choline group at the end of polymer chains causes in PTOBDME-choline a decrease in the thermal stability range compared to precursor PTOBDME. A 5% weight loss is observed for PTOBDME-choline at 230°C, while PTOBDME loses 5% weight at about 280°C. The thermal stability of PTOBEE-Choline is similar to that of polyester PTOBEE. PTOBEE-choline has 5% weight loss at 281°C, and PTOBEE at 280°C (see **Figures 6** and **7**). In the thermal stability curve of PTOBDME-choline, the first degradation step observed at 230°C is followed by two other weight loss step at 280 and 448°C. Two decomposition steps are observed at 280 and 466°C in PTOBEE-choline.

In the DSC experiment of PTOBDME-choline, performed at 10°C/min, **Figure 6(b)**, a glass transition can be observed at 58.2°C, in the first heating run, and a weak endothermic peak at 99.5°C is interpreted as due to the first order transition from crystal phase to liquid crystal state. An exothermic peak at 171.2°C is also observed which is not explained, but the beginning of a second endothermic peak at 200°C can be attributed to fusion to the isotropic. In the cooling process, two exothermic peaks at 155°C and at 175 are observed, probably associated to crystal formation. In the second heating, a very broad endothermic peak at 100.2°C is observed again associated to the transition to liquid crystal mesophase.

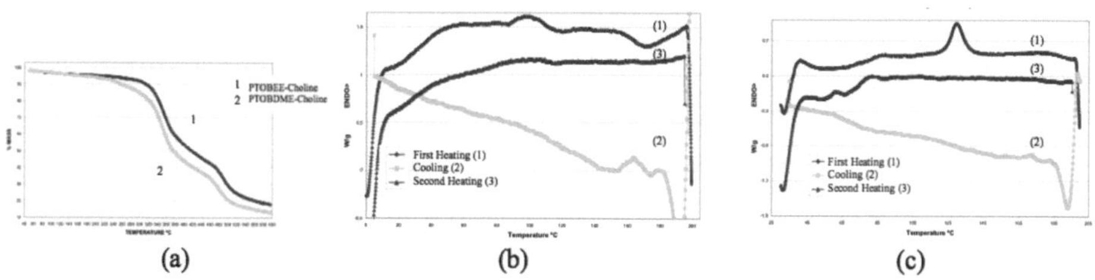

Figure 6. (a) Thermogravimetry of PTOBDME-choline and PTOBEE-choline; (b) DSC analysis of PTOBDME-choline and (c) DSC of PTOBEE-choline, both at 10°C/min.

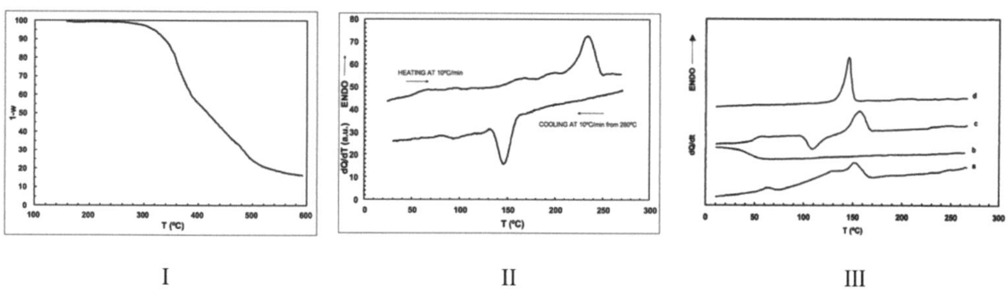

Figure 7. (I) Thermogravimetry of precursor PTOBDME; (II) DSC analysis of PTOBDME; (III) DSC of PTOBEE: (a) first heating process of the original sample, (b) subsequent cooling down, (c) second heating process; and (d) DSC of (—) PTOBEE. First heating run of the original sample. All at 10°C/min.

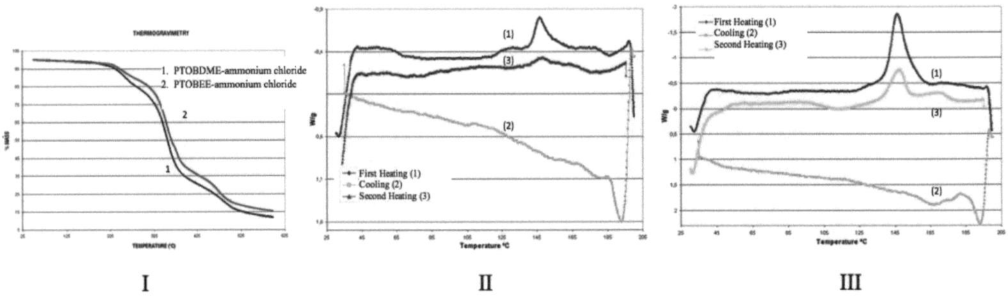

Figure 8. (I) Thermogravimetric curve of PTOBDME-ammonium and PTOBEE-ammonium; (II) DSC analysis of PTOBDME-ammonium; (III) and PTOBEE-ammonium chloride. All at 10°C/min of.

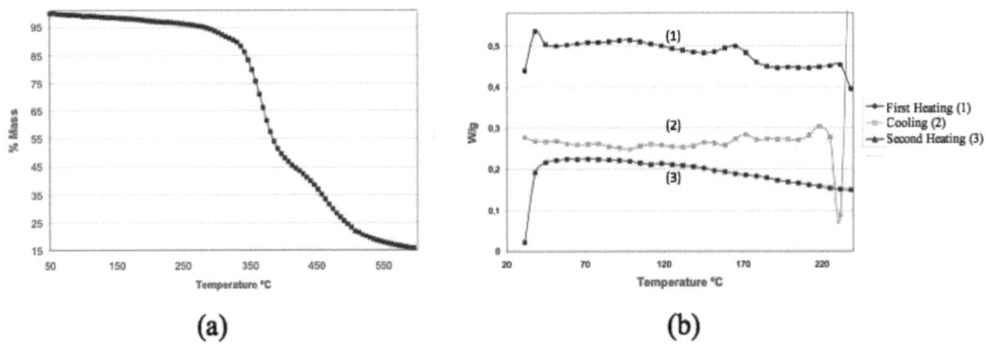

Figure 9. (a) Thermogravimetric curve of PTOBUME-amide; (b) DSC analysis of PTOBUME-amide.

In the DSC experiment of PTOBEE-choline at (10°C/min), **Figure 6(c)**, a glass transition can be observed at 60°C, and an endothermic peak at 130.2°C is attributed to the transition crystal to liquid crystal. A decreasing of baseline from 183.7°C to the end of heating was also observed in the first heating run due to a nonconcluded endothermic process or to the beginning of degradation to the polymer. A broad exothermic peak observed the cooling around 145°C would correspond to a crystallization from the mesophase state. In the second heating, only two glass transitions can be observed at 65 and 85°C.

The presence of ammonium chloride group at the end in the polymer chains, in **Figure 8-I**, produces a decrease of the thermal stability range compared to precursor polyesters. At 278°C, PTOBEE-ammonium chloride loses 10% weight and PTOBDME-ammonium chloride at 260° C, while precursor PTOBDME and PTOBEE at 310°C. In the thermal stability curve of the ammonium-polymers, the first degradation step observed at 228 and 230°C, respectively, was not observed in PTOBEE and PTOBDME. The two next inflexion points at 310 and 311°C and 466 and 471°C were equivalents to the observed in the precursor polyesters, which would indicate the same type of decomposition to principal core of the chain.

In the DSC experiment of PTOBDME-ammonium chloride, at 10°C/min, **Figure 8-II**, a very broad exothermic peak centered at 96.8°C, is observed in the first heating, associated to low enthalpy value, which can be attributed to crystal to crystal transitions, involving molecular reordering between crystalline phases. An endothermic peak at 146.9°C is interpreted due to the transition to liquid crystal mesophase; finally, an exothermic peak at 186.8°C is observed. In the cooling run, very weak exothermic peaks at 154.4 and at 104.1°C were observed due crystallization process. In the second heating, a broad exothermic peak centered at 75.2°C, an endothermic peak at 149.1°C, and finally, an exothermic peak at 179.8°C were observed again.

The DSC experiment of PTOBEE ammonium choride, at 10°C/min, **Figure 8-III**, shows in the first heating run a broad exothermic peak centered at 69.1°C, and a very strong endothermic peak at 146.2°C due to the fusion transition from crystalline phase to liquid crystal mesophase, and finally, a weak endothermic peak at 173.3°C, perhaps due to a partial fusion to isotropic. During the cooling, an exothermic peak appeared at 166°C would correspond to a crystallization from the mesophase state, and in the second heating, the broad exothermic peak observed in the first heating was observed to higher temperature centered at 114.8°C; the two endothermic peaks were again observed at 147.6 and 170.1°C.

The thermogravimetric curve and the DSC analysis of PTOBUME-amide are given in **Figure 9**. At 265°C, it loses 5% weight. At 340°C, a first decomposition step begins, followed by another three at 400, 450, and 510°C. In the first heating of the DSC, an endothermic peak is observed at 160°C interpreted as the transition to the mesophase state. In the cooling run, several week exothermic peaks could be associated to crystal formation processes.

6. Optical characterization

6.1. Optical activity of PTOBDME-ammonium and PTOBEE-ammonium

As in the polyester precursors PTOBEE-ammonium chloride and PTOBDME-ammonium chloride presented an unexpected optical activity and chiral morphology, although they were synthesized starting from equimolar quantities of TOBC and the racemic mixture of the corresponding glycol. The obtained chirality has been evaluated by optical rotatory dispersion, in **Figure 10**, the values of optical activity are given as $[\alpha]^{25°C}$, at different wavelengths. **Table 9** shows the measured values.

In the optical characterization of precursor cholesteric liquid crystal polyesters [1, 3], even an increase of chirality was observed for a second fraction of the polymer, obtained by precipitation, after days of reaction of the liquors mother with respect to the initial first fraction of the polymer. The optical activity of PTOBDME-choline, PTOBEE-choline, PTOBUME-amide and PTOBEE-amide, has not been studied at the end of the present article but will be reported in the future.

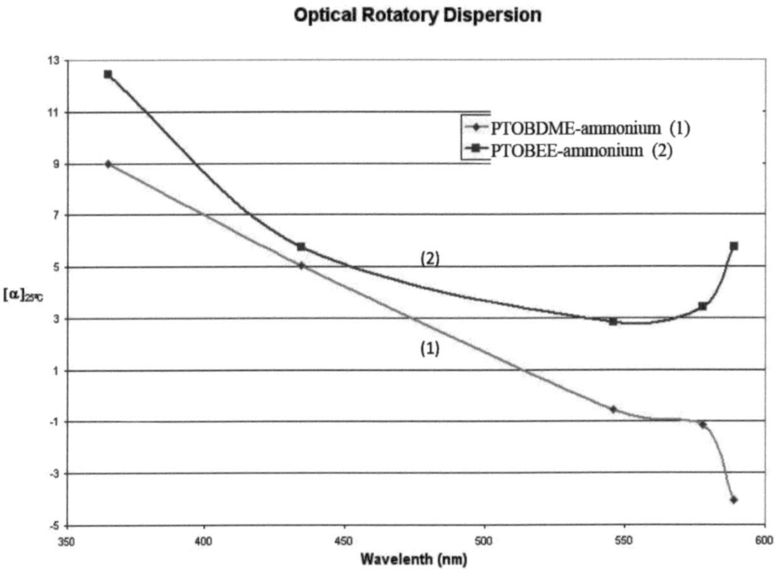

Figure 10. Optical activity of PTOBDME-ammonium chloride and PTOBEE-ammonium choride. Expressed as $[\alpha]^{25°C}$ in DMSO-d_6 at different wavelengths.

Polymers (0.2 g/100 ml in DMSO)	Hg (365 nm)	Hg (435 nm)	Hg (546 nm)	Hg (578 nm)	Na (589 nm)
PTOBEE-ammonium chloride	+12.44°	+5.75°	+2.85°	+3.44°	+5.75°
PTOBDME-ammonium chloride	+9.00	+5.06	–0.55	–1.15	–4.04

Table 9. Optical activity of PTOBDME-ammonium and PTOBEE-ammonium, expressed by optical rotatory dispersion.

7. Conclusions

The synthetic methods of six new multifunctional cationic cholesteric liquid crystal polymers designed as PTOBDME-choline $[(C_{34}H_{36}O_8)_n\text{–}C_5H_{13}N]$; PTOBEE-choline $[(C_{26}H_{20}O_8)_n\text{–}C_5H_{13}N]$; PTOBDME-ammonium $[(C_{34}H_{36}O_8)_n\text{–}C_5H_{13}N]$; PTOBEE-ammonium $[(C_{26}H_{20}O_8)_n\text{–}C_5H_{13}N]$; PTOBUME-amide $(C_{33}H_{33}O_9N)_n$ and PTOBEE-amide $(C_{26}H_{19}O_9N)_n$ are given and their characterization by ^{1}H, ^{13}C-NMR, COSY, and HSQC is reported.

The NMR analysis let us to conclude that the enantiomeric polymer chains present stereo regular head-tail, isotactic structure, explained in terms of the higher reactivity of the primary hydroxyl group in the glycol, with respect to the secondary one, through the polycondensation reaction.

According to our previous experience, each enantiomer, with two independent sets of signals observed by ^{1}H and ^{13}C-NMR, differentiated with apostrophe (') and without it ( ), could be attributed to two diastereomeric conformers: *gg* and *gt*, related with two possible staggered conformations, of the torsion along the chemical bond containing the asymmetric carbon atom in the spacer, along the copolymer backbone, with two possible helical screw sense of the polymer chain and in all the studied polymers. Chirality in racemic PTOBDME was proposed to be due to the kinetic resolution of a preferable helical diastereomer, such as *Sgt*, with respect to the possible four forms, while the R/S ratio of asymmetric carbon atoms remained 50:50.

The presence of choline group or ammonium chloride groups at the end of polymer chains causes in precursor polyesters a decrease in their thermal stability range. PTOBDME-choline losses 5% weight at 230°C (PTOBDME at 280°C). The thermal stability of PTOBEE-choline is similar to that precursor PTOBEE, with 5% weight loss at 281°C.

At 260°C, PTOBDME-ammonium loses 10% weight and PTOBEE-ammonium at 278°C (precursor PTOBDME and PTOBEE at 310°C).

All the synthetized cationic liquid crystal polymers show in DSC an endothermic peak assigned to the first order transition from crystalline phase to liquid crystal mesophase: PTOBDME-choline at 99.5°C; PTOBEE-choline at 130.2°C; PTOBDME-ammonium at 146.9°C; and PTOBEE-ammonium at 146.2°C.

At 265°C, PTOBUME-amide loses 5% weight. At 340°C, it has a first decomposition step, followed by another three at 400, 450, and 510°C. In the DSC first heating, it shows the endothermic peak due to the mesophase transition at 160°C.

Optical ORD values are provided for the second fractions of PTOBDME-ammonium and PTOBEE-ammonium.

Acknowledgements

The author thanks Dr. Javier Sanguino Otero for his valuable help during the development of this Project. She also thanks the financial support obtained in the Project "Nuevos vectores no virales basados en polímero cristal-líquido colestérico (PCLC) y su uso para transfección génica". PTR1995-0760-OP.

Author details

Mercedes Pérez Méndez

Address all correspondence to: perezmendez@ictp.csic.es

Instituto de Ciencia y Tecnología de Polímeros (ICTP), CSIC, Madrid, Spain

References

[1] Pérez Méndez M, Sanguino Otero J. Cholesteric Liquid-Crystal Copolyester, Poly[oxycarbonyl-1,4-phenylene-oxy-1,4 terephthaloyl-oxy-1,4-phenylene-carbonyloxy

(1,2-dodecane)] $[C_{34}H_{36}O_8]_n$, Synthesized from Racemic Materials: Kinetics, Structure and Optical Characterization. International Journal of Engineering Research and Applications (IJERA). July 2015;**5**(7, Part-2):48-62. ISSN: 2248-9622, http://www.ijera.com/papers/Vol5_issue7/Part%20-%202/H57024862.pdf

[2] Fayos J, Sanchez-Cortes S, Marco C, Pérez-Méndez M. Journal of Macromolecular Science Part B Physics.Conformational analysis and molecular modeling of cholesteric liquid-crystal polyesters based on XRD, Raman and transition thermal analysis. Journal of Macromolecular Science Part B Physics. 2001;**40**:553-576

[3] Perez-Mendez M, Marsal R, Garrido L, Martin-Pastor M. Self-Association and Stereoselectivity in a Chiral Liquid-Crystal Colesteric Polymer Formed under Achiral Conditions. Macromolecules. 2003;**36**:8049-8055

[4] (a) Pérez-Méndez M, Marco C. New synthesis, thermal properties and texture of cholesteric poly[ethyl ethylene 4,4′-(terephthaloyldioxy)dibenzoate]. Acta Polymerica. 1997;**48**:502-506. (b) Pérez-Méndez M, Marco Rocha C. Process for obtaining cholesteric liquid crystals by stereoselective recrystallization. Patents: EP1004650-A, WO9831771-A, WO9831771-A1, AU9854863-A, ES2125818-A1, US6165382-A, MX9906732-A1, JP2001513827-W, AU739076-B, DE69824182-E

[5] (a) Bilibin AY, Ten'kovtsev AV, Piraner ON, Skorokhodov SS. Synthesis of highmolecular weight liquid crystal polyesters based on a polycondensation mesogenic monomer. Polymer Science U.S.S.R. 1984;**26**(12):2882-2890. (b) Bilibin AY, Skorokhodov SS. Rational path of the synthesis of liquid-crystalline highmolecular weight polyesters and their properties in solution. Macromolecular Symposia. 1989;**29**:9-23

[6] Pérez Méndez M, Rodríguez Martínez D, Fayos Alcañíz J. Structure of non-viral vectors based on cholesteric liquid-crystal polymers by SAXS. International Journal of Advancement in Engineering Technology, Management and Applied Science (IJAETMAS). 2016;**03**(11):27-41. ISSN 2349-3224, http://www.ijaetmas.com/wp-content/uploads/2016/11/IJ16M1031.pdf

[7] Pérez Méndez M, Hammouda B. SAXS and SANS investigation of synthetic cholesteric liquid-crystal polymers for biomedical applications. Journal of Materials Science and Engineering B. 2013;**3**(2):104-115

[8] Pérez Méndez M, Rodriguez Martinez D, King SM. pH-induced size changes in solutions of cholesteric liquid crystal polymers studied by SANS. Journal of Physics: Conference Series. Dynamics of Molecules and Materials-II. 2014;**554**(012011):1-11. IOP Publishing

[9] (a) Cobas JC, Sardina FJ, Concepts in Magenetic Resonance. Nuclear magnetic resonance data processing. MestRe-C: A software package for desktop computers. Concepts in Magenetic Resonance. 2003;**19**:80-96; (b) MestReNova 11.0.4, Mestrelab Research SL, Santiago de Compostela, Spain, www.mestrelab.com, 2017

[10] ChemDraw Professional, Version 15.1.0.144 (PerkinElmer Informatics 1985-2016)

[11] Knapp S, Levorse AT. Synthesis and Reactions of Iodo Lactams. Journal of Organic Chemistry; **53**:4006-4014

[12] Rani S, Vankar YD. An efficient one step dihydroxylation of 1,2-glycals with oxone in acetone. Tetrahedron Letters. 2003;**44**:907-909

[13] Corey PF, Ward FE. Buffered potassium peroxymonosulfate-acetone epoxidation of .Alpha.,.Beta.-unsaturated acids. The Journal of Organic Chemistry. 1986;**51**(10):1925-1926

[14] Zhu W, Ford WT. Oxidation of alkenes with aqueous potassium peroxymonosulfate and no organic solvent. The Journal of Organic Chemistry. 1991;**56**(25):7022-7026

[15] Ella-Menye JR, Sharma V, Wang G. New synthesis of chiral 1,3-oxazinan-2-ones from carbohydrate derivatives. The Journal of Organic Chemistry. 2005;**70**(2):463-469

Chapter 3

From a Chiral Molecule to Blue Phases

Emine Kemiklioglu

Additional information is available at the end of the chapter

http://dx.doi.org/10.5772/intechopen.70555

Abstract

Chiral molecules play an important role in a wide range from biological structures of plants and animals to chemical systems and liquid crystal display technologies. These molecules were used in different research fields due to their opaqueness and iridescent colors changes as a function of the variation in temperature after their discovery by Lehman in 1889. The iridescent colors and different optical textures of cholesterol make it attractive for the new study field of cholesteric liquid crystals. The direction of the cholesteric liquid crystals generates a periodic helical structure depending on the chirality of molecules. This helical structure might be right or left handed configuration and it is very sensitive to the external conditions, such as chiral dopant concentration and temperature. The variation in a helical structure, which was induced by these external conditions, had a great attraction for the scientists working on the chirality in liquid crystals and their applications. This chapter will provide a general introduction not only about the chirality in nature and its application in liquid crystals, especially in blue phases but also about the trends in the stabilization of blue phases and the investigation of their electro-optical properties for advanced applications in display, photonic devices.

Keywords: chiral molecule, chirality, cholesteric liquid crystal, blue phase liquid crystal, Bragg reflection, polymer stabilization

1. Introduction

Chiral molecules play an important role in a wide range from biological structures of plants and animals to chemical systems and liquid crystal display technologies. The chiral molecules as liquid crystals in the form of cholesterol in the biological substances were discovered by the biologist Friedel in 1922 [1, 2]. After cholesterol was extracted from plants, cholesteryl esters were obtained by treating cholesterol with fatty acids and Reinitzer observed that cholesterol has two different melting points during heating from the crystalline phase or upon cooling from the isotropic liquid [3]. The cholesteryl esters were examined with the help of a polarizing optical microscope by chemist Reinitzer in 1888 [3], biologists Planer in 1861 [4] and physicist Lehman in 1889 [5] and they noticed that these esters showed an opaqueness and iridescent colors with the changes in temperature. Therefore, the structure and optical behavior of cholesterol were explained with the contribution of biologist, physicist, chemist and the iridescent colors and different optical textures of cholesterol make it attractive for the new study field of cholesteric liquid crystals. The cholesteric liquid crystals are generally derivatives of the cholesterol which exhibit in organic compounds of elongated molecules (like nematic liquid crystal) without mirror symmetry [6–8]. They can be formed using pure chiral compounds or the mixture of the chiral and achiral compounds [9–13]. The direction of the cholesteric liquid crystals generates a periodic helical structure depending on the chirality of molecules. This helical structure might be right or left handed configuration and it is very sensitive to the external conditions, such as chiral dopant concentration and temperature [14]. The variation in a helical structure, which was induced by these external conditions, had a great attraction for the scientists working on the chirality in liquid crystals and their applications.

Moreover, recent studies showed that the reduction of the pitch of the helical structure of the cholesteric liquid crystal by adding chiral dopants generates different phases, such as blue phase [15, 16]. Blue phases are mesophases with double-twisted cylinders of cholesteric liquid crystals and they come into existence in a self-organized three-dimensional (3D) structures in the narrow temperature range between the cholesteric and isotropic phases [3–17]. Recent studies showed the trends in the stabilization of blue phase in order to expand its narrow temperature range using different stabilization methods, such as photopolymerization [18–22], nanoparticles doping [23–27], polymer-modified carbon nanotubes (CNTs) [28, 29]. Specifically, the temperature range of blue phase was broadened up to 60 K by using a polymerized polymer network, called as the polymer-stabilized blue phase (PSBP) [18] whereas blue phase was stabilized over a range of about 50°C by using a mixture nematic bimesogenic liquid crystals [30]. Furthermore, blue phase has some advantageous in the display applications due to its outstanding electro-optical properties. Blue phases have field-induced birefringence (Kerr effect) and their response time is in the level of submillisecond. Additionally, blue phases do not need any surface modification which leads simplicity in the fabrication process and they have wide and symmetric viewing angle.

This chapter will be focused on the stabilization and electro-optical properties of blue phases and their potentials for advanced applications in display as well as photonic devices [18–22, 31, 32]. The chapter concludes with the studies related to the recent novel studies on the encapsulation of blue phases [33], the stabilization of the encapsulated blue phases [34] and polymerization-induced polymer-stabilized blue phase [35–37].

2. Cholesteric liquid crystals

2.1. Chirality

The word of chirality originates from Greek as a meaning of hand and chirality was discovered by Lord Kelvin in 1894 [38]. It was described as a property of a molecule that cannot be superimposed over its own mirror image. The main feature that gives rise to chirality at a molecular scale is the presence of an asymmetric ally handed carbon atom. A chiral molecule typically has a carbon atom in the center of the molecule surrounded by four different substituents and these molecules cannot be transformed into their mirror image by rotations [39–41]. Moreover, all these different groups are not in the same plane, although the positions of these groups form the corners of a tetrahedron with a central carbon atom (**Figure 1**). The configuration of these molecules can be classified into two different group, such as 'R'(for Latin rectus, right) or 'S'(for Latin sinister, left). In that case, handedness can be defined as right hand or left hand for a chiral molecule [42]. In order to define handedness of the molecule, the four groups are arranged in a priority list. Groups with the higher atomic number take precedence of the groups of lower atomic number. Each chiral center is labeled as R or S related to the priority of the substituents of the molecule based on their atomic numbers. To determine the handedness of the molecule, first the chiral center is determined according to the lowest priority of the four substituents. If the priority of the other three substituents decreases in clockwise direction, it is called R (right handed), if it decreases in counterclockwise direction, it is called S (left handed). **Table 1** summarizes a list for the enantiomers of chiral materials. Moreover, an enantiomer can be named by its ability to rotate the plane of plane-polarized light (+/−). The enantiomer is labeled (+), if it rotates light in a clockwise direction. If it rotates the light counterclockwise direction, it is labeled as (−). Liquid crystals may have multiple chiral centers with handedness and configuration. Moreover, the chirality of atoms can be detected by optical experiments [43], which shows that the broken mirror symmetry in stable atoms as a function of absorption of light.

Chiral molecules may give rise to an intrinsic helical structure of the director in liquid crystals inducing chirality [44–49]. The liquid crystal state is a mesophase between solid and liquid which is characterized by the alignment of rod-like molecules which has two aromatic rings

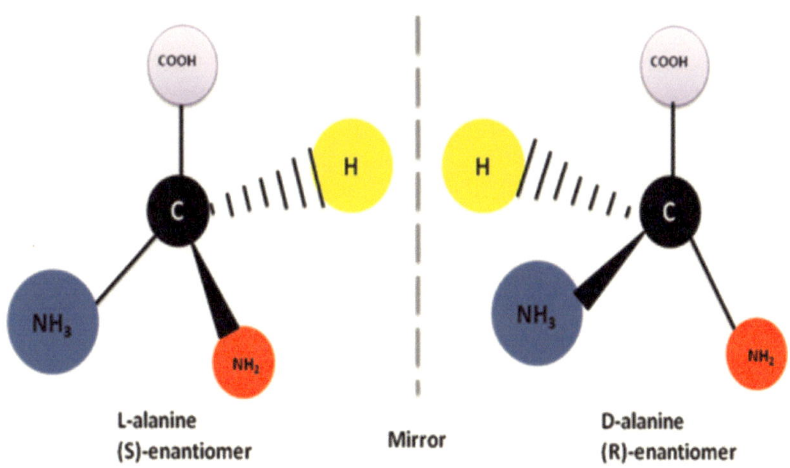

Figure 1. Simulation of a chiral molecule which cannot be superimposed with its mirror image [53].

Name of an enantiomer	Chemical structure of an enantiomer	Properties of enantiomers
(R) Limonene	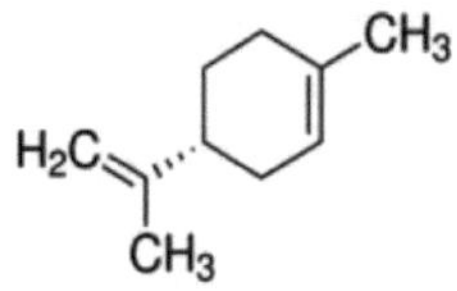	Orange smell
(S) Limonene	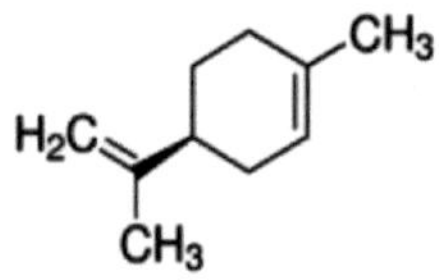	Lemon smell
(S) Asparagine	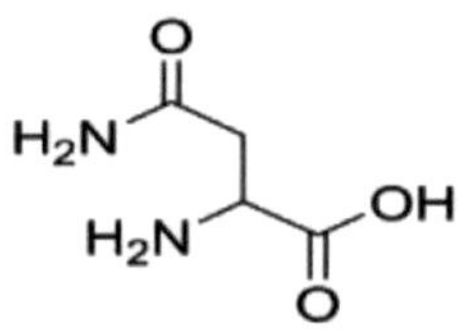	Sweet taste
(R) Asparagine	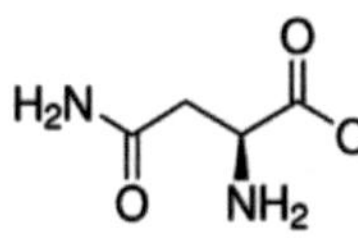	Bitter taste
Pseudoephedrine	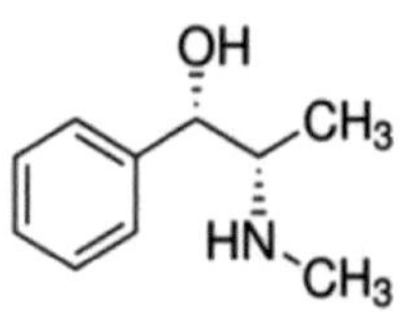	Active ingredients in over-the-counter nasal decongestants
(S) Carvone	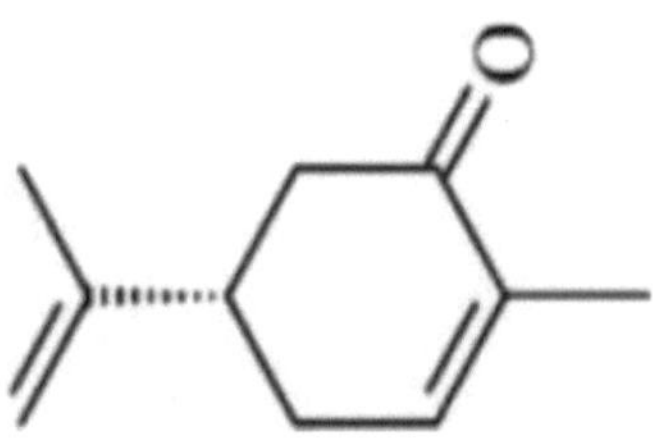	Caraway flavor
(R) Carvone	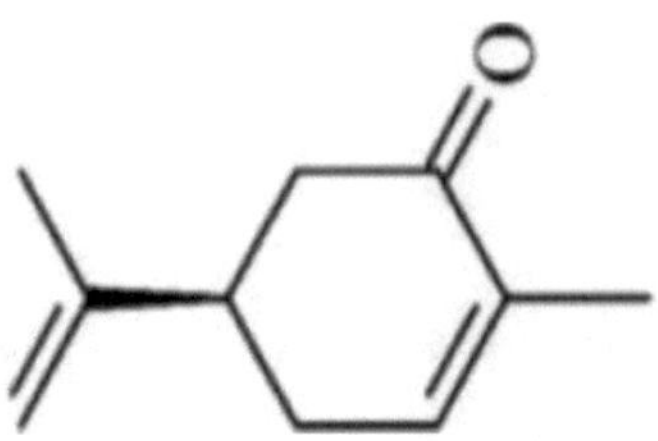	Spearmint flavor

http://dx.doi.org/10.5772/intechopen.70555

Name of an enantiomer	Chemical structure of an enantiomer	Properties of enantiomers
(S) Ibuprofen	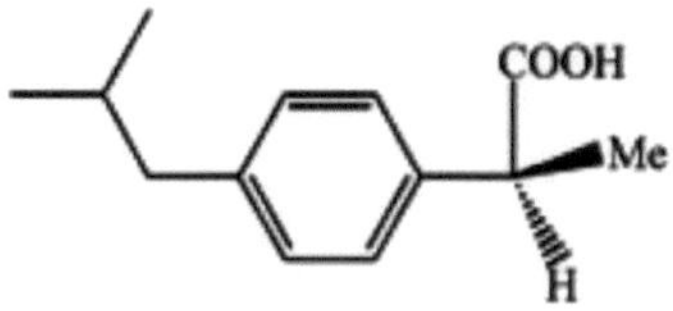	The over-the-counter painkiller
(R) Thalidomid	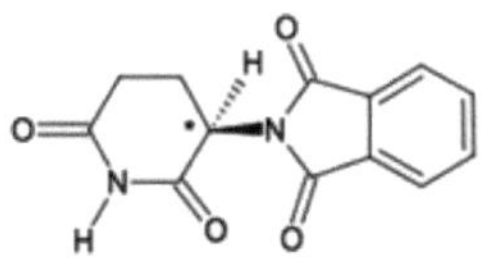	Sedative and antinausea
(S) Thalidomid	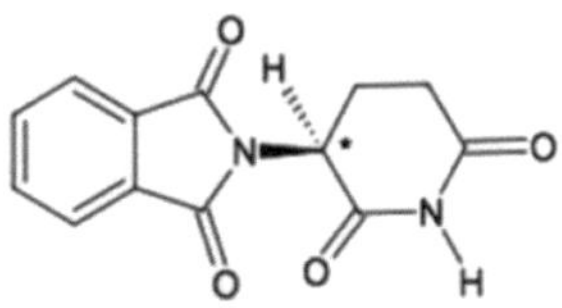	Teratogen
Cloramphenicol	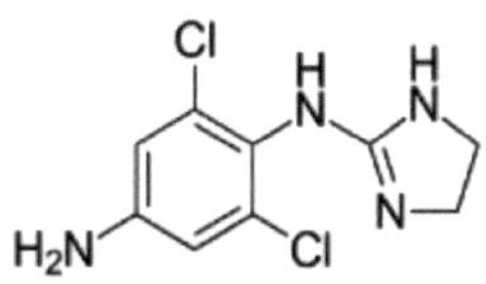	It is an antibiotic and it works to treat serious infections induced by certain bacteria
Levomethamphetamine	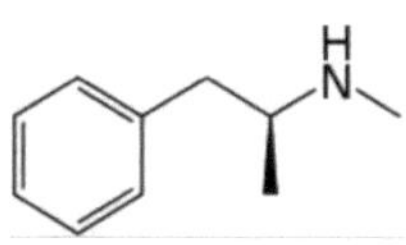	Active ingredients in over-the-counter nasal decongestants

Table 1. Enantiomers and their properties.

with the aliphatic chains. These rod-like shaped molecules are usually formed liquid crystal materials by aligning along a certain direction and this certain direction forms a helical structure with the addition of chiral molecules. In that case, the liquid crystal phase is called chiral nematic (cholesteric) phase which is one of the several additional phases in the temperature range between the crystalline and the isotropic liquid state (**Figure 2**). Moreover, chirality can be induced in smectic and columnar phases which are a quasi-long range positional order in less than three dimensions. In contrast to thermotropic liquid crystals, chiral columnar liquid crystals are formed by amphiphilic molecules in lyotropic liquid crystals. However, these amphiphilic molecules can be arranged in an anisotropic structure as in thermotropic liquid crystals, when the concentration of these molecules in a solvent is enough. Recently, Takezoe et al. induced a molecular chirality in the bent-core molecules which do not have molecular handednesses. They successfully induced a chirality in these achiral molecules based on the packing of the bent-core molecules [15, 17].

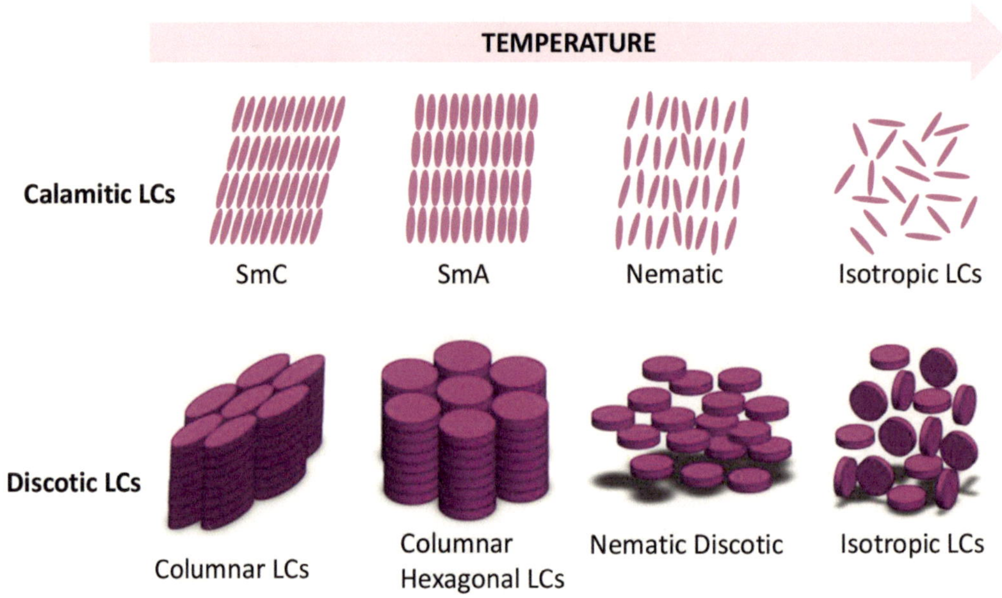

Figure 2. Phase sequence in thermotropic liquid crystals.

2.2. Cholesteric liquid crystals

The cholesteric phase is a mesophase which exhibits between the smectic and isotropic phases in thermotropic liquid crystals. Chiral nematic liquid crystals are a type of liquid crystal which has a helical structure based on the molecular chirality of its components (**Figure 3**). This phase can be formed using the chiral dopants in an achiral nematic forms new chiral materials with specific helical pitches (**Figure 4**) [50–52].

Cholesteric liquid crystals arrange within layer without any positional ordering in the layer whereas the director axis rotates with the layers as shown in **Figure 5**. The rotation of the director axis is periodic and its full rotation of 360° is called the pitch, p. The value of the pitch may change as a function of enantiomeric excess in an ideal mixture containing chiral and racemic.

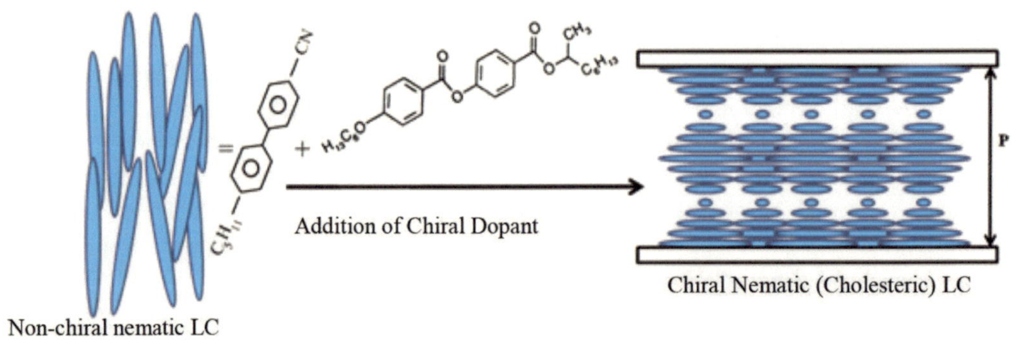

Figure 3. Simulation of formation of the helical structure of a chiral nematic (cholesteric) phase.

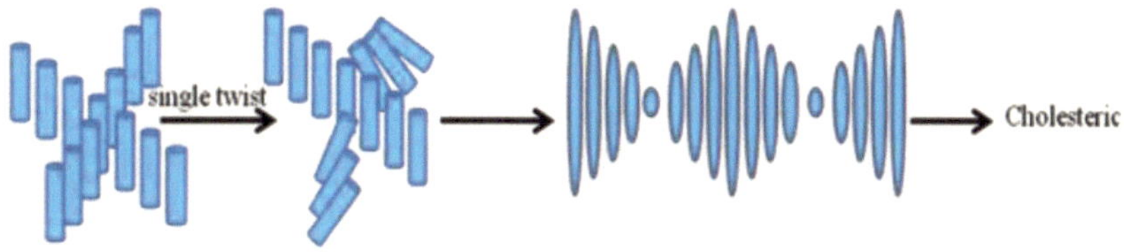

Figure 4. Formation of helical chiral nematic phase using chiral dopant.

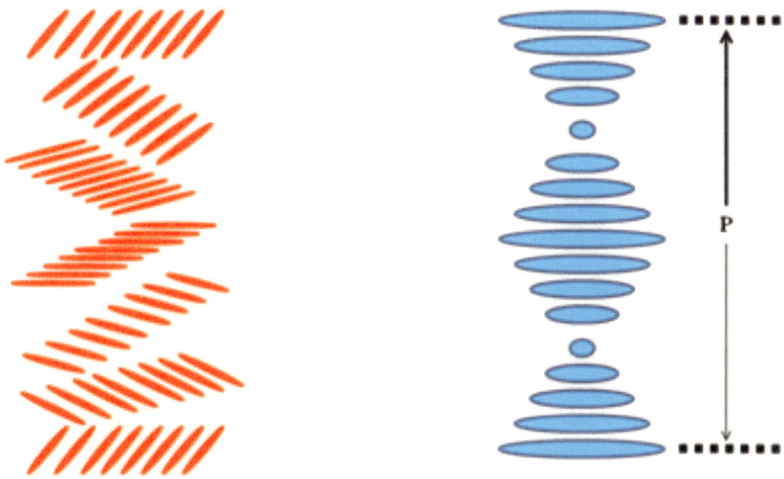

Figure 5. An illustration of cholesteric liquid crystal with a pitch p. Pitch plays an important role in the reflection of the wavelength of the incident light, as a result of the periodic structure of cholesteric liquid crystals [53].

Pitch plays an important role in the reflection of the wavelength of the incident light, as a result of the periodic structure of cholesteric liquid crystals [53]. Cholesteric liquid crystals have the ability to reflect a handedness of circularly polarized light when the pitch has the same wavelength of visible light [53]. The light will be circularly reflected if it is the same handedness as that of the cholesteric liquid crystal, whereas it will be circularly transmitted with opposite handedness as that of the cholesteric liquid crystal [44]. This selective reflection of circularly polarized light exhibits an iridescent color depending on the angular deviation. This property of selective refraction may practically be used in the application of liquid crystals, such as thermometers, polarizing mirrors, refractive electro-optic displays and optical storage [53–58].

Furthermore, chirality in liquid crystals can be described related to inverse of the pitch of the material and a shorter helical pitch has a higher chirality. The normalized reciprocal of the pitch (p^{-1}) is described as the helical twisting power (HTP) of a molecule and it can be defined as the chiral dopant's ability to induce helicity in the molecule [17].

$$HTP = 1/c * p \tag{1}$$

where p is helical pitch in microns and c is concentration of chiral dopant in the cholesteric liquid crystal mixture.

However, q_o is the pitch in the ground state given by Eq. (2) and it is induced in nematic liquid crystal where there are long-range distortions

$$q_o = 2\pi/p \tag{2}$$

and the pitch introduces in a scalar quantity of the free energy of cholesteric phases [44]:

$$F = \frac{1}{2}\left[K_{11}(\nabla.n)^2 + K_{22}\left(n.\nabla xn + q_o\right)^2 + K_{33}(nx\nabla xn)^2\right] \tag{3}$$

where n is the director, K_{11} is splay elastic constant, K_{22} twist elastic constant and K_{33} is bend elastic constant. When the chirality of a material is high enough, in other words the pitch of the molecule is around 100 nm, another phase becomes energetically favorable, which is called blue phase with self-organized three-dimensional double twist structure [17].

3. Blue phases

Blue phases were first observed in 1888 by Reinitzer who noticed a brief hazy blue color that exhibited in the narrow temperature range between the chiral nematic (cholesteric) and the isotropic phases [17]. Blue phases are locally isotropic fluids. Moreover, the molecules are self-organized and complex three-dimensional (3D) structures and characterized by crystallographic space group symmetry in this kind of liquid crystal phase. The blue phases are generated by double-twisted cylinders separated by defect lines (**Figure 6**). Effectively, blue phase is classified by the network of the defect line and three network states are known as BPI, II and III as a function of increasing temperature. The Bravais lattice is body-centered and simple cubic for BPI and BPII, respectively, as shown in **Figure 7** [59]. The BPI and BPII have soft, frequently coagulating platelet-small domains in a size of micrometer to submillimeter. The lattice constant which is around 100 nm depends on the radius of double-twisted helix and photonic band. This constant is mostly in the blue wavelength range and has the same order of magnitude as the cholesteric pitch. Additionally, the BPIII is called 'blue fog' since it has a cloudy and an amorphous appearance.

Moreover, Bragg scattering of BP which is the characteristics of the selective reflections of BPs have been comprehensively investigated since 1980 [59, 60]. Because of the exhibition of the BPs in a narrow temperature range, studies on BPs have been a challenge to the experimentalists

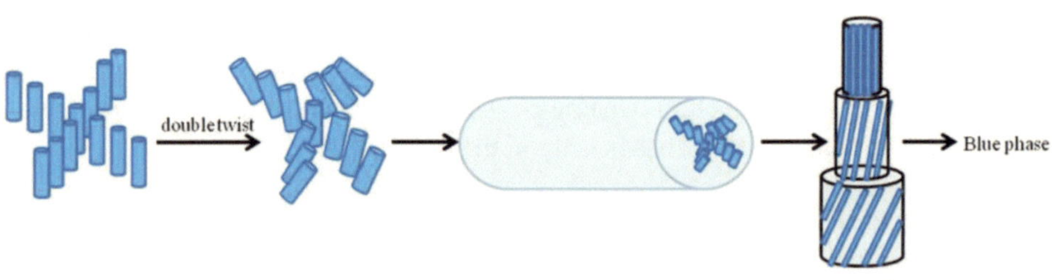

Figure 6. The simulation of a double twist cylinder structure of blue phase.

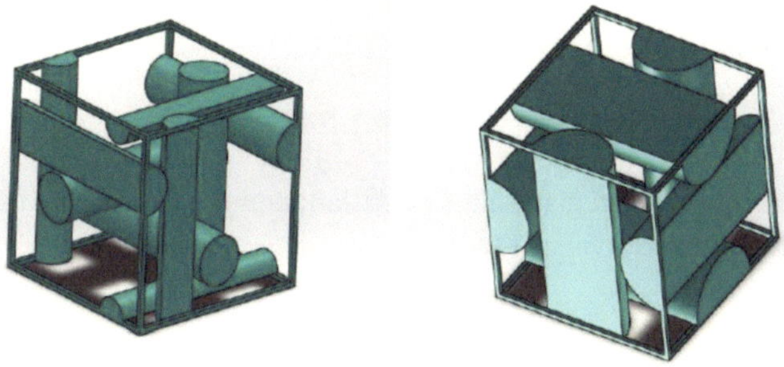

Figure 7. Blue phase structures in (a) BPI and (b) BPII.

[61, 62]. Recently, it became possible to observe BPs in wide temperature range via the stabilization method. The thermodynamic stability of BPs which were composed of chiral nematic liquid crystal with a low chirality have been predicted using Landau theory [63] and it was proved that the instability of cholesteric phase increases at the temperature near the transition point depending on the increment in the planar helix structure. Alternatively, the temperature of the blue phase liquid crystal can be determined by the help of Meiboom's defect model depending on the Oseen-Frank elasticity equation [64]. The presence of the defect lines is essential for entity of the lattice structure in blue phase liquid crystals and the energy cost of the defects should be low enough to stabilize the entire phase for narrow range temperatures. Moreover, the free energy per unit length for the disclination line in BPs can be described as in Eq. (4):

$$F_{discl} = F_{core} + F_{int}F_{surf} + F_{el}, \qquad (4)$$

where F_{discl} is the total free energy per unit length of the disclination, F_{el} is the elastic energy related the defect, F_{surf} is the free energy at the disclination surface and F_{int} is the energy related to melting of area to the isotropic core. For blue phase double-twist cylinder lattices, the free energy calculations of Meiboom et al. comprise F_{core} as the only temperature-dependent term:

$$F_{core} = \alpha(T_{iso} - T)\ \pi R_o^2 \qquad (5)$$

where T_{iso} is the isotropic transition temperature, R_o is the defect core radius size and the difference in free energies of the isotropic and ordered phases at temperature T is represented by $\alpha(T_{iso}-T)$.

The surface energy at the interface between core and cholesteric is characterized by a surface tension, σ (Eq. (6));

$$F_{interface} = 2\sigma\pi R_o \qquad (6)$$

F_{surf} can be turned into a surface integral and it is negligible and ignored, since surface terms do not scale competitively with the bulk terms. In that case the interior surface of the disclination must be taken into consideration and the solution covers the energy per unit length of the disclination line (Eq. (7)):

$$F_{surf} = -\pi(K_{22} + K_{24}) = -\pi K \tag{7}$$

F_{el} is the elastic energy where K is the elastic constant and R_{max} is the radius of the double twist cylinder and R_o is the defect core radius.

$$F_{el} = \frac{1}{4}\pi K ln\left(\frac{R_{max}}{R_o}\right) \tag{8}$$

According to Eq. (8), one parameter in this equation must minimize the energy of cost of the disclination line in BPs to expand the BP temperature range. It is expected to move the isotropic particles, such as nanoparticles or monomers, towards isotropic areas of liquid crystals in order to minimize the core energy. The addition of these nanoparticles into an isotropic phase of sample and cooling to the BP give rise to an aggregation of these nanoparticles in the defect lines. These nanoparticles will interrupt any inclination towards orientational order inside the core when temperatures decreased into the blue phase. However, the surface energy at the interface between core and cholesteric was assumed zero during the energy minimization of the system [53].

3.1. Stabilization of blue phases

Blue phase liquid crystals have a great potential for various applications due to their electro-optical properties, such as fast response time, wide and symmetric viewing angle and lack of requirement of any surface alignment layer. However, BPs have limited usage in the practical applications because of their narrow temperature range [18, 65, 66]. Recently, two independently reported methods to expand the BP temperature range have a great attraction to blue phase materials, which have become a hot topic of comprehensive research in exploiting applications in new optics, photonics and information displays based on the outstanding electro-optical properties of BPLCs. The first reported approach uses a tiny amount of monomer for polymerization and it has been reported polymer stabilization which helps to expand the BP temperature range to more than 60 K including room temperature with an ultrafast response time [18]. Kikuchi et al. [67–69] developed a technique to extent the BP temperature range with a polymerized polymer network, denoted as the polymer-stabilized blue phase (PSBP). The synchrotron small angle X-ray scattering measurements exhibited that polymers are selectively concentrated in the disclination cores and a remarkably unique accumulation structure in the PSBP [70]. This result evidently conforms the mechanism of the stabilizing effect of BPI originating from the immobilization of the disclination in the blue phase by polymers. The first method proposed that the polymer network which is concentrated not only in the isotropic defect core but also in the disclination core of BP causes an increment in the temperature range of BP. Therefore, cross-linked network of the polymer which was produced by the process of in-situ polymerization blocked the molecular reorientation of liquid crystal directors [18]. The latter approach reported the usage of the nematic bimesogenic liquid crystal mixtures to stabilize the defect structures in the blue phase. This method provided a successfully polymerization of BPs in a wide temperature range [71] which is more than 50°C [30]. Recently, Coles' group reported a study about the stabilization benefitted from the flexoelectric coupling between polar order and curvature of the director. Unfortunately, the report does not address the question of whether BPI appears at 16.5°C on heating from the smectic phase. Therefore, thermodynamic stability of BPI were not clearly indicated [53, 72, 73].

Alternatively, Yoshizawa et al. [74] successfully manage to extent the temperature range of the BPs more than 10°C using chiral T-shaped compounds. Yelamaggad et al. [75] were able to stabilize the BP more than 20°C using chemically linked bent core molecules.

Wang et al. [76] both introduced BPs in a wide temperature range using ZnS nanoparticles and showed the stability of the cubic structures against the electrical field. Recent studies on BPs with an broadened temperature range make them more attractive for applications because of some specific electro-optical (E-O) properties of BPs, such as fast response time [31], wide viewing angle and also any surface treatments are not necessary for the BPs. Moreover, Kemiklioglu et al. stabilized the cholesteric blue phases using polymerizable silicon-based nanoparticles to expand the temperature range of BPS. They showed that these polymerizable nanoparticles help to modify the interfacial properties of disclination cores broadening the blue phase temperature range and also the polymer concentration plays an important role in the thermodynamic stability of modulated liquid crystal blue phases. They also reported inorganic polymer leads to significant reduction in the switching voltage from about 140 to 40 V in corresponding device as a result of the low surface energy property of the inorganic polymers [72, 73, 77, 20].

3.2. Carbon nanotubes doping for the stabilization of blue phases

Nanoparticles [63, 64, 30, 72, 73] and polymer-modified carbon nanotubes (CNTs) [67] are emerging as new classes of nanoscaled materials and have become the subjects of extensive research because of their potential in improving the mechanical, electrical and thermal features of composite materials. Recently, with new approaches, such as doping MgO [68], ZnS [69] and CNTs into LCs, it has been possible to overcome the limitations of the transition temperature range and physical properties of LCs [53].

Carbon nanotubes are not only anisotropic but also metallic or semiconducting nanoparticles based on the diameter and helicity of the carbon rings [78]. Moreover, they categorized into two different morphologies namely, single-wall carbon nanotubes (SWCNTs) and multi-wall carbon nanotubes (MWCNTs). Moreover, the typical length of SWCNT which changes from submicron to microns is an important parameter for the determination of tensile strength of SWCNTs since they show exceptional tensile strength depending on their high aspect ratio and rigidity. In addition, the diameter in the range from 0.5 to 2 nm leads a high aspect ratio of tubes [53, 79]. Besides, MWCNTs show the similar electronic behaviors with those of SWCNTs due to weak coupling between cylinders of them.

Carbon nanotubes have became an important research topic for the liquid crystal scientists after their discovery by Iijima in 1991 [80] because of the extraordinary electrical properties and strong interactions of the CNTs with the mesogenic units of liquid crystals [81]. Recently, different groups have reported studies on the alignment and characterization of CNTs in nematic liquid crystals [82–86] as well as on the dielectric [87–89] and electro-optical properties [90, 91] of CNTs [93]. Different textures of CNTs were observed when the nematic LC droplets were embedded in a polymer matrix medium [53, 80, 92]. However, carbon nanotubes have been studied in blue phase liquid crystals to expand the temperature range of the blue phases depending on thermal stability of carbon nanotubes. Another group studied CNT-doped polymer-stabilized blue phase (PSBP) LC. The CNT-doped PSBP mixtures exhibit a good thermal stability in a wide BP temperature range which is more than 40°C. They reported that BP temperature range and electro-optical properties, such as Kerr constant, switching voltages and response times of the PSBP LCs were able to improved when a mixture of monomer and BPLCs doped with CNT [93].

Moreover, the electro-optical behaviors of liquid crystals were studied as a function of the addition of carbon nanotubes. Several liquid crystal textures were observed in the CNT-doped liquid crystal mixture based on the field-induced movement of CNTs inside nematic liquid crystal by applying a high electric field [94–96]. Furthermore, it has also been demonstrated that the rising time of CNT-doped nematic liquid crystals leads a decrement in the threshold voltage both of the twisted nematic and in-plane switching cells with the dispersion of a very small concentration of CNT dopant [90, 91, 97, 98]. The addition of CNTs in the optical controlled birefringence cells lead to a fast response time due to the increment in anchoring energy of the alignment layer by CNT doping [53, 99].

3.3. Polymer stabilization of blue phases

Although BPLCs have the potential for various applications because of their electro-optical properties, such as fast response time, wide and symmetric viewing angle, the narrow temperature range of BPs is still one of the main limitations for their practical applications [18, 65, 66]. Therefore, stabilization of blue phase (PSBP) liquid crystals via polymerization have been studied commonly for two decades [18, 19, 32, 100–105] because of their great potential for use in display devices or as an optical modulator.

Recently, many studies on the increment in the BP temperature range via polymer stabilization methods have drawn attention to blue phase materials, which have thereafter become a subject of extensive research in exploiting applications in new optics, photonics and information displays because of the outstanding electro-optical properties of BPLCs [53]. One of these studies include a method which uses a small amount of polymer for polymerization that is phase-separated to the defects of the blue phase based on the concept of the polymer-stabilized liquid crystal the orientation of liquid crystal directors can be stabilized by a crosslinked network dispersed in a liquid crystal [18]. Moreover, the polymer network plays a fundamental role in the increment of temperature range, causing the thermodynamic stabilization of BP [18, 19, 32, 100]. Furthermore, another study explored that the stabilization of cholesteric blue phases using polymerizable silicon-based nanoparticles to modify the interfacial properties of disclination cores and broaden the blue phase temperature range. This study showed that the polymer concentration has an important effect on the thermodynamic stability behaviors of modulated liquid crystal blue phases. There was a significant reduction in the switching voltage of the device as a result of the low surface energy property of the inorganic polymer used in the experiments. This low surface energy leads a significant decrement in the switching voltage from 140 to 40 V [19, 20, 72, 73]. The significant reduction in the switching voltage and widening of blue phase temperature range are useful for new electro-optical applications [22, 53]. Besides, Coles et al. reported that they managed to expand the blue phase temperature range to over 50°C by using the nematic bimesogenic liquid crystal mixtures to stabilize the defect structures of the blue phase [30]. However, a three-dimensional polymer network is formed by the reaction of benzoyl radicals with the double bonds of the diacrylate monomer through a chain reaction [53, 101]. Furthermore, molecular mobility of the network polymers obtained in the PSBP affects the stability of PSBP [53, 102]. Additionally, the electro-optic properties can be improved due to the variation of the flexibility of the molecule, the length of the rigid core and the polymerizable functional group of reactive monomer [53, 103]. All these studies showed that either thermally polymerizable [101] or photopolymerizable [18, 19, 32, 100–105] monomers can be used in the polymer stabilization of BP liquid crystals and PSBP liquid crystals have some advantages [31, 104–111]. PSBP liquid crystals become an attractive material as a next generation display technology [109–111] since, these materials have submillisecond response time, and wide viewing angle and also they do not need any surface alignment [53]. On the other hand, high operating voltage, and a low-contrast ratio due to residual birefringence and hysteresis [107, 108] are some disadvantages of PSBP liquid crystals which limits the wide-spread applications of them. There are two common approaches in order to overcome the issue of high operating voltage, there are two common approaches. One of these approaches is using a small electrode gap to produce a strong electric field [53, 110, 111]. The second approach is electric-field-induced birefringence known as the Kerr effect [53, 71]. Kerr effect is an electro-optical behavior of blue phases. Blue phase liquid crystals not have birefringence in the absence of the electric field and Kerr effect can be explained as the induced birefringence occurrence in the presence of the electrical field. PSBP liquid crystal with large Kerr constant was reported by Kikuchi et al. [107] and Wu et al. [111, 112–114]. Kerr effect can be calculated using Eq. (9)

$$\Delta n(E) = \lambda K E^2 \tag{9}$$

where K is the Kerr constant, λ is the probe wavelength, and E (= V/l, where V is the applied voltage and l is the distance between electrodes) is the applied electric field. Induced birefringence is linearly proportional to E^2, where E is the electric field and this linear relationship is valid only in the low field region [53, 111, 112].

3.4. Polymer dispersion of blue phases

Polymer-dispersed liquid crystals (PDLCs) have become the topic of considerable interest during the last decades, because of their potential applications such a smart windows, flexible displays, projection displays and holographic gratings [115–119]. The PDLC films have been widely studied as a candidate for the large area display because of the simplification of the preparation process and because their light transmittance is higher than conventional LCs in the absence of polarizer by the reason of their light scattering nature [120–126]. PDLC films are a mixed phase of micron-sized liquid crystal droplets, which are randomly dispersed inside a polymer matrix [127]. In general, the polymer weight concentration is between 30 and 60%

[127]. However, it was known four different methods for the fabrication of PDLCs. The first one of these four methods is encapsulation (emulsification) which includes the liquid crystal inside an aqueous solution of film-forming polymer. The second one is called the solvent-induced phase separation (SIPS) which is used to dissolve the liquid crystal and thermoplastic polymer and forms a single phase by evaporating the solvent at a certain rate. The third method is known as thermally induced phase separation (TIPS) which includes the heating of the liquid crystal and thermoplastic polymer to obtain a single phase. After heating procedure, the liquid crystal phase separates into droplets with the cooling of mixture at a controlled rate. The fourth method is polymerization-induced phase separation (PIPS). This method contains the liquid crystal that is dissolved into the monomer [93] and uses ultraviolet radiation to initiate the free radical polymerization of monomers [127]. One of the main advantages of this method is the possibility to form a composite directly between two glass substrates coated with indium-tin-oxide (ITO) without any requirement of laminating procedure. The above methods produce a wide size distribution of liquid crystal domain size [127]. PDLCs are operated based on the micron-sized LC droplet dispersion inside the polymer matrix and the scattering performance of the PDLC film is determined by the LC droplet size. The operation principle of the PDLC films—electrically switchable between light scattering and transparent states or vice versa depends on the refractive indices matching between guest and host materials [128, 129]. The PDLC films normally seems milky since the randomly orientation of LC molecules inside the droplets causes a scattered light at zero voltage. As a function of an applying a voltage across the PDLC film, the LC directors align in the direction parallel to the applied field. Due to matching in indices of refraction between polymer and liquid crystal molecules under the electric field, PDLC film becomes transparent at normal viewing direction. Additionally, H-PDLC, which is the another type of PDLC, includes liquid crystal droplets smaller than that of PDLC [130] and they are staged in varying planes in accordance with the polymer. There are two modes of H-PDLCs, which are called transmissive and reflective. In the transmissive mode, diffraction occurs by an applying voltage and light is reflected in the absence of electric field. In the reflective mode, light is reflected in the absence of electric field, with the applying voltage it transmits through the display.

Polymer-dispersed liquid crystals (PDLCs) which are a class of important electro-optical (E-O) materials comprise of dispersions of micron-sized LC droplets inside a polymer matrix which were discovered by Fergason [127, 130]. The PDLC films were fabricated by using different methods, such as solvent evaporation, thermal induction or polymerization-induced phase separation [131]. The first method used in the preparation of PDLC films is the encapsulation (emulsification) of the liquid crystal inside an aqueous solution of film-forming polymer [130]. After water evaporated at a certain rate to induce phase separation, the film is laminated between two conductive electrode coated substrates. The second method which is called the solvent-induced phase separation (SIPS) includes solvent which is used to dissolve the liquid crystal and thermoplastic polymer and create a single phase. The certain solvent evaporation rate induces the phase separation. The third method is the thermally induced phase separation (TIPS). This method uses liquid crystal and thermoplastic polymer. These two phases are heated to obtain a melting and then mixed to form a single phase. Liquid crystal phase separates into droplets with cooling of the mixture at a controlled rate. The fourth method is known as polymerization-induced phase separation (PIPS) containing the liquid crystal, monomer and a small amount of catalyst. After exposing the prepolymer mixture to an external stimulus, for example, light or heat, the monomer gels into a polymer matrix and liquid crystal phase separates into droplets.

The operation principle of the PDLC films based on the electrical switching between light scattering and transparent states due to index matching between guest and host materials [115, 132, 127]. The PDLC films normally appear milky and scatter incident ambient light because the LC molecules orient randomly inside of the droplets in the absence of voltage. With the applying a voltage across the PDLC film, the LC directors align in the direction parallel to the applied field. As a result of index matching between polymer and LC molecules in the presence of the electric field, the PDLC film becomes transparent when viewed along the normal direction. These PDLC films have significant advantages for electro-optical device applications, since PDLCs do not require any polarizers also PDLCs have the property of high light transmittance [127]. A number of reports have appeared recently suggested application areas for PDLCs ranging from switchable light modulators [71, 133], smart windows [127], information displays [134] and holographically formed optical elements and devices [127–135].

The electro-optic properties of PDLC devices, such as displays and smart windows can be improved by using BPLCs. The polymer dispersed or encapsulated blue phase liquid crystal films have many advantages when compared to that of polymer dispersed or encapsulated nematic liquid crystals [33–35, 53]. One of these advantages of BPLCs is field-induced birefringence due to their submillisecond response time, which is at least one order of magnitude faster than the present nematic LC-based displays [53]. BPLCs do not require any surface alignment layer; thus, the device fabrication process is greatly simplified [53]. Another significant advantage of BPLCs is their wide and symmetric viewing angle due to the fact that their 'voltage off' state is optically isotropic and the 'voltage on' state forms multidomain structures [53, 108, 109]. Moreover, BPLC can be a substantial candidate for polymer encapsulated LC films due to their fast switching properties [53]. Due to all these advantages of BPLCs, polymer encapsulated blue phase liquid crystal films are a strong candidate for the next generation of displays and spatial light modulators due to their optical properties [22, 31, 32] and also these films have potential for advanced applications in displays and photonic devices [53].

3.5. Polymer encapsulated blue phase liquid crystals

Encapsulation is one of the major methods used in the fabrication of PDLC films [53]. This method includes emulsion-based PDLC films which are formed of small liquid crystal droplets inside the aqueous solution of water soluble polymer [136, 137] or a colloidal suspension of a water insoluble polymer [130, 138]. Moreover, the energy is required to form the encapsulated droplets. This required energy which generally arises from the chemical potential of components or from mechanical devices break-up the droplets. The emulsion system is obtained by high shear; for example, by ultrasonication or high-pressure homogenizers and the rate of solidification and polymer solubility play a role in the yield. Microspheres formed by rapid solidification of the polymer may give a higher yield due to encapsulation of some of the soluble fractions in the matrix [53, 140, 141]. Besides, the emulsion is coated on a conductive substrate before allowing the water to evaporate to produce thin polymer films containing liquid crystal droplets dispersed in a matrix. Evaporation process is one of the key factors for the droplet shape deformation which affects the alignment of LC inside the film cavities, which has a significant effect on the physical properties of PDLC films. Droplets become spherical or oblate in the polymer film [71, 120, 127, 132–145] depending on the evaporation process. However, the size distribution of the liquid crystal droplets in the emulsion can be modified by the preparation process and materials used to produce the emulsion, for example, the stirring time and speed, viscosities of polymer and liquid crystal. With increase in time of mixing in an ultrasonic cleaner, the droplet size of emulsion decreases [53, 146]. Moreover, surfactant type and concentration are the other key factors which affect the size, stability, and polydispersity of the droplets. The droplet size and polydispersity index decreased with increase in surfactant concentration [53]. The size and size distribution of encapsulated LC droplets can have a significant effect on the electro-optical properties of the films. Large area applicability of the emulsion system enhances the range of useful applications of PDLCs, ranging from switchable light modulators [71, 133], smart Windows [127] and information displays [134], as well as holographically formed optical elements and devices [135, 137, 142–144]. In PDLC systems, LC droplets are dispersed in a polymer film and these LCs can be oriented in the polymer droplets leading a switching from scattering to transparent states or vice versa with an applying electric field. There is a mismatching of refractive indices in the field-off state and these refractive indices of LC and the polymer match in the field-on state [53, 115, 127, 132–134, 146]. This phenomenon gives the electro-optical performance of the corresponding device. This electro-optical performance of displays and smart windows can be improved by replacing the nematic LCs with BPLCs, and the dispersed or encapsulated BPLCs leads a development in the original optical and E-O properties with an external electical field [53].

A recent study has appeared recently demonstrating polymer-encapsulated blue phase (PEBP) liquid crystal films were prepared via solvent evaporation-induced phase separation of a mixture of blue phase liquid crystal (BPLC) and polymer latex [33]. It was observed that the PEBP films induced the birefringence between crossed polarizers at low switching voltage and with fast response time. PEBP samples generated considerably large Kerr constants, in the range of $1.83*10^8$–$20*10^8$ V^{-2} m (at 633 nm), which are about 10 times higher than those of the reported PSBPs [139]. Therefore, PEBP liquid crystal films are strong candidates for next-generation displays as a result of the outstanding E-O properties of blue phases [33].

3.6. Polymer stabilization of polymer encapsulated blue phase liquid crystals

A novel report has appeared recently suggesting the stabilization of encapsulated BPLC droplets is useful technique to expand their wide temperature range and improve the E-O properties of PDLCs using BPLCs [33]. Besides, polymer stabilization is one of the most effective methods to expand the thermal stability of BPLCs. However, liquid crystal/polymer composites can be classified into two distinct groups as polymer-dispersed liquid crystal (PDLC) and polymer-stabilized liquid crystal (PSLC). Both PDLC and PSLC methods are usually operated between a transparent state and an opaque state [53]. In the PDLC systems, droplets of liquid crystal are dispersed in a polymer film, which can be switched from scattering state to transparent state or vice versa with an applied electric field. In the reported study, composite materials of polymer-dispersed blue phase (PDBP) LCs were studied by combining PDLC films and BPLCs using two preparation methods, that is, polymer encapsulation and polymer stabilization in order to fabricate polymer encapsulated–polymer-stabilized blue-phase (PEPSBP) LC droplets[33]. Encapsulated droplets were stabilized via the polymerization of reactive monomers after they were produced in a polyvinyl alcohol solution by emulsification. It was reported that polymer stabilized droplets caused an expansion of the BP temperature range from 53 to below 0°C. Moreover, this study concluded that low switching voltage and fast response time based on the decrement in the interfacial energy of polymer encapsulated and stabilized BPLC droplets. Furthermore, stabilization and positions of droplets in the aqueous phase have a great attraction for the researchers [142–146].

Kemiklioglu et al. firstly demonstrated the polymer-dispersed blue-phase liquid-crystal films between two indium-tin-oxide-coated conductive substrates by switching between light scattering and transparent states with applying an electric field across the films [35].

They experimentally investigated the photoinitiator effect on the electro-optical properties of the polymer-dispersed blue phase liquid crystals as well as the ratio between the crosslinking agent and the monomer. They showed that the increasing monomer concentration reduces the switching voltage of the corresponding device. Moreover, the increment of the monomer concentration in the polymer-dispersed blue phase liquid crystal samples leads an increment in the contrast ratio. All these significant improvements in the electro-optical properties of polymer-dispersed blue-phase liquid crystal devices are promising for new electro-optical applications [35].

Author details

Emine Kemiklioglu

Address all correspondence to: emine.kemiklioglu@cbu.edu.tr

Department of Bioengineering, Faculty of Engineering, Manisa Celal Bayar University, Manisa, Turkey

References

[1] Friedel G. Annales de Physique. 1922;**18**:273

[2] Streyer L. Chapter 10, Introduction to Biological Membranes Biochemistry. New York: W. H. Freeman; 1975

[3] Reinitzer F. Monatshefte fuer Chemie. 1888;**9**:421

[4] Planer P. Liebigs Annalen. 1861;**118**:25

[5] Lehmann O. Uber flissende Kristalle. The Journal of Physical Chemistry. 1889;**4**:462

[6] Chandrasekhar S. Liquid Crystals. Cambridge: Cambridge University Press; 1992

[7] Beliakov VA. Optics of Complex Structured Periodic Media. New York: Springer-Verlag; 1992

[8] de Gennes PG, and Prost J. The Physics of Liquid Crystals, Clarendon Press, Oxford, 1993

[9] Chilaya G. Induction of chirality in nematic phases. Revue de Physique Appliquée. 1981;**16**:193

[10] Solladie G, Zimmermann RG. Angewandte Chemie. 1985;**96**:335

[11] Gottarelli G, Hibert M, Samori B, Solladie G, Spada GP, Zimmermann R. Journal of the American Chemical Society. 1983;**105**:7318

[12] Chilaya GS, Lisetski LN. Soviet Physics Uspekhi. 1981;**24**:496

[13] Chilaya GS, Lisetski LN. Molecular Crystals and Liquid Crystals. 1986;**140**:243

[14] Heppke G, LoÈtzch D, Oestrecher F. Zeitschrift für Naturforschung. 1986;**42a**:279

[15] Gorecka E, Cepic M, Mieczkowski J, Nakata M, Takezoe H, Zeks B. Physical Review E. 2003;**67**:061704:1-061704:5

[16] Nakata M, Takanishi Y, Watanabe J, Takezoe H. Physical Review E. 2003;**68**:041710:1-041710:6

[17] Crooker P. In: Kitzerow H-S, Bahr C, editors. Chirality in Liquid Crystals. New York: Springer; 2001

[18] Kikuchi H, Yokota M, Hisakado Y, Yang H, Kajiyama T. Nature Materials. 2001;**1**:64

[19] Hisakado Y, Kikuchi H, Nagamura T, Kajiyama T. Advanced Materials. 2005;**17**:96

[20] Haseba Y, Kikuchi H, Nagamura T, Kajiyama T. Advanced Materials. 2005;**17**:2311

[21] Iwata T, Suzuki K, Amaya N, Higuchi H, Masunaga H, Sasaki S, Kikuchi H. Macromolecules. 2009;**42**:2002

[22] Kemiklioglu E, Hwang JY, Chien L-C. Physical Review E. 2014;**89**:042502

[23] Zhang T, Ge J, Hu Y, Yin Y. Nano Letters. 2007;**7**:3203

[24] Banerjee SS, Chen DH. Chemistry of Materials. 2007;**19**:6345

[25] Suda M, Kameyama N, Suzuki M, Kawamura N, Einaga Y. Angewandte Chemie, International Edition. 2007;**47**:160

[26] Rangappa D, Naka T, Kondo A, Ishii M, Kobayashi T, Adschiri T. Journal of the American Chemical Society. 2007;**129**:11061

[27] Taguchi M, Takami S, Adschiri T, Nakane T, Sato K, Naka T. CrystEngComm. 2841;**13**:2011

[28] Shaffter MSP, Windle AH. Advanced Materials. 1999;**11**:937

[29] Velasco-Santos C, Martinez-Hernandez AL, Fisher F, Rouff RS, Castano VM. Journal of Physics D: Applied Physics. 2003;**36**:1423

[30] Castles F, Morris SM, Terentjev EM, Coles HJ. Physical Review Letters. 2010;**104**:157801

[31] Gardiner D, Morris SM, Castles F, Qasim MM, Kim W-S, Choi SS, Par HJ, Chung I-J, Coles HJ. Applied Physics Letters. 2011;**98**:263508

[32] Hussain Z, Masutani A, Danner D, Pleis F, Hollfelder N, Nelles G, Kilickiran P. Journal of Applied Physics. 2011;**109**:114513

[33] Kemiklioglu E, Chien L-C. Applied Physics Express. 2014;**7**:091701

[34] Kemiklioglu E, Chien L-C. Liquid Crystals. 2017;**44**:722

[35] Kemiklioglu E, Chien L-C. EPJ E-Soft Matter & Biological Physics. 2017;**40**:11524

[36] Martinez-Gonzalez JA, Zhou Y, Rahimi M, Bukusoglu E, Abbott NL, de Pablo JJ. PNAS. 2015;**112**:13195

[37] Bukusoglu E, Wang X, Martinez-Gonzalez JA, de Pablo JJ, Abbott NL. Advanced Materials. 2015;**27**:6897

[38] Pasteur L, Hebd CR. Sean Academy of Sciences Paris. 1848;**26**:535

[39] Eliel EL, Wilen SH, Mander LN. Stereochemistry of Organic Compounds. New York: Wiley; 1994

[40] Nasipuri D. Stereochemistry of Organic Compounds: Principles and Applications. New Delhi: Wiley Eastern; 1991

[41] Collins AN, Sheldrake GN, Crosby J, editors. Chirality in Industry: The Commercial Manufacture and Applications of Optically Active Compounds. Chichester: Wiley; 1992

[42] Hegstrom RA, Kondepudi DK. Scientific American. January 1990;**1990**:98

[43] Bouchiat M-A, Pottier L. Science. 1986;**234**:1203

[44] de Gennes PG, Prost J. The Physics of Liquid Crystals, 2nd, Oxford University Press, Oxford, 1993

[45] Chandrasekhar S. Liquid Crystals. 2nd ed. Cambridge: Cambridge University Press; 1992

[46] Demus D, Goodby J, Gray GW, Spiess H-W, Vill V, editors. Handbook of Liquid Crystals. Vol. 1-3. Weinheim: Wiley-VCH; 1998

[47] Elston S, Sambles R, editors. The Optics of Thermotropic Liquid Crystals. Taylor & Francis; London, 1998

[48] Goodby JW. Journal of Materials Chemistry. 1991;**1**:307-318

[49] Goodby JW, Slaney AJ, Booth CJ, Nishiyama I, Vuijk JD, Styring P, Toyne KJ. Molecular Crystals and Liquid Crystals. 1994;**243**:231-298

[50] Pelzl G, Diele S, Weissflog W. Advanced Materials. 1999;**11**:707-724

[51] Takezoe H, Takanishi Y. Japanese Journal of Applied Physics. 2006;**45**:597-625

[52] Link DR, Natale G, Shao R, Maclennan JE, Clark NA, Korblova E, Walba DM. Science. 1997;**278**:1924-1927

[53] Kemiklioglu E. Polymer stabilized and dispersed blue phases [Dissertation]. USA: Kent State University; 2014

[54] Yang DK, Doane JW, Yaniv Z, Glasser J. Applied Physics Letters. 1994;**64**:1905

[55] Petri A, BraÈuchle C, Leigeber H, Miller A, Witzel H-P, Kreuzer F-H. Liquid Crystals. 1993;**15**:113

[56] Palfy-Muhoray P, Singer KD. Optics & Photonics News. Sept 1995;**1995**:17

[57] Schlichting W, Faris S, Li L, Fan B, Kralik J, Haag J, Lu Z. Molecular Crystals and Liquid Crystals. 1997;**301**:771

[58] Lu SY, Chien LC. Optics Letters. 2010;**35**:562

[59] Etchegoin P. Physical Review E. 2000;**62**:1435

[60] Barbet-Massin R, Cladis P, Pieranski P. Physical Review A. 1984;**30**:1161

[61] Lehmann O. Zeitschrift für Physikalische Chemie. 1906;**56**:750

[62] Gray GW. Journal of the Chemical Society. 1956;**3733**

[63] Wright DC, Mermin ND. Reviews of Modern Physics. 1989;**61**:385

[64] Meiboom S, Sethna JP, Anderson WP, Brinkman WF. Physical Review Letters. 1981; **46**:1216

[65] Coles HJ, Pivnenko MN. Nature. 2005;**436**:997

[66] Choi H, Higuchi H, Ogawa Y, Kikuchi H. Applied Physics Letters. 2012;**101**:131904

[67] Gerber PR. Molecular Crystals and Liquid Crystals. 1985;**116**:197

[68] Alexander GP, Yeomans JM. Physical Review Letters. 2007;**99**:067801

[69] Alexander GP, Yeomans JM, Zumer S. Faraday Discussions. 2010;**144**:159

[70] Yoshizawa A, Sato M, Rokunohe J. Journal of Materials Chemistry. 2005;**15**:3285

[71] Yelamaggad CV, Shashikala IS, Liao G, Shankar Rao DS, Prasad SK, Li Q, Jakli A. Chemistry of Materials. 2006;**18**:6100

[72] Wang L, He W, Xiao X, Meng F, Zhang Y, Yang P, Wang L, Xiao J, Yang H, Lu Y. Small. 2189;**8**:2012

[73] Szleifer I, Rozen RY. Polymer. 2005;**46**:7803

[74] Dierking I, Scalia G, Morales P. Journal of Applied Physics. 2005;**97**:044309

[75] Iijima S. Nature. 1991;**354**:56

[76] Jeong SJ, Sureshkumar P, Jeong KU, Srivastava AK, Lee SH, Jeong SH, Lee YH, Lu R, Wu ST. Optics Express. 2007;**15**:11698

[77] Lagerwall J, Scalia G, Haluska M, Dettlaff-Wegliskowska U, Roth S, Giesselmann F. Advanced Materials. 2007;**19**:359

[78] Trushkevych O, Collings N, Hasan T, Scardaci V, Ferrari AC, Wilkinson TD, Crossland WA, Milne WI, Geng J, Johnson BFG, Macaulay S. Journal of Physics D. 2008;**41**:125106

[79] Basu R, Iannacchione GS. Journal of Applied Physics. 2008;**104**:114107

[80] Javadian S, Dalir N, Kakemam J. Liquid Crystals. 2017. DOI: 10.1080/02678292.2016.1278051

[81] Lisetski LN, Minenko SS, Samoilov AN, Lebovk NI. Journal of Molecular Liquids. 2017;**235**:90

[82] Huang C-Y, Hu CY, Pan HC, Lo KY. Japanese Journal of Applied Physics. 2005;**44**:8077

[83] Lee W, Wang CY, Shih YC. Applied Physics Letters. 2004;**85**:513

[84] Peterson MSE, Georgiev G, Atherton TJ, Cebe P. Liquid Crystals. 2017. DOI: 10.1080/02678292.2017.1346212

[85] Huang C-Y, Pan HC, Hsiem CT. Brief communication. Japanese Journal of Applied Physics. 2006;**45**:6392

[86] Chan C, Crawford G, Gao Y, Hurt R, Jian K, Li H, Sheldon B, Sousa M, Yang N. Carbon. 2440;**43**:2005

[87] Baik I-S, Jeon SY, Jeong SJ, Lee SH, An KH, Jeong SH, Lee YH. Journal of Applied Physics. 2006;**100**:074306

[88] Kemiklioglu E, Chien LC. Liquid Crystals. 2016;**43**:1067

[89] Jeon SY, Park KA, Baik I-S, Jeong SJ, Jeong SH, An KH, Lee SH, Lee YH. Nano. 2007;**2**:41

[90] Srivastava AK, Jeong SJ, Lee MH, Lee SH, Jeong SH, Lee YH. Journal of Applied Physics. 2007;**102**. DOI: 043503

[91] Chen HY, Lee W. Optical Review. 2005;**12**:223

[92] Lee W, Shih YC. Journal of the Society for Information Display. 2005;**13**:743

[93] Lu SY, Chien LC. Optics Express. 2008;**16**:12777

[94] Hur S-T, Gim M-J, Yoo HJ, Choi S-W, Takezoe H. Soft Matter. 2011;**7**:8800

[95] Hikmet RAM. Journal of Applied Physics. 1990;**68**:4406

[96] Mizunumuna T, Oo TN, Nagano Y, Ma H, Haseba Y, Higucki H, Okumura Y, Kikuchi H. Optical Materials Express. 2011;**1**:1561

[97] Kikuchi H, Hisakado Y, Uchida K, Nagamura T, Kajiyama T. Proceedings of SPIE. 2004; **5518**:182

[98] Crawford GP, Zumer S. Liquid Crystals in Complex Geometries. Taylor and Francis Ltd.; London, 1996

[99] Rao L, Ge Z, Wu ST. Journal of Display Technology. 2010;**6**:115

[100] Lin YH, Chen HS, Lin HC, Tsou YS, Hsu HK, Li WY. Applied Physics Letters. 2010; **96**:113505

[101] Iwata T, Takaoka T, Suzuki K, Amaya N, Higuchi H, Kikuchi H. Molecular Crystals and Liquid Crystals. 2007;**470**:11

[102] Choi SW, Yamamoto SI, Haseba Y, Higuchi H, Kikuchi H. Applied Physics Letters. 2008;**92**:043119

[103] Ge Z, Gauza S, Jiao M, Xianyu H, Wu ST. Applied Physics Letters. 2009;**94**:101104

[104] Ge Z, Rao L, Gauza S, Wu ST. Journal of Display Technology. 2009;**5**:250

[105] Rao L, Ge Z, Wu ST, Lee SH. Applied Physics Letters. 2009;**95**:231101

[106] Jiao M, Li Y, Wu ST. Applied Physics Letters. 2010;**96**:011102

[107] Chen KM, Gauza S, Xianyu H, Wu ST. Journal of Display Technology. 2010;**6**:49

[108] Yan J, Cheng HC, Gauza S, Li Y, Jiao M, Rao L, Wu ST. Applied Physics Letters. 2010;**96**:071105

[109] Yoon S, Kim M, Kim MS, Kang BG, Kim MK, Srivastava AK, Lee SH, Ge Z, Rao L, Gauza S, Wu ST. Liquid Crystals. 2010;**37**:201

[110] Doane JW, Vaz NA, Wu BG, Zumer S. Applied Physics Letters. 1986;**48**:269

[111] Drazic PS. Liquid Crystal Dispersions. Singapore: World Scientific; 1995

[112] Ono H, Kawamura T, Frias NM, Kitamura K. Journal of Applied Physics. 2000;**88**:3853

[113] Bowley CC, Crawford GP. Applied Physics Letters. 2235;**76**:2000

[114] Sutherland RL, Tondiglia VP, Natarajan LV, Bunning TJ, Adams WW. Applied Physics Letters. 1994;**64**:1074

[115] Wu BG, Erdmann JH, Doane JW. Liquid Crystals. 1989;**5**:1453

[116] Jain SC, Thakur RS, Lakshmikumar ST. Journal of Applied Physics. 1993;**73**:3744

[117] Choi C-H, Kim B-K, Kajiyama T. Molecular Crystals and Liquid Crystals. 1994;**247**:303

[118] Amundson K, Blaaderen A, Wiltzius P. Physical Review E: Statistical Physics, Plasmas, Fluids, and Related Interdisciplinary Topics. 1997;**55**:1646

[119] Kalkar AK, Kunte VV, Deshpande AA. Journal of Applied Polymer Science. 1999;**74**:3485

[120] Mormile P, Musto P, Petti L, Ragosta G, Villano P. Applied Physics B: Lasers and Optics. 2000;**70**:249

[121] Fergason JL. Journal of the Society for Information Display. 1985;**16**:68

[122] Doane JW, Golemme A, West JL, Jr Whitehead JB, Wu B-G. Molecular Crystals and Liquid Crystals. 1988;**165**:46

[123] Kashima M, Cao H, Liu H, Meng Q, Wang D, Li F, Yang H. Liquid Crystals. 2010;**37**:339

[124] Whitehead JB, Gill NL. Proceedings of SPIE. 2005;**5741**:12

[125] Drzaic PS. Liquid Crystal Dispersions. Singapore, World Scientific Publishing, Chapter 4; 1995

[126] Vaz NA. Proceedings of SPIE. 1989;**1080**:2

[127] Drzaic PS. Journal of Applied Physics. 1986;**60**:2142

[128] Cupelli D, Nicoletta FP, Manfredi S, Vivacqua M, Formoso P, De Filpo G, Chidichimo G. Solar Energy Materials & Solar Cells. 2009;**93**:2008

[129] White TJ, Natarajan LV, Tondiglia VP, Lloyd PF, Bunning TJ, Guymon CA. Polymer. 2007;**48**:5979

[130] Hsiao V, Lu C, He GS, Pan M, Cartwright AN, Prasad PN. Optics Express. 2005;**13**:3787

[131] Landfester K. Macromolecules. 1999;**32**:2679

[132] Izquierdo P, Esquena J, Tadros Th. F., Dederen C, Garcia-Celma MJ, Azemar N, Solans C. Langmuir. 2012;**18**:26

[133] Maschke U, Coqueret X, Benmouna M. Macromolecular Rapid Communications. 2002; **23**:159

[134] Wu ST, Yang DK. Reflective Liquid Crystal Displays. Wiley; England, 2001

[135] Cheoi YC, Yang DK. Applied Physics Letters. 2011;**98**:023502

[136] Drzaic PS. Liquid Crystals. 1988;**3**:1543

[137] Fergason JL. US Patent, 4, 1984, 435 047

[138] Bunning TJ, Natarajan LV, Tondiglia VP, Sutherland RL, Vezie DL, Adams WW. Polymer. 1995;**36**:2699

[139] Bunning TJ, Natarajan LV, Tondiglia VP, Sutherland RL. Annual Review of Materials Science. 2010;**30**:83

[140] Klosowicz SJ, Aleksander M. Opto-Electronics Review. 2004;**12**:305

[141] Serbutoviez C, Kloosterboer JG, Boots HMJ, Paulissen FAMA, Touwslager FJ. Liquid Crystals. 1997;**22**:145

[142] Qi J, Crawford GP. Displays. 2004;**25**:177

[143] Wang X, Bukusoglu E, Abbott NL. Chemistry of Materials. 2017;**29**:53

[144] Bukusoglu E, Wang X, Zhou Y, Martínez-González JA, Rahimi M, Wang Q, de Pablo JJ, Abbott NL. Soft Matter. 2016;**12**:8781

[145] Martínez-González JA, Zhou Y, Rahimi M, Bukusoglu E, Abbott, NL, de Pablo JJ. PNAS. 2015;**112**:13195

[146] Bukusoglu E, Wang X, Martínez-González JA, de Pablo JJ, Abbott NL. Advanced Materials. 2015;**27**:6892

Chapter 4

Phase Transition effect on the Parametric Instability of Liquid Crystals

Martin Hernández Contreras

Additional information is available at the end of the chapter

http://dx.doi.org/10.5772/intechopen.70240

Abstract

We review advances in the last few years on the study of the Faraday instability onset on thermotropic liquid crystals of nematic and smectic A types under external magnetic fields which have been investigated with a linear stability theory. Especially, we show that thermal phase transition effects on nematics of finite thickness samples produce an enhanced response to the instability as a function of the frequency of Shaker's movement. The linear stability theory has successfully been used before to study dynamical processes on surfaces of complex fluids. Consequently, in Section 1, we show its extension to the study of the instability in the nematics, which set the theoretical framework for its further application to smectics or other anisotropic fluids such as lyotropic liquid crystals. We present the dispersion relationships of both liquids and its dependence on interfacial elastic parameters governing the surface elastic responses to external perturbations, to the sample size and their bulk viscosities. Finally, we point out the importance of following both experimental and theoretical analysis of various effects that needs to be incorporated into this model for the quantitative understanding of the hydrodynamics behavior of surface phenomena in liquid crystals.

Keywords: liquid crystals, parametric instability, surface hydrodynamics, phase transition, nonlinear waves, complex fluids

1. Introduction

The Faraday wave instability emerges as a macroscopic nonlinear behavior of the dynamics at interfaces of different liquids and vapor [1–22]. It appears when the vessel containing the liquid is vibrated vertically with a given acceleration until the quiescent equilibrium interface develops unstable surface waves. It has been observed in Newtonian fluids [6], but their most interesting realizations occur in complex fluids where their viscoelastic responses are present due to different time scales associated with the molecular relaxation processes [8, 16]. Therefore, it is important to determine how the onset of the Faraday instability is determined by the underlying bulk fluid elasticity. The basic study of this phenomenon poses challenges to hydrodynamic and statistical theories that need to be adapted or developed to explain the onset of the instability. The description of the Faraday instability has recently motivated the advance of new experiments [23–26] and theoretical approaches for anisotropic fluids such as liquid crystals [27–40]. Also, it seems that its comprehensive experimental and theoretical investigation may have a major impact on the development of sensor technologies based on interfaces with biological or chemical components of practical interest. Absorbed molecules at the interface provide a surface coverage with absorbed molecular species that can present activity. Their impact on the instability has not well been understood yet. Liquid crystals remain as an ideal complex fluid where a controlled fine tuning of the cohesive energy of the absorbed molecules of interest and the nematogen's bulk average orientations imposed by surface treatment of their anchoring energy [41–46] can be experimentally reached. Reports on birefringent experiments on a lyotropic suspension of fd virus describe the effect of bulk microrheology on the surface wave [23]. The hysteresis of the wave amplitude under the harmonic external driving acceleration, and how the imposed perturbation shear lowers the viscosity for increasing driving impulse were observed. Such a rheological response of the liquid crystal led to a hydrodynamically induced transition from isotropic to nematic phase change, which produces the formation of patches at the deformed crest of the interface. In this chapter, we review recent work on the Faraday instability on thermotropic liquid crystals, and the effect that a thermal phase transition experienced by a nematic liquid crystal toward its isotropic state has on the instability onset. Thus, we review our understanding of thermal phase transitions that produce enhanced response on dynamical properties at the interface of thermotropic liquid crystals. The liquid crystal is subjected to vertical vibrations of the container which induce hydrodynamics instability on its surface. Temperature variations produce phase changes on liquid crystals [47]. We present our results on the liquid crystal phase change effect in the dynamics of the Faraday wave instability [39]. We focus our discussion on this coupled phenomenon on a hydrodynamic level of description based on the Navier-Stokes equation for the field velocity response of the liquid crystal. Our presentation incorporates the constitutive equation for taking into account properly the heat transfer into the liquid crystal which drives the phase transition. Also, the significant effects of various elastic parameters such as surface tension, bending modulus, and interfacial elasticity of the interface on the sustained wave are discussed. Further discussion shows how those elastic parameters determine the onset of the hydrodynamic instability through the critical acceleration of the surface wave, which is temperature dependent when the liquid crystal experiences a phase change. To set the theoretical framework, in Section 2 the phenomenological free energy of layers and surface deformations and its dependence on the elastic parameters are introduced. In Section 2.1, the hydrodynamic level of description of a model nematic liquid crystal is made. This section includes the mean field viscous shear stress tensor of the liquid crystal and the corresponding boundary conditions. In Section 2.2, we discuss the case of model nematic with nematogens aligned perpendicular to an external magnetic field but parallel to the surface. Such a model represents an isotropic liquid crystal case. We further present an analysis of the critical acceleration as a function of temperature variation from nematic up to the phase transition to the isotropic liquid crystal phase in Section 2.3. In Section 2.4, we analyze the dispersion relationship as a function of all elastic parameters for a specific temperature. We then present in Section 2.5 the dispersion relation of an isotropic liquid. In Section 3, we discuss the occurrence of a parametric instability in smectic A liquids. In Sections 3.1 and 3.2, the finite thickness layer dispersion relationships for sustained Faraday surface waves in two configurations of the director on the magnetic field that orient the nematogens are provided. In Section 4, we discuss the experimental results in the literature on the phase transition effect on Faraday waves due to changes in particle concentration in a lyotropic liquid crystal of fd virus. In Section 5, a conclusion paragraph is provided. Finally, there is a list of the most relevant and updated list of references.

2. Thermotropic nematic liquid crystal layers

We consider a finite thickness layer of depth L of nematic fluid with its molecules been oriented parallel to the liquid-air interface in the X-axis direction by an external magnetic field H as shown in **Figure 1**.

The vector position giving the local elastic response of the interface has components $\zeta(x, y, t)$ and $\xi(x, y, t)$ to a normal and in-plane perturbations. Those deformation fields are determined by a surface tension γ and interfacial shear elasticity ε produced by adsorbed surfactants at the interface which lower the surface tension. Perturbations of the interface produce coupling of vertical and lateral deformation whose strength is given by a parameter λ. A splay module K determines the curvature distortion of the nematogen's layers deformation, and a compressibility modulus B yields its compression. The magnetic field orients the nematogens as they are characterized by a magnetic susceptibility χ_a. Thus, the elastic-free energy of layers distortion is given by [48]

$$F_{\text{bulk}} = \frac{1}{2}\int d^3 r \left\{B(\partial_x u)^2 + K\left(\partial^2{}_y u + \partial^2{}_z u\right)^2 + \chi_a H^2\left[\left(\partial_y u\right)^2 + \left(\partial_z u\right)^2\right]\right\}. \quad (1)$$

whereas the interface elastic-free energy may be written approximately as [49]

$$F_{surface} = \frac{1}{2}\int d^2 r \left\{ \begin{array}{l} \gamma\left((\partial_x\zeta)^2 + \left(\partial_y\zeta\right)^2\right) + \varepsilon\left((\partial_x\xi)^2 + \left(\partial_y\xi\right)^2\right) + \kappa'\left(\partial^2{}_x\zeta + \partial^2{}_y\zeta\right)^2 \\ -2\lambda(\partial_x\xi)\left(\partial^2{}_x\zeta + \partial^2{}_y\zeta\right)^2 \end{array} \right\}. \quad (2)$$

The molecular field $u(x, y, z, t)$ takes into account the bulk elastic deformation of the stack of layers and which is caused by the acceleration, thermal fluctuations, and for smectics, also due to the movement (permeation) of molecules between layers $\dot{u} - v_x = \Lambda_p h$, where Λ_p, h is the permeation length and molecular field, respectively, $\partial_\beta := \partial/\partial_\beta$, $\beta = x, y, z$. If we consider the Faraday wave on a liquid crystal that supports a monolayer, then the surface elastic response is affected by surface tension, dilational, and coupling modulus as given in Eq. (2). Crilly et al. [50] have shown using photon correlation spectroscopy experiments that the thermotropic phase transition on monoglyceride monolayers that originate from the little molecular area fluctuations can be accurately detected. They found that two surface elastic moduli can explain the shear normal to the surface $\gamma = \gamma_0 - \pi + i\omega\gamma'$ and the other governs dilational distortion in

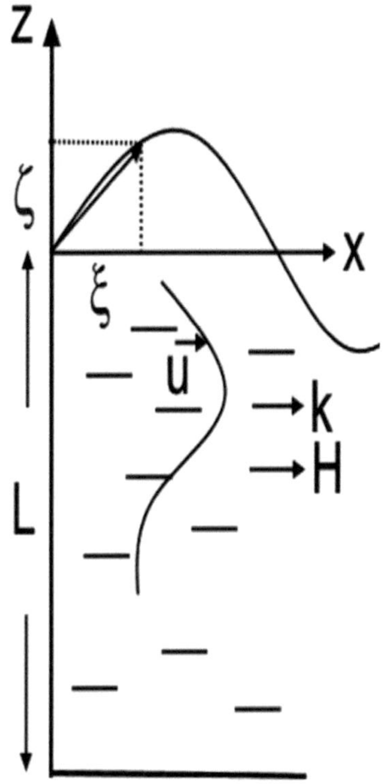

Figure 1. Interface of nematic liquid crystal layer of thickness L and air. A magnetic field **H** orients the nematogens parallel to the wave vector **k** which lies along the X-axis. Gravity modulation acts in the Z-axis direction. A vector position with components (ξ, ζ) designates the inplane and normal elastic deformation of the interface, respectively.

the interfacial plane $\varepsilon=\varepsilon_0+i\omega\varepsilon'$. Both are complex quantities, with $i^2=-1$ and γ', ε' viscous coefficients of a monolayer of surfactants with pressure π. γ_0, ε_0 are unperturbed values of elastic parameters without the monolayer. The imaginary parts are dissipative processes. However, Crilly et al. [50] showed that only the real parts of these moduli determine completely the thermotropic phase change of the surfactant monolayer, thus, our expression in Eq. (2) is valid to incorporate surface viscoelastic properties. The third elastic parameter λ that couples normal and tangential elasticity seems not to be detected experimenlly yet. The glycerol monooleate (GMO) monolayer has a phase transition temperature about 15.5°**C**where a lipid undergos an all-trans state at low temperature to a gauche conformation of the lipid (pointing out the chain-melting transition) for temperatures greater than 15.5°**C**. We note, however, that the liquid crystal methoxy benzylidine butyl aniline (MBBA) has a transition temperature of 45°**C** [51]; therefore, one might expect a theory prediction for the effect of thermotropic phase changes of lipid monolayers on the Faraday waves to occur prior to the phase transition in the nematic-isotropic liquid. Such a study is possible to perform either experimentally and theoretically. From the experimental viewpoint, Crilly et al. [50] measured the elastic properties of the monolayer and provided results similar to those of **Figures 2** and **3** (see **Figure 1** of [50]), both well below and above the transition temperature as it is required in the modeling approach that is presented in this review. However, the determination of the

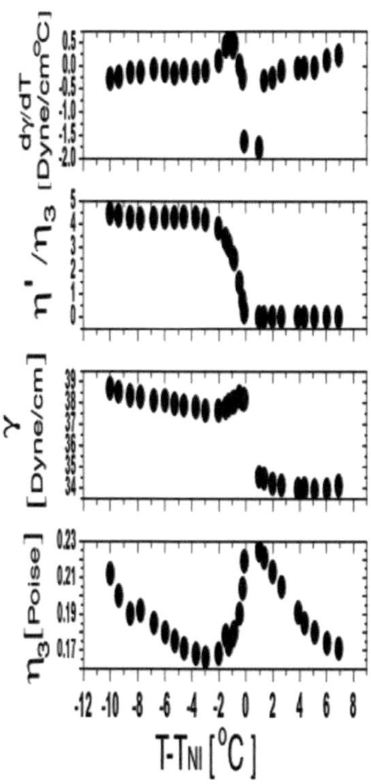

Figure 2. From bottom up are the interpolated values of viscosities η_3, surface tension γ, viscosity η'/η_3 of MBBA as a function of the nematic-isotropic phase transition temperature. These values were interpolated from the experimental studies of [54].

elastic properties of GOM monolayer and their variation as a function of temperature as in **Figures 1–3** of paper [50], when the monolayer is deposited on surfaces of liquid crystals, are still lacking, and their complete experimental measurement as a function of temperature would be quite valuable. Such experimental studies would pave the way for models which can be extended to understand thermotropic phase changes on monolayers and their impact on liquid crystal surface dynamics. Presently, one can resort to several useful models that study the surface hydrodynamics such as parametric instability, thermal capillary waves on polymer solutions and gels [13, 52], and coarse-grained effective field theories [27] and atomistic simulation techniques [37] and lattice hydrodynamics [17]. These methods have been demonstrated to provide a rich description of the effects of phase transition at interfaces of liquid crystals. From the experimental side, there is an accumulated knowledge of prototype systems that are now well characterized with tested experimental techniques. Those studies provide accurate information on static (elastic parameters [50, 53]). Also, the time-dependent properties (power spectrum of the intensity of scattered light) at interfaces of isotropic and simple liquids supporting surfactant monolayers [54] have been reported. Those studies need to be extended to include anisotropic liquids. An example of two such comprehensive experimental studies was performed by Langevin [54] in the 1970s where a series of systematic experiments on the variation of viscous and elastic parameters of some liquid crystals

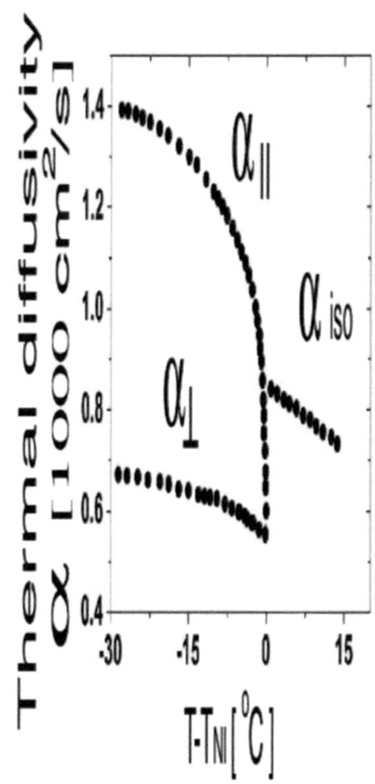

Figure 3. Interpolation of the entropy of deformation that contributes to Marangoni flow of nematogens during the thermal phase transition of MBBA from nematic to isotropic phase as a function of transition temperature. These values were interpolated from the experimental studies of [59].

experiencing thermotropic phase change were reported, and the thorough experiments by Earnshaw [55] on surfaces of water supporting monolayers of surfactants. These researches provided tabulated comprehensive data of the elastic parameters and viscous properties on bulk and in the plane of the interface in a wide range of temperatures and not only at a single specific temperature of interest. Our phenomenological description of the elastic behavior of the interface given in Eq. (2) relies on models developed for polymer solution surfaces with absorbed non-active surfactants [49]. They are simple models that do not incorporate the anchoring energy of nematogens on the liquid crystal side toward the lipid monolayer and its strength of variation. A more reliable model that takes into account surface elasticity and the modification of anchoring energy of nematogens due to absorption of lipids at surfaces was developed in [34–36], and the application of this method to study theoretically the main surface modes of thermal waves at surfactant-laden liquid-liquid crystal interfaces was made in [33]. In reference [33], it was predicted that the thermal wave has a dispersion law that involves energy dissipation through anisotropic coefficients due to compression of the surfactant layer. And a new relaxation mode of the director appears due to the boundary condition of anchoring at the interface in a perpendicular direction to the plane.

The experimental elucidation of thermotropic phase changes in monolayers of lipids deposited on top of liquid crystals and its impact on interface-laden of liquid crystals is an open topic to be investigated yet. A possible program for that task would involve the determination of the variation of the surface area of mono- or bilayers of surfactants like GMO which may be supported in isotropic, nematic, or smectic phases of liquid crystals. Moreover, the surface tension and dilational modulus versus temperature in an ample range around the critical temperature have been measured. Experimental techniques of photon correlation spectroscopy and surface quasielastic light scattering have been used in this context for isotropic fluids. Moreover, the entropy of formation of the film (see **Figure 2** in Ref. [50] for temperature gradient of surface tension), which is related to the Marangoni effect of diffusion of molecules in the surface, should be measured correspondingly. Such experiments might observe the phase transition in the film and also determine the surface viscosities ignored in our model of Eq. (2). It is expected that the viscosities will help in clarifying the nature of the monolayer phases [54–56]. A comprehensive experimental determination of the bulk Leslie viscosities of the most known liquid crystals in an ample interval of temperatures near their critical phase transition is scarce [53]. Their values can lead to the prediction of enhanced macroscopic hydrodynamic response on the parametric surface dynamics of liquid crystals [39, 40], or polymer melts [57]. A similar experimental program may be set up for studying lyotropic liquid crystals to determine the viscosities and elastic parameters as a function of temperature. Those thermodynamic parameters enter as inputs in models like the one developed in this section. As an alternative to experimental measurements of bulk and interfacial viscoelastic properties of liquid crystals, one can also resort to molecular dynamics simulation. This computational technique allows the determination of the bulk and surface elastic parameters for certain molecular models of liquid crystals where the nematogens are under pairwise forces of the Weeks-Chandler-Andersen type, and the particles of ellipsoidal shape constitute the colloidal suspensions. From the practical point of view, it is known that the viscoelastic behavior of films of surfactants has very practical consequences, for instance, the electrical properties of GOM films depend on the formation of pores arising from significant fluctuations in the area near the phase transition [37, 38].

2.1. Parallel magnetic field to wave vector and along the *X*-axis

In this section, we present the hydrodynamic description of the velocity field of the center of mass of an infinitesimal volume element containing nematogens and its corresponding boundary conditions on a fluid-air interface [39]. We consider a nematic layer in contact with air, which has an equilibrium interface that is located at position $z=0$ and has a depth L and infinite lateral extension in order to avoid viscous boundary layer effects due to the wall of the container and also beneth the surface as it is shown in **Figure 1**. This model can be generalized to liquid-liquid interfaces that separate two phases of a liquid crystal. To ignore the finite size effect of the vessel, we assume that it has an infinite lateral extension. The fluid is subjected to vertical cosine-like vibration with an acceleration $g(t)=g-a\cos(\omega t)$, where a is the acceleration of the shaker in a reference frame fixed to the container. ω is the angular frequency of the oscillation and g being the gravitational acceleration. Because the director of the nematogens remains fixed by the magnetic field, their time variation is neglected in the hydrodynamic equations. Under these conditions, a spatial ripple at the interface is generated which propagates symmetrically in the *X*- and *Y*-axis directions. Therefore, it is sufficient to consider that the surface wave propagates with wave vector $\boldsymbol{k}$ in the *x*-coordinate only and independently of coordinate *y*. The static magnetic field H keeps the director n=(1,0,0) of the nematogens in the *X*-axis direction oriented. Langevin calculated that thermal fluctuations produce periodic distortions of the nematic director with frequencies of strength $\omega_{undulation} = (Kk^2 + \chi_a H \cdot H)/\eta \approx 9.9$ Hz for MBBA nematic liquid with viscosity η. Both the frequency of the mechanical excitation and that of inertial effects $\omega_{inertia} = \eta/\rho L^2 = 10^6 \omega_{undulation}$ are larger than the elastic frequency of director variations. ρ is the density of the nematic. Thus, the nematic director dynamics is neglected in the hydrodynamic description of the generated surface wave. Therefore, the governing equation of the fluid velocity v about the quiescent state of rest is provided by the linearized Navier-Stokes equation which uses a reference frame that moves attached to the container [58]

$$\rho \frac{\partial V}{\partial t} = \nabla \cdot \sigma \tag{3}$$

The total stress tensor of the liquid $\sigma = -p\mathbf{I} + \sigma' + \sigma^r - \rho g(t)\widehat{\mathbf{e}}_\mathbf{z}\widehat{\mathbf{e}}_\mathbf{z}$ has dissipative contributions σ' due to the bulk viscous response and an elastic part σ^r given by the different elastic parameters in the free energy (Eq. (2)). Here, the unit vector $\widehat{\mathbf{e}}_\mathbf{z}$ is directed along the *Z*-axis, and the hydrostatic fluid pressure is p. The unit matrix $\mathbf{I}_{\beta,\delta} = 1$ if $\beta = \delta$ and zero otherwise. For nematic liquids

$$\sigma'_{ij} = \eta' n_i n_j V_{lo} n_l n_o + 2\eta_2 V_{ij} + 2(\eta_3 - \eta_2)(n_i n_l V_{lj} + n_j n_l V_{li}) \tag{4}$$

where the shear viscosities η_2, η_3, and η' are defined in terms of the Leslie coefficients as $\eta_3 = \frac{\alpha_2 + \alpha_5}{2} - \alpha_2\gamma_2/(2\gamma_1)$, $\eta_3 = \frac{\alpha_4}{2}$, $\eta' = \alpha_1 + \gamma_2^2/\gamma_1$, where $\gamma_1 = \alpha_3 - \alpha_2$, and $\gamma_2 = \alpha_3 + \alpha_2$. The strain rate $V_{ij} = (\partial_i V_j + \partial_j V_i)/2$ has components $i, j, l, o = x, y, z$. For frequencies that are much less than the first sound frequency of the liquid, it holds the incompressible condition

$$\nabla \cdot V = 0. \tag{5}$$

In the experiment, a constant and vertical gradient of temperature that produces Marangoni flow with the rate of change of local temperature *T* variation given by the linearized heat diffusion equation [39] is applied from top to the bottom of the vessel

$$\frac{\partial T}{\partial t} = AV_z + \alpha(\partial_x^2 T + \partial_z^2 T) \tag{6}$$

where $A = -3^\circ$C/mm is the temperature gradient per unit length with the heating occurring from the air side. At the bottom solid wall $T = 0^\circ$C. The balance of the normal and tangential shear stresses at the interface $z = 0$ defines the boundary conditions to be

$$\sigma_{zz} = 0, \sigma_{xz} = 0, \sigma_{yz} = 0 \tag{7}$$

Normal interface displacements are balanced by the elastic force obtained from Eq. (2)

$$\sigma^r_{zz} = f_z = \gamma\left(\partial_x^2 + \partial_y^2\right)\zeta + \lambda\left(\partial_x^2 + \partial_y^2\right)\xi. \tag{8}$$

The tangential forces at the interface result from the Marangoni instability due to surface tension variation with temperature, and the in-plane elastic deformation which are obtained from Eq. (2) as

$$\sigma^r_{xz} = f_x = \frac{d\gamma}{dT}[\partial_x T - A\partial_x\zeta] - \varepsilon\left(\partial_x^2 + \partial_y^2\right)\xi + \lambda\left(\partial_x^2 + \partial_y^2\right)\zeta, \sigma^r_{yz} = 0. \tag{9}$$

In this last equation, elongational deformation ε and the coupling with normal deformation through the elastic parameter λ are included. This is the case when one needs to study interfaces that support mono- and bilayers of surface-active surfactants. Because the normal displacement is small compared with the wavelength, one can approximate the velocity field with the rate of surface deformations through the components of the vector that locates the interface positions as follows (see **Figure 1**):

$$\partial_t\zeta = V_z, \quad \partial_t\xi = V_x \text{ at } z = 0, \kappa\partial_z T = 0. \tag{10}$$

In the last equality, κ is the thermal conductivity of a thermally insolated surface with fixed flux. At the botton of the container $z = -L$, there is no slip of fluid

$$V = 0 \tag{11}$$

moreover, there is no penetration of the wall

$$\partial_z V_z = 0. \tag{12}$$

In the application that follows below, we do not consider the effect of the coefficients ε and λ and will be made zero. From the first two conditions of Eq. (7) and with help of Eq. (8), we obtain

$$\eta_3\left[\nabla_\perp^2 - \partial_z^2\right]V_z = \frac{d\gamma}{dT}\partial_x^2[T - A\zeta] \tag{13}$$

From the second identity of Eq. (7), and from Eq. (8) together with Eq. (9), we obtain

$$\left[\rho\partial_t - \eta_2\nabla^2\right]\nabla^2 V_z = (\eta_3 - \eta_2)\nabla^2\left[\partial_x\partial_z V_x - \partial_x^2 V_z\right] + \eta'\partial_x^2\partial_x\partial_z V_x. \tag{14}$$

Now taking the divergence with the gradient operator $\nabla_\perp := (\partial_x, \partial_y)$ of the Navier-Stokes (Eq. (3)) moreover, using the divergence of Eq. (4) yields

$$\nabla_\perp^2 p = \left[\rho\partial_t - \eta_2\nabla^2 - (\eta_3 - \eta_2)\partial_x^2\right]\partial_z V_z + (\eta_3 - \eta_2)\left[\nabla_\perp^2 + \nabla^2\right]\partial_x V_x + \eta'\partial_x^2\partial_x V_x \tag{15}$$

We now replace this expression in the z component of the total stress tensor σ and take its Fourier transform $\tilde{u} = \int d^2 r e^{i\mathbf{k}\cdot\mathbf{r}} u$, $i^2 = -1$ with the result (from now on, we will not use $\tilde{u}$ for any transformed function but just simply u)

$$\left[\partial_t + (\nu_3 + 2\nu_2)k^2 - \nu_2\partial_z^2\right]\partial_z V_z = -(\nu_3 - \nu_2)\left[-2ik^3 + ik\partial_z^2\right]V_x + \nu' ik^3 V_x - g(t)k^2\zeta - \frac{\gamma}{\rho}k^4\zeta. \quad (16)$$

As a consequence, the Fourier transformed form of Eq. (13) is

$$\nu_3\left[k^2 + \partial_z^2\right]V_z = \frac{k^2}{\rho}\frac{d\gamma}{dT}[T - A\zeta], \quad (17)$$

whereas Eq. (14) takes the form

$$\left[\partial_t - \nu_2\left(\partial_z^2 - k^2\right)\right]\left(\partial_z^2 - k^2\right)V_z = (\nu_3 - \nu_2)\left(-\partial_z^2 + k^2\right)\left[ik\partial_z V_x + k^2 V_z\right] + \nu' ik^3\partial_z V_x, \quad (18)$$

With $\upsilon_j := \frac{\eta_j}{\rho}, j = 2, 3, \nu' := \eta'/\rho$. Similarly, Eq. (6) becomes

$$\frac{\partial T}{\partial t} = AV_z + \alpha\left(\partial_z^2 T - k^2 T\right), \quad (19)$$

The acceleration $g(t)$ is a function with period $2\pi/\omega$, and according to Floquet theory the solution of Eqs. (5) and (16)–(19) is the superposition $\zeta(t) = \sum_{n=-\infty}^{\infty}\zeta_n\, exp\,(\mu_n t)$ [6]

where the modes $\mu_n = s + i(n + \alpha_r)\omega$. The quantities ζ_n, s, α_r are real numbers. There are two types of waves Harmonic with frequency equaling the external forcing frequency and determined by $\alpha_r = 0$,with ζ_n equaling its complex conjugate and subharmonic with $\alpha_r = 1/2$, with $\zeta_n = \zeta_{n-1}^*$. The velocity V_z also has a similar Floquet expansion that is substituted in Eqs. (5) and (18) providing the equation for each component of the velocity amplitude V_{zn}

$$\left[\partial_z^4 + b_n\partial_z^2 + c_n\right]V_{zn}(z) = 0, \quad (20)$$

where $b_n = -\left[\frac{\mu_n}{\nu_3} + k^3\left(2 + \frac{\nu'}{\nu_3}\right)\right], c_n = k^2\left[\frac{\mu_n}{\nu_3} + k^2\right]$. A trial solution of Eq. (20) in the form $v_{zn}(z) \sim e^{m(k)z}, C = m^2$ simplifies Eq. (20) to $C^2 + b_n C + c_n = 0$. We denote the two possible modes $C_j := m_j^2, j = 1, 2$ which lead to the solution of Eq. (20) in the form

$$V_{zn}(z) = P_n\cosh(zm_1) + Q_n\sinh(zm_1) + R_n\cosh(zm_2) + S_n\sinh(zm_2). \quad (21)$$

The coefficients P_n, Q_n, R_n, S_n in this solution are derived by the substitution of Eq. (21) in Eqs. (10)–(12) (we recall $\xi = \lambda = 0$ here). Finally, the use of Eq. (21) in Eq. (16) yields the surface amplitude of deformation ζ_n

$$M_n\zeta_n = a(\zeta_{n-1} + \zeta_{n+1}) \quad (22)$$

with

$$M_n = \frac{2}{k}\left\{\begin{array}{l} w_0^2 + \frac{m_1}{k}\left[\mu_n + k^2\left(3\nu_3 + \nu'\right) - \nu_3 m_1^2\right]\frac{Q_n}{\zeta_n} + \\ + \frac{m_2}{k}\left[\mu_n + k^2\left(3\nu_3 + \nu'\right) - \nu_3 m_2^2\right]\frac{S_n}{\zeta_n} \end{array}\right\} \equiv \frac{2}{k}D_n, \quad (23)$$

moreover, the capillary frequency $w_0^2 := gk + \frac{\gamma k^3}{\rho}$. The coefficients are

$$Q_n = \frac{\zeta_n \mu_n A_1}{\left(m_1^2 - m_2^2\right)\left(1 + \frac{k^2\alpha}{\mu_n}\right) de}$$

$$A_1 = -m_2\left(k^2 + m_1^2\right) - m_2\left[k^4\left(\frac{A}{\mu_n \eta_3}\frac{d\gamma}{dT} + 1\right) + k^2 m_1^2\right]\frac{\alpha}{\mu_n} + \left\{k^2 + m_2^2 + \left[k^4\left(\frac{A}{\mu_n \eta_3}\frac{d\gamma}{dT} + 1\right) + k^2 m_2^2\right]\frac{\alpha}{\mu_n}\right\}\cdot \left(m_2 \cosh(Lm_1)\cosh(Lm_2) - m_1 \sinh(Lm_1)\sinh(Lm_2)\right),$$

$$S_n = \frac{\zeta_n \mu_n B_1}{\left(m_1^2 - m_2^2\right)\left(1 + \frac{k^2\alpha}{\mu_n}\right) de}$$

$$de := -m_2 \cosh(Lm_2)\sinh(Lm_1) + m_1 \cosh(Lm_1)\sinh(Lm_2),$$

$$B_1 = -m_1\left(k^2 + m_2^2\right) - m_1\left[k^4\left(\frac{A}{\mu_n \eta_3}\frac{d\gamma}{dT} + 1\right) + k^2 m_2^2\right]\frac{\alpha}{\mu_n} + \left\{k^2 + m_1^2 + \left[k^4\left(\frac{A}{\mu_n \eta_3}\frac{d\gamma}{dT} + 1\right) + k^2 m_1^2\right]\frac{\alpha}{\mu_n}\right\}\cdot \left(m_1 \cosh(Lm_1)\cosh(Lm_2) - m_2 \sinh(Lm_1)\sinh(Lm_2)\right),$$

$$P_n = -\frac{\zeta_n \mu_n}{m_1^2 - m_2^2}\left[k^2 + m_2^2 + k^4 \frac{A}{\mu_n \eta_3}\frac{d\gamma}{dT}\frac{(\alpha/\mu_n)}{\left(1 + k^2\frac{\alpha}{\mu_n}\right)}\right],$$
$$R_n = \frac{\zeta_n \mu_n}{m_1^2 - m_2^2}\left[k^2 + m_1^2 + k^4 \frac{A}{\mu_n \eta_3}\frac{d\gamma}{dT}\frac{(\alpha/\mu_n)}{\left(1 + k^2\frac{\alpha}{\mu_n}\right)}\right], \tag{24}$$

Eq. (22) is the first of our most significant results. It permits the calculation of the wave amplitude modes ζ_n as dictated by the viscoelastic properties of the nematic liquid and the hydrodynamic fluid velocity. It does not depend on adjustable free parameters and includes the Marangoni number $\boldsymbol{Ma} = (A/\mu_n\eta_3)d\gamma/dT$ that takes into account the thermal instability induced by the heating process applied to the nematic layer. All the material parameters that appear in Eq. (22) have already been reported by the experimental work of other authors [58] and that we are going to use further in the subsequent text. In contrast for a system of semi-infinite thickness, we obtained

$$M_n^\infty = \frac{2}{k}\left(w_0^2 - \frac{\mu_n}{k(m_1 + m_2)\left(1 + k^2\frac{\alpha}{\mu_n}\right)}\left\{\begin{array}{c}\left[\mu_n + k^2\left(3\nu_3 + \nu'\right)\right]\left(k^2 - m_1 m_2\right) \\ -\nu_3 k^2\left(m_1^2 + m_2^2 + m_1 m_2\right) - \nu_3 m_1^2 m_2^2 \\ + \frac{\alpha}{(m_1 - m_2)^2}\left[\begin{array}{c} A_2\left(m_1 m_2 + \nu_3 m_1 m_2^3 + B_3 m_2^2 - \nu_3 m_2^4\right) \\ +B_2\left(m_1 m_2 + \nu_3 m_2 m_1^3 + B_3 m_1^2 - \nu_3 m_1^4\right)\end{array}\right]\end{array}\right\}\right) \equiv \frac{2}{k} D_n^\infty\left(k, \mu_n = s + i(\alpha_r + n)\omega\right), \tag{25}$$

where the coefficients

$A_2 = k^4\left(\frac{A}{\mu_n\eta_3}\frac{d\gamma}{dT}+1\right)+k^2m_1^2$, $B_2 = k^4\left(\frac{A}{\mu_n\eta_3}\frac{d\gamma}{dT}+1\right)+k^2m_2^2$ and $B_3=\mu_n+(3\nu_3+\nu')k^2$.

2.2. Magnetic field in the *Y*-axis direction with wave vector along the *X*-axis

In this section, we consider a nematic liquid layer of thickness L with the average director of nematogens **n**=(0,1,0) oriented in the Y-axis direction by the magnetic field and the wave propagates with wave number $\boldsymbol{k}$ in the X-axis. Thus, there is no coupling of the director with the flux. The hydrodynamic Navier-Stokes equation is the same as an isotropic liquid with a single viscosity $\eta:=\eta_3=\eta_2$. The components of the viscous stress tensor in Eq. (4) now read [53]

$$\sigma'_{xx}=2\eta\partial_x V_x,\ \sigma'_{zz}=2\eta\partial_z V_z,\ \sigma'_{zx}=\eta(\partial_x V_z+\partial_z V_x). \tag{26}$$

Moreover, the forces normal to the interface and in the plane are, respectively, the same as in Eqs. (8)–(9) of Section 4. The boundary conditions are the same as in Eqs. (10)–(12), and the heat diffusion equation (Eq. (13)) is still valid. Consequently, the same method of section **a** results in the following eigenvalue equation for the mode amplitude of the wave:

$$M_n^i\zeta_n = a(\zeta_{n-1}+\zeta_{n+1}), \tag{27}$$

with

$$M_n^i=\frac{2}{k}\left\{w_0^2+\nu\left(q_n^2+k^2\right)\frac{Q_n^i}{\zeta_n}\right\}+4\nu q_n\frac{S_n^i}{\zeta_n}, \tag{28}$$

and

$$Q_n^i=\zeta_n\left(\begin{array}{l}\nu q_n k^2\left\{-2\left(1+k^2\frac{\alpha}{\mu_n}\right)+\frac{A}{\nu\eta}\frac{d\gamma}{dT}\frac{1}{k^2-q_n^2}\left[1-q_n^2\frac{\nu}{\mu_n}+(\nu+\alpha)\frac{k^2}{\mu_n}\right]\right\}+\\ \nu\left\{-\left(k^2+q_n^2\right)\left(1+k^2\frac{\alpha}{\mu_n}\right)+\frac{A}{\nu\eta}\frac{d\gamma}{dT}\frac{k^2}{k^2-q_n^2}\left[1-q_n^2\frac{\nu}{\mu_n}+(\nu+\alpha)\frac{k^2}{\mu_n}\right]\right\}\\ \times\left[-q_n\cosh(kL)\cosh\left(q_nL\right)+k\sinh(kL)\sinh\left(q_nL\right)\right]\end{array}\right)$$
$$\times\left\{\left(1+k^2\tfrac{\alpha}{\mu_n}\right)\left[q_n\sinh(kL)\cosh\left(q_nL\right)-k\cosh(kL)\sinh\left(q_nL\right)\right]\right\}^{-1} \tag{29}$$

2.3. Thermal phase transition effect on surface dynamics

In this section, we study the critical acceleration, and wave number of the Faraday wave at the interface of nematic liquid crystal and air as a function of temperature. We use the real material parameters reported in the literature of nematic methoxy benzylidine butyl aniline liquid crystal that experiences a thermal nematic-isotropic phase transition [58]. We now explain how we calculated the wave properties just mentioned. These properties a_c, k_c need to be determined in the nematic phase and then in the isotropic phase of the liquid. Notice that the critical parameters are obtained in separate plots of the driving acceleration a for the onset of Faraday waves versus wave number. The lowest value of $a = a_c$ in the lower branch in that plot for a given k yields their critical values, so we need to first see this plot, the so-called instability tongue because of its shape, to know what it is the corresponding minimum value of a and associated wave number. One of such plots is given in **Figure 4** of the neutral stability curve.

This picture was obtained using Eq. (27) for a semi-infinite medium of nematic and temperature of $T_{NI} - T = 3^{\circ}$C from the nematic-isotropic transition temperature and the experimental data for MBBA $\eta_3 = 0.0163$ Pa s, $\gamma = 0.03853$ N/m, $\rho = 1.03881 \times 10^3$ kg/m^3 [58]. The main wave sustained by the interface is of subharmonic type.

For the pure nematic state, we first solved numerically Eqs. (22)–(24) with $s = 0, \alpha_r = 1/2$ for $n = 22$ modes using the real materials data of the nematic MBBA obtained from interpolations of the experimental data which are provided in **Figures 2** [58] and 3 [59]. Experiments with laser light scattering from the interface of MBBA performed by Langevin [54, 58] show that the nematic state has two viscosities $\eta^{'}, \eta_3$, the surface tension α and the interface structural entropy $d\gamma/dT$ as the only necessary data to characterize fully the phase transition. MBBA has the critical temperature of $T_c \approx 45^{\circ}$C, and above this temperature the anisotropy of the viscosity disappears. This was confirmed experimentally by Langevin who found that in the isotropic state of the liquid the intensity of scattered light is well characterized by a single bulk viscosity $\eta = \eta_3 = \eta^{'}$ and the same elastic parameters above. In both cases, the density of the liquid crystal was the same and equal to the constant value $\rho = 1.03881 \times 10^{-3}$ kg/m^3 as required by the

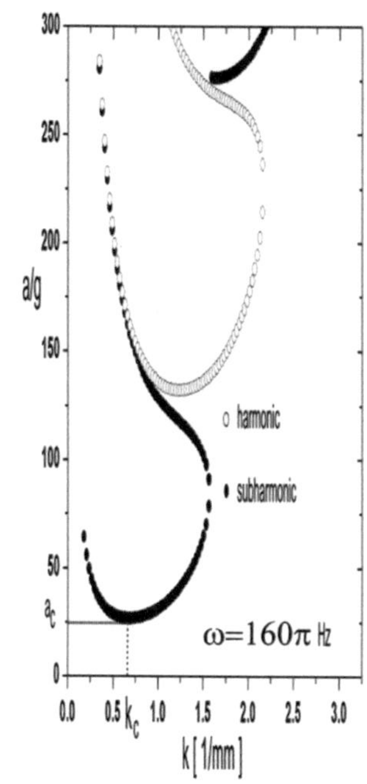

Figure 4. Neutral stability curve of Faraday waves of subharmonic type for a semi-infinite layer of nematic in contact with an air interface. The use was made of the real material parameters of MBBA $\eta_3 = 0.0163$ Pa s, $\gamma = 0.03853$ N/m, $\rho = 1.03881 \times 10^3$ kg/m^3, and temperature $T_{NI} - T = 3^{\circ}$C from the phase transition.

incompressibility condition (Eq. (5)). We used a liquid layer of thickness $L = 4.5$ mm. Thus, for the isotropic phase we used the hydrodynamic equations which are similar to those of a simple liquid derived in this section. The critical parameters were numerically calculated from Eqs. (27)–(29) for the isotropic branch. The result of such a procedure is depicted in **Figure 5**.

Figure 5 shows the transition of the main sustained waves which are of subharmonic type, from low up to high temperatures across the critical temperature. One can observe that there is a significant variation of (a_c, k_c) due to their discontinuous behavior at the critical temperature of phase transition. **Figure 5a** and **b** depict k_c, and **Figure 5c** yields a_c, versus the nematic-isotropic transition temperature $T - T_c$ at two values of the external frequency ω. For $\omega = 20\pi$ Hz, the plots of **Figure 5a** and **c** with symbol O correspond to the inclusion of Marangoni flow. For a higher frequency $\omega = 40\pi$ Hz we used symbol • in **Figure 5b** and **c** and Marangoni flow is included too. However, in **Figure 5** the plots with symbol ⋆ do not include Marangoni flow. We used the viscosities and surface tension in all the ranges of temperature variations as shown in **Figures 2** and **3**. When the Marangoni number is included $Ma = (A/\mu_n \eta_3) d\gamma/dT$ in Eqs. (22)–(24) and (27)–(29), the critical acceleration and wave number (a_c, k_c) roughly coincide with their values when $Ma = 0$ is neglected. The same results are obtained if the anistotropy of the thermal diffusivity is taken as either $\alpha_{\|}$ or $\alpha_{\perp}$; therefore in the plot of **Figure 5** only $\alpha_{\|}$ for the nematic side and α_{iso} for the isotropic branch are used. On the other hand, from **Figures 2** and **3** we can observe that the vertically applied temperature gradient of $A = -3^{\circ}$C/mm on top of MBBA produces major changes in the magnitude of viscosities, surface tension, and on thermal diffusivity. However, the single parameter, the gradient of surface tension $\frac{d\gamma}{dT}$, which is related to Marangoni flow, does not affect the Faraday wave. This phenomenon occurs for instance at $\omega = 40$Hz in **Figure 5**. We note from **Figure 5** that (a_c, k_c) display the same discontinuous behavior as the viscous and elastic parameters do across the thermal phase transition temperature. However, a different mechanism for discontinuity of these critical parameters was measured experimentally by Huber et al. [57] during the surface freezing of a polymer monolayer made of tretraconazole melt. They found that a decrease of temperature leads to the formation of a monolayer at the interface of polymer-air that changed the surface tension but without an observable quantitative change of bulk viscosity. As a consequence, a high flow velocity gradient close to the surface at a characteristic temperature during the cooling down process from the high temperature regime is produced. Such an effect was accurately confirmed with a Faraday wave calculation by those authors and it is similar to our results here. We note that an exact match of the experimentally measured power spectrum of scattered light by thermal fluctuations of the nematic-air interface with a theoretical calculation ignoring Marangoni flow and for constant temperature suggests that a monolayer of nematogens does not form at the interface of MBBA. Unlike the MBBA liquid, the critical parameters of the Faraday instability at the interface of the liquid-vapor of CO_2 [6] do not display discontinuous behavior as in the liquid crystal study.

2.4. Dispersion relation of an MBBA liquid layer with the inclusion of Marangoni flow

In **Figure 6**, the dispersion relation of the real MBBA (**Figure 6a**) together with an ideal model of a nematic (**Figure 6b** and **c**) as calculated from Eqs. (22)–(24). The real MBBA liquid

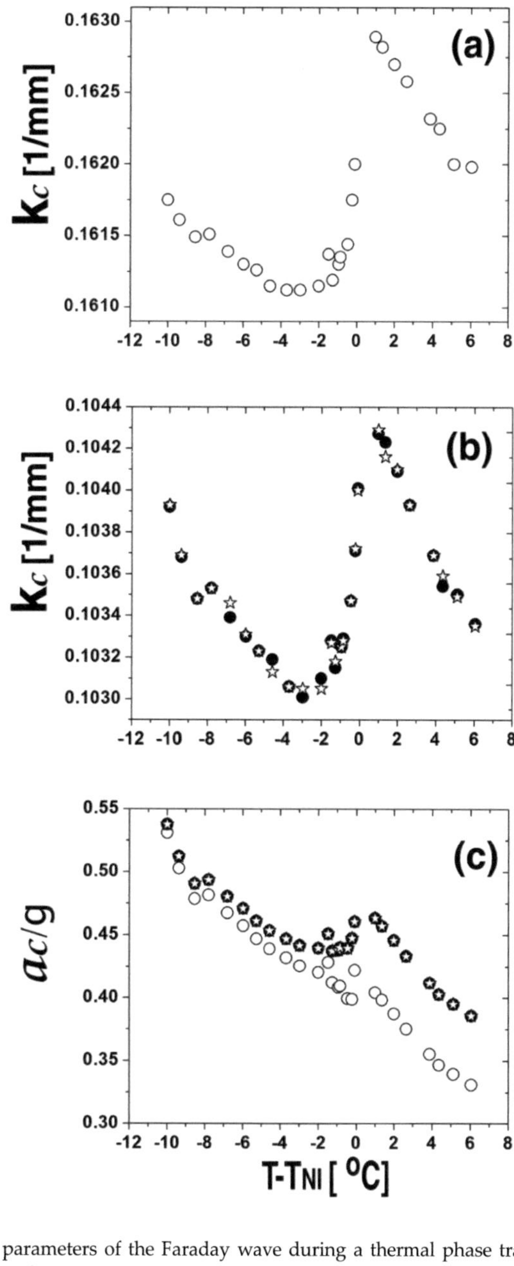

Figure 5. Calculated critical parameters of the Faraday wave during a thermal phase transition of nematic MBBA. In **Figure 5**, the critical wave number k_c (5a, 5b) and acceleration a_c (5c) as a function of the transition temperature are plotted. Two cases of frequency of excitation were used, $\omega=20\pi$ Hz with symbol O, (5a, 5c), $\omega=40\pi$ Hz symbol • and white stars (5b, 5c). The nematic layer depth is $L=4.5$ mm. Symbols O and • include Marangoni flow, and it is neglected in the plot with a star symbol. The material parameters used across the thermal phase transition are those given in **Figures 2** and **3**.

(**Figure 6a**) presents a minimum of $a_c\approx0.455$ g at oscillaton frequency $\omega=55\pi$ Hz for a layer thickness of $L=4.5$ mm. In this picture, the Marangoni flow was included, whereas the ideal model of nematic has a higher viscosity $\eta_3=0.163$ Pa than the real one, and its thickness is

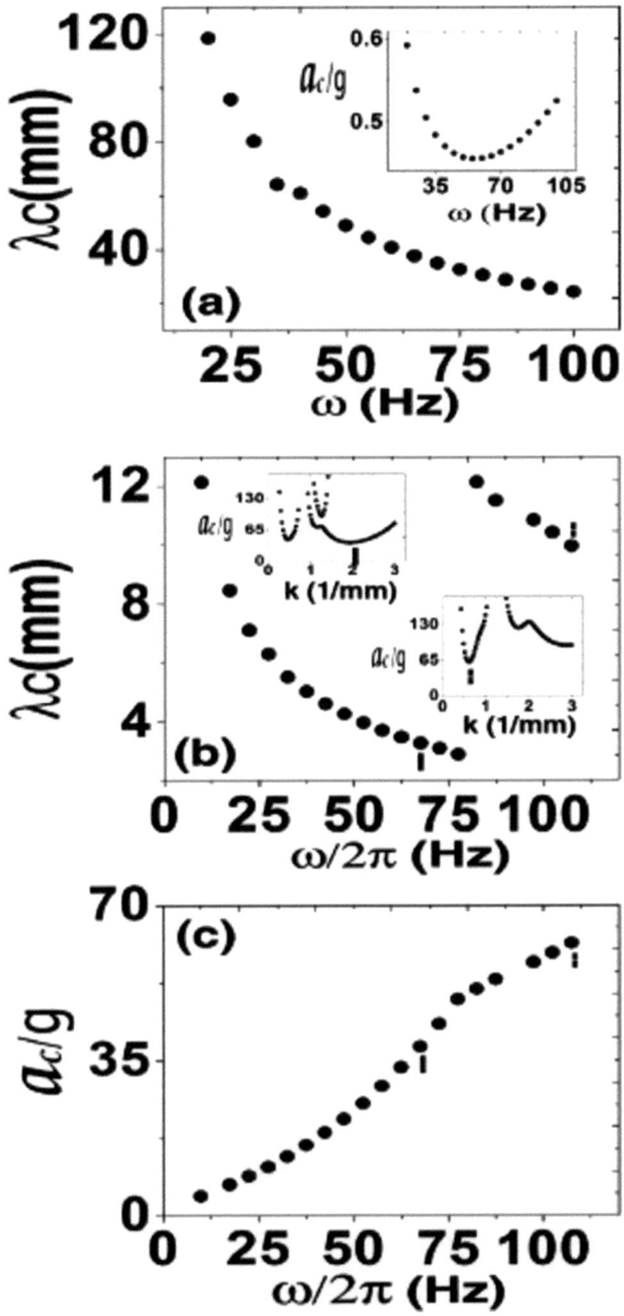

Figure 6. Dispersion relation of subharmonic waves of nematic MBBA at the fixed transition temperature $T_{NI} - T$=3°C for layer thickness of L=4.5 mm (6a) and L=2.5mm for (6b, 6c). Inset in **Figure 6a** depicts the critical acceleration calculated using the real parameters of MBBA and inclusion of Marangoni flow was made. Pictures in **Figure 6b** and **c** are for a model of nematic with 10 times the viscosity of the real nematic and neglect Marangoni flow and its critical acceleration is given in 6c. The left and right vertical lines in the insets of panels b and c depict the critical accelerations $\frac{a_c}{g} = 38.12, 51.13$ which occur at wave vectors k_c=1.9284 , 0.6312 mm^{-1} and frequencies ω=135π , 215π Hz, respectively.

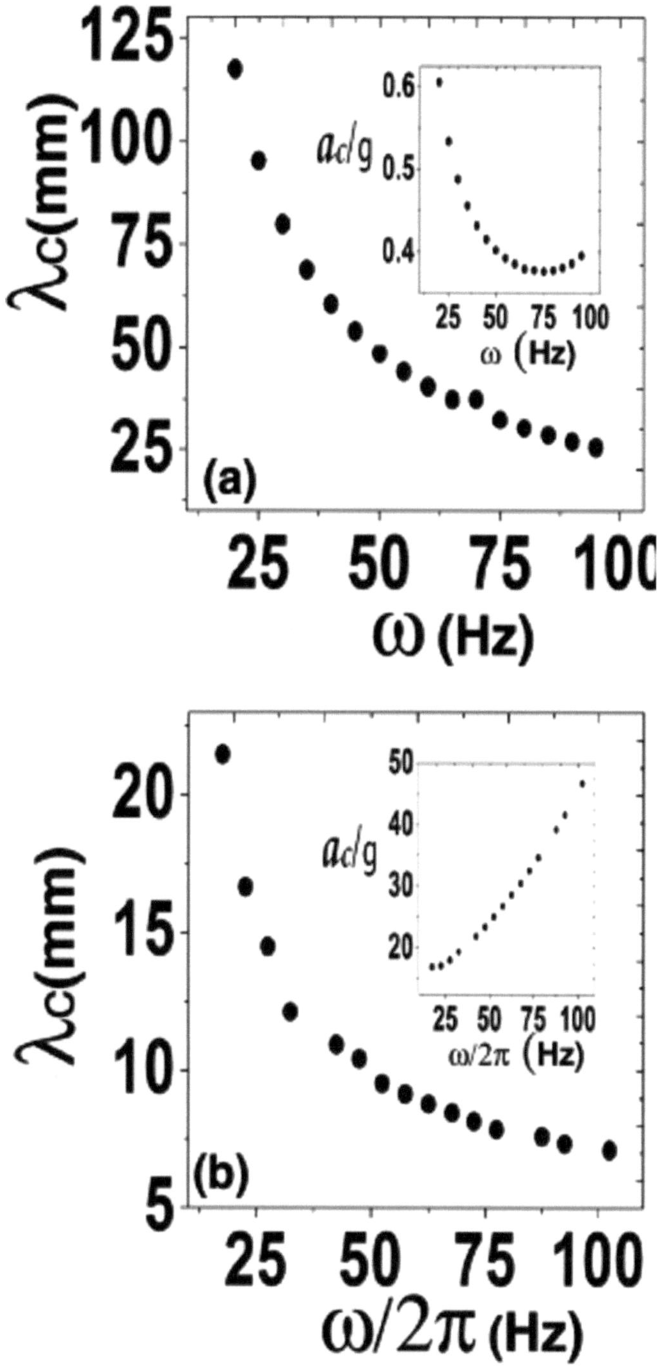

Figure 7. Dispersion relation of subharmonic waves of the isotropic nematic for low nematic depths $L = 4.5, 2.5$ mm, **Figure 7a** and **b**, respectively. Other parameters are as in **Figure 6**. **Figure 7a** includes Marangoni number and uses real data of MBBA, whereas **Figure 7b** is for a model nematic with 10 times the real viscosity with no Marangoni flux. The insets correspond to the critical acceleration versus frequency.

almost half the size of the real MBBA $L=2.5$ mm without Marangoni instability. The ideal model shows a monotonously increasing critical acceleration versus frequency (**Figure 6b** and **c**). A similar minimum value of a_c was detected in silicon oil and water system. From **Figure 6b** and **c**, it can be seen that the harmonic type wave length λ_c versus frequency shows a bifurcation. First, λ_c decreases continuously as the frequency starts growing until a value $\omega \approx 165\pi$ Hz is reached and suddenly the wave length makes a jump to a new value of the same magnitude $\lambda_c=12.12$ mm; it has at the starting value of the stimulus excitation $\omega \approx 20\pi$ Hz. The inset in **Figure 6b** shows this transition in a regime of critical acceleration depicted with the small vertical lines. The discontinuities in wave length have been documented also to appear in viscoelastic fluids, in Newtonian liquids, and for silicon water oil systems.

2.5. Dispersion relation of isotropic liquid crystal layer with the inclusion of Marangoni flow

In this case, the dispersion relation is calculated from Eqs. (27)–(29). **Figure 7** presents the dispersion relation, whereas the insets depict the curve of the critical acceleration as a function of frequency for two systems; one with a layer thickness of $L=4.5$ mm, **Figure 7a** that includes Marangoni flow, and the second one for $L=2.5$ mm with $Ma=0$ as shown in **Figure 7b**. The same pattern of behavior as in **Figure 6** is obtained.

3. Thermotropic smectic A liquid crystal layers

3.1. Smectic order parameter, wave vector, and magnetic field directed along the X-axis direction

In this section, we discuss the Faraday instability in smectic A liquid crystal layers [40]. We consider a smectic liquid crystal of average thickness L and infinite lateral extension in contact with a vapor. The nematogens are oriented by an external magnetic field in the X-axis as depicted in **Figure 8**.

The stack of layers deformation is given by Eq. (1), whereas the elastic response of the interface is given by Eq. (2). The governing hydrodynamic equations of the velocity, viscous stress tensor, and the boundary conditions were reported in Refs. [40, 53]. Following their use and with the help of the methods of Section 2, we derived the following recursive equation of the amplitude of deformation ζ_n:

$$M_n\zeta_n = a(\zeta_{n-1} + \zeta_{n+1}) \tag{30}$$

$$M_n = \frac{2}{k}\left\{ w_0^2 + \left\{ \frac{\left[Bk^2 + KS_1^2 + Kk^2S_1 + \chi_a H^2 S_1\right]}{\rho} + \left[\mu_n + \lambda_p Bk^2 + \lambda_p\left(KS_1^2 - \chi_a H^2 S_1\right)\right]\left[\mu_n + k^2\left(3\nu_3 + \nu'\right) - \nu_3 S_1\right] \right\} \frac{(A_1 + B_1)}{i\zeta_n} \right\},$$

where

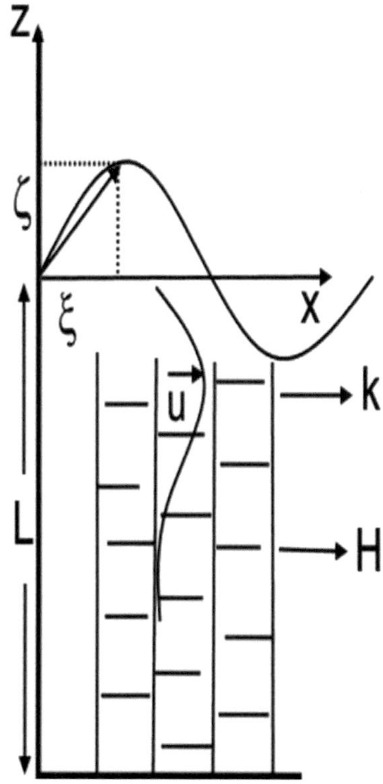

Figure 8. Model of the deformation of a smectic A liquid crystal layer of thickness *L* due to the external perturbation. The liquid crystal has a free interface with air. The nematogens are oriented by a magnetic field **H** in the *X*-axis direction and parallel to the interface surface. The velocity flow given by the wave vector ***k*** is parallel to the external field.

$$A_1 = \frac{\zeta_n i e^{L\sqrt{S_1}}}{2kR_n\sqrt{S_1}}\left(-1+\coth(L\sqrt{S_1})\right)[\mu_n S_1 + k^2(-1+e^{L\sqrt{S_1}})R_n]$$

$$B_1 = \frac{\zeta_n i}{kR_n\sqrt{S_1}(-1+e^{L\sqrt{S_1}})}[\mu_n S_1 e^{L\sqrt{S_1}} - k^2(-1+e^{L\sqrt{S_1}})R_n]$$

$$R_n = \lambda_p BK^2 + \lambda_p KS_1^2 - \lambda_p \chi_a H^2 S_1 + \mu_n, \; S_1 = k^2 d_n / c_n,$$

$$c_n = \frac{\mu^2{}_n}{\lambda_p K v_3} + k^2\left(\frac{\mu_n}{\lambda_1^2 v_3} + \frac{\mu_n}{\xi_H^2 v_3} + \frac{2\mu_n}{\lambda_p K} + \frac{v'\mu_n}{\lambda_p K v_3} + \frac{\kappa_s{}^2}{\lambda_1^2}\right) + k^4\left(\frac{1}{\xi_H^2} + \frac{2}{\lambda_1^2} + \frac{v'}{v_3\lambda_1^2}\right)$$

$$d_n = \frac{\mu^2{}_n}{\lambda_p K v_3} + k^2\left(\frac{\mu_n}{\lambda_1^2 v_3} + \frac{\mu_n}{\lambda_p K}\right) + \frac{k^4}{\lambda_1^2}, \; \xi_H = \sqrt{K/\chi_a}/H, \; \lambda_1 = \sqrt{K/B},$$

$$\lambda_p = \frac{1}{\rho v_3 \kappa_s{}^2}, \; v' = \frac{\eta'}{\rho}, \; \kappa_s \approx 100\overset{o}{A}.$$

In this eigenvalue equation, we ignored the elongational elasticity and coupling between in-plane and normal elastic deformations. A plot of the Faraday stability curve is given in **Figure 9**

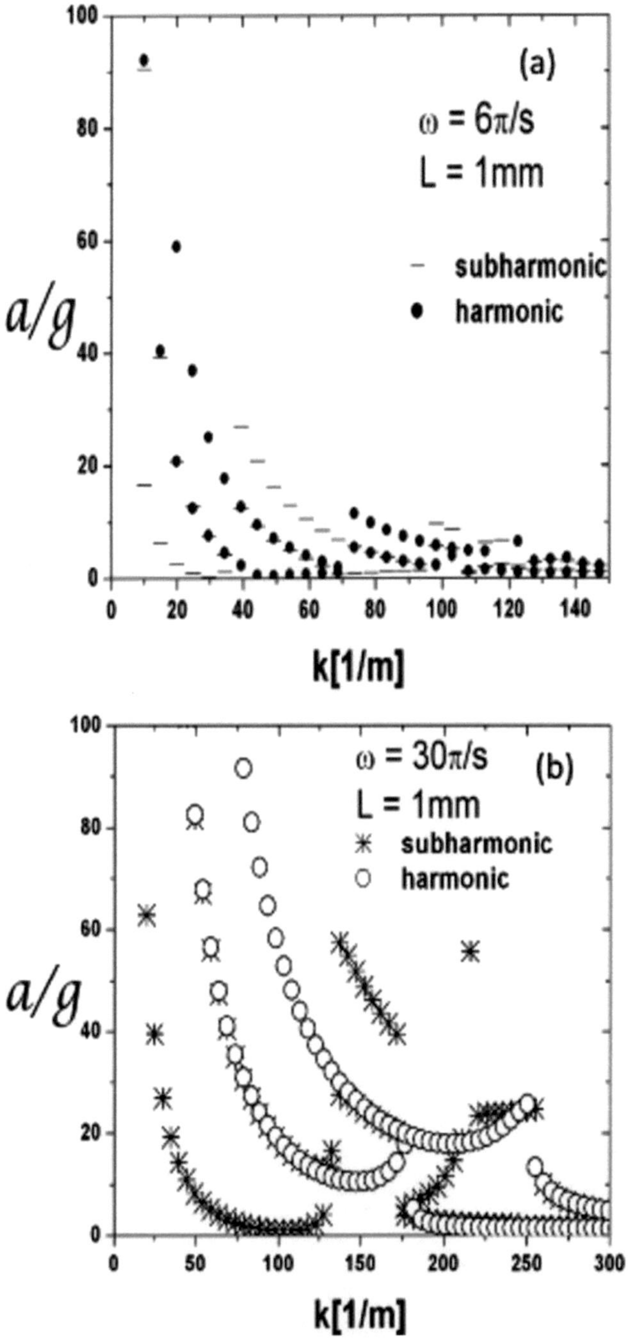

Figure 9. Plot of the acceleration versus wave number response of the surface wave sustained at the interface of a smectic A liquid-air for two frequencies and fixed room temperature. The configuration of the nematogens is given in **Figure 8**. The material parameters are provided in Section 3.1 and are typical of smectic A.

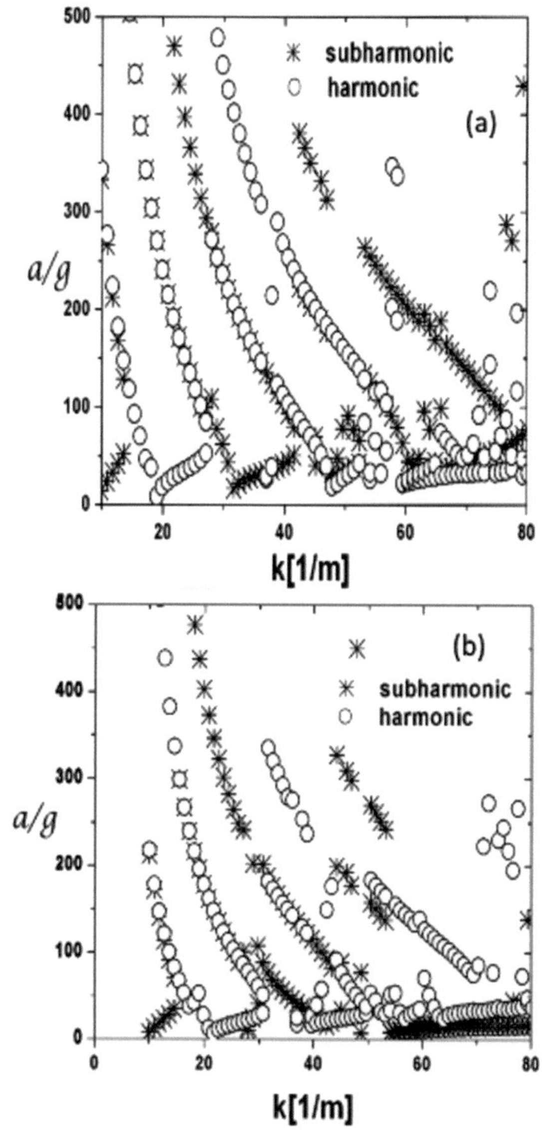

Figure 10. Neutral stability curve of Faraday wave at the interface of Sm A and air. There is no applied magnetic field. The nematogens are oriented in layers parallel to the interface and the flow velocity is in the *X*-axis direction. The material parameters used are given in the main text. Note that subharmonic waves are plotted with symbol *, whereas harmonic waves are depicted with ∘. This plot and that of **Figure 9** show that small depth Sm A liquid crystal layers can sustain parametric waves for excitation frequencies of typical experiments.

where the driving acceleration is plotted regarding the wave number. This picture shows us that parametric instability can be generated in shallow layers of smectic liquid crystals when it is excited with low frequencies. The magnitude of the normalized acceleration a/g falls well in the range of resolution of accessible experimental techniques that have been used before to study other fluids. The material parameters used to make this plot correspond to typical smectic liquids: $K=10^{-11}$ N, $B=10^6$ N/m^2, $L=10^{-4}$m, $\lambda_p=10^{-14}$ m^4/Ns, $\rho=10^3$ kg/m^3, $\eta_3=1$ P, $\eta_3=\eta_2$, $\eta'=10^{-2}\eta_3$, $\chi_a=10^{-8}$ kg/(m s^2 G), $H=3000$ G, the magnetic field that keeps fixed the orientation of nematogens, and $\gamma=0.033$ N/m. In particular, **Figure 9a** points out that the excitation frequency $a/g\approx 0$ for the low frequency $\omega=6\pi$ Hz, that is, infinitesimal accelerations, can excite subharmonic waves in a similar manner as it occurs in inviscid ideal fluids.

3.2. Smectic A layers parallel to the surface and no magnetic field

In this case, we derived the modes of the surface amplitude of deformation ζ_n that satisfies the eigenvalue Eqs. (27)–(29) with the function

$$M_n = \frac{2}{k}\left\{ w_0^2 + \left(\frac{Bk\sqrt{S_1}}{\rho} + \frac{\sqrt{S_1}}{k}\mu_n\left[\mu_n + k^2\left(3\nu_3 + \nu'\right) - \nu_3 S_1\right]\right)\coth\left(L\sqrt{S_1}\right)\right\}. \qquad (31)$$

The curly brackets in the above expression for M_n with the value $n=0$ yield the well now dispersion relationship of thermal capillary waves reported by other authors. In **Figure 10**, we depict the stability curve of external acceleration versus wave number at a fixed frequency of oscillation for the given material parameters; $K=10^{-10}$ N, $B=10^4$ N/m^2, layer thicknesses $L=0.03, 0.05$ m, frequency $\omega=18\pi$ Hz, $\lambda_p=10^{-14}$ m^4/(N s), $\rho=10^3$ kg/m^3, and viscosities and surface tension as in Section 3.1 earlier. We can observe that parametric surface waves also can evolve in this configuration of smectic A liquid. In a forthcoming manuscript, we will evaluate this parametric instability when the smectic phase can develop from the nematic phase as a thermal transition.

4. Lyotropic liquid crystal

In the previous section, we investigated how the bulk microstructure of the liquid crystal can modify the parametric instability through a thermal phase change. Now, we describe phase changes produced by particle volume fraction variations. Using birefringent measurements, Ballesta et al. [23] demonstrated a hydrodynamic phase change from isotropic to the nematic ordering of particles in a colloidal suspension. For the first time in the Faraday instability that occurs at the interface of air-colloidal suspension made of fd virus, they found that Faraday waves induce local nematic ordering of the nematogens in the wave crest when the colloid concentration is increased and close to the isotropic-nematic critical concentration. Such regions of nematic ordering become more permanent as the concentration is raised, and finally large areas of stable nematic patches that follow the wave flow are developed. This phenomenon was interpreted as a change in the local viscosity from its unperturbed value and decreases as a function of shear generated by the surface movement, which may reach high values of 100 Hz. A consequence of such shear thinning of viscosity is the appearance of hysteresis in the amplitude of the normal direction of the surface deformation as a function of the driving acceleration. Their analysis of the hysteretic behavior of the wave amplitude required them to use the Cross model of viscosity for bulk Newtonian fluids. Whereas for interpreting the intensity of observed birefringent experiments, it was necessary to use the viscosity, and order parameter of a model of rod-like suspension of particles in the isotropic phase [60]. A recent complete numerical simulation of the hydrodynamic equations governing the Faraday waves was developed by Perinet et al. [61] for a system formed by two immiscible fluids with the supporting fluid forming a shallow layer smaller than its boundary layer. These authors confirmed a hysteresis of the amplitude of the surface deformation as a function of the driving acceleration. They conclude that the wave amplitude bifurcates into two different waves. The hysteresis of the lower amplitude wave is attributed to a change of the shear stress in the fluid that results from variations in the fluid flow that produce a balance of hydrostatic and lubrication stresses.

Unlike lyotropic liquids, a successful continuum mean field model for thermotropic nematics has allowed the description of the rheology of the bulk of confined layers of nematics under oscillatory shear. Such a model has predicted shear thinning of the viscosity as a function of the oscillation frequency of the imposed shear. The extension of this study to understand the shear thinning effect on viscosity in the experiments of Ref. [23] seems feasible. A different perspective is obtained with computer simulations as those made by Germano et al. [38]. These researches have shown through molecular dynamics simulations on a molecular model of bulk ellipsoids with pair interaction of Weeks-Chandler-Andersen type as they say "nematic fluids may adopt inhomogeneous steady states under shear flow" [38]. Thus, shear flow modifies the molecular ordering in the liquid crystal producing changes in macroscopic viscosities like shear thinning and thickening and shear banding similarly as that found by Ballesta in their experiment of fd virus suspension under parametric instability. However, those simulations cannot be compared directly with the experiments of Ballesta [23]. Germano et al. [38] were interested in the capillary waves spectrum at the interface formed during the transition from nematic to isotropic. However, they did compare their simulations with a theory of Landau-De Gennes type for the free energy of the interface which incorporates an average director parallel to the interface. They found isotropic capillary waves that propagate at long wavelengths governed by the macroscopic surface tension. At short wavelength, however, the surface tension becomes anisotropic and depends on the wave vector. In a recent series of papers by Popa-Nita et al. [27–31], they developed a Landau-De Gennes theory to describe the capillary waves originating from thermal fluctuations, and at the interface of a ternary mixture of liquid crystal, colloid, and impurities. They considered both homeotropic (perpendicular to the interface) and also the variation of the nematic director. As in the Germano et al. method, Popa-Nita uses a free energy of the liquid crystal that predicts the bulk phase diagram, and additionally a Cahn-Hilliard equation was incorporated for taking into account the diffusion of impurities and colloids inside the liquid crystal. For such a mixture, they predicted the surface tension to decrease with the presence of colloids, whereas the impurities enhance its strength. Also, the temperature of the bulk phase transition is lowered on the pure liquid crystal nematic-isotropic transition temperature. The interfaces so formed experience thermal fluctuations. With the help of this approach, Popa-Nita were able to find that there are two regions of propagating capillary waves as it was also observed by Germano et al. in their simulation work. In the first region of long wavelengths, there is dissipation produced by shear flow and the ternary mixture behaves like an isotropic fluid which can be described by a single effective bulk viscosity. The hydrodynamic equations of the velocity field underlying that dispersion relation depends on the respective viscosity for each of the two phases formed which are separated by a sharp interface. With appropriate boundary conditions on each thermodynamic phase, the dispersion relation of the capillary waves was predicted. The generated wave depends on the average effective constant surface tension of the nematic and isotropic interface. The propagating ripple depends on one viscosity and the compression and bending modulus of the surface. The second region corresponds to a diffuse gap of particles close to the interface and corresponds to low values of wavenumber. This wave is dominated by the relaxation of the order parameter and the surface tension which is dependent on the density variation within the diffuse zone and the inhomogeneous distribution of nematogens inside it. The boundary conditions consist of the matching of the velocity field inside the diffuse zone with that from the bulk isotropic and nematic regions. The theoretical model of Popa-Nita [27–31] might be useful to study Faraday waves in interfaces of phase-separated regions of liquid crystals as a function of the concentration of particles. Presently, the phase transition on bulk phases of liquid crystals constitutes a large body of knowledge [62], but its effect on the dynamical responses of parametric waves on the interfaces so formed in the transition is still an open subject of research.

5. Conclusion

We reviewed recent results underlying the hydrodynamics description of Faraday waves under a thermal phase transition in thermotropic nematic liquid crystals. The numerical evaluation of the effect of phase change on the critical acceleration at the onset of the instability points out its pertinent experimental observation with birefringence or surface light-scattering techniques. Consequently, other liquid crystals can be studied with this theoretical approach. In Section 4, one such experimental example of a lyotropic liquid crystal of fd virus was mentioned. Also, a correction to the conceptual framework of Sections 2 and 3 to include the effect of variations of volume fraction of particles that can lead to a phase transition can be considered in this case.

Acknowledgements

The author acknowledges the General Coordination of Information and Communications Technologies (CGSTIC) at CINVESTAV for providing HPC resources on the Hybrid Supercomputer "Xiuhcoatl," which has contributed to the research results reported in this paper.

Author details

Martin Hernández Contreras

Address all correspondence to: marther@fis.cinvestev.mx

Physics Department, Center for Research and Advanced Studies of the National Polytechnic Institute, CD Mexico, Mexico

References

[1] Christiansen B, Alstrom P, Levinsen T. Ordered capillary-wave states: Quasicrystals, hexagons, and radial waves. Physical Review Letters. 1992;**68**:2157-2160. DOI: 10.1103/PhysRevLett.68.2157

[2] Edwards WS, Fauve S. Patterns and quasi-patterns in the Faraday experiment. Journal of Fluid Mechanics. 1994;**278**:123-148. DOI: 10.1017/S0022112094003642

[3] Kudrolli A, Pier P, Gollub JP. Superlattice patterns in surface waves. Physica D. 1998;**123**: 99-111. DOI: 10.1016/S0167-2789(98)00115-8

[4] Arbell H, Fineberg J. Spatial and temporal dynamics of two interacting modes in parametrically driven surface waves. Physical Review Letters. 1998;**81**:4384-4387. DOI: 10.1103/PhysRevLett.81.4384

[5] Benjamin T, Ursell F. The stability of the plane free surface of a liquid in vertical periodic motion. Proceedings of the Royal Society A. 1954;**225**:505-501. DOI: 10.1098/rspa.1954.0218

[6] Kumar K, Tuckerman LS. Parametric instability of the interface of two fluids. Journal of Fluid Mechanics. 1994;**279**:49-68. DOI: 10.1017/S0022112094003812

[7] Benchhoefer J, Ego V, Manneville S, Johnson B. An experimental study of the onset of parametrically pumped surface waves in viscous fluids. Journal of Fluid Mechanics. 1995;**288**:325-350. DOI: 10.1017/S0022112095001169

[8] Müller HW, Zimmermann W. Faraday instability in a linear viscoelastic fluid. Europhysics Letters. 1999;**83**:169-174. DOI: 10.1209/epl/i1999-00142-5

[9] Wagner C, Müller HW, Knorr K. Faraday waves on a viscoelastic liquid. Physical Review Letters. 1999;**83**:308-311. DOI: 10.1103/PhysRevLett.83.308

[10] Kumar S. Vibration-induced interfacial instabilities in viscoelastic fluids. Physical Review E. 2002;**65**:026305. DOI: 10.1103/PhysRevE.65.026305

[11] Kumar S, Fauve S. Faraday instability with a polymer solution. The European Physical Journal B. 1999;**9**:175-178. DOI: 10.1007/s100510050753

[12] Kumar S. Parametrically driven surface waves in viscoelastic liquids. Physics of Fluids. 1999;**11**:1970-1981. DOI: 10.1063/1.870061

[13] Ovando-Vazquez C, Vazquez Rodriguez O, Hernandez-Contreras M. Faraday waves on finite thickness smectic a liquid crystal and polymer gel materials. AIP Conference Proceedings. 2008;**1077**:135-141. DOI: 10.1063/1.3040251

[14] Ballesta P, Manneville S. Signature of elasticity in the Faraday instability. Physical Review E. 2005;**71**:026308. DOI: 10.1103/PhysRevE.71.026308

[15] Müller HW. Parametrically driven surface waves on viscous ferrofluids. Physical Review E. 1998;**58**:6199-6205. DOI: 10.1103/PhysRevE.58.6199

[16] Mekhonoshin VV, Lange A. Faraday instability on viscous ferrofluids in a horizontal magnetic field: Oblique rolls of arbitrary orientation. Physical Review E. 2002;**65**:061509. DOI: 10.1103/PhysRevE.65.061509

[17] Perinet N, Juric D, Tuckerman LS. Numerical simulation of Faraday waves. Journal of Fluid Mechanics. 2009;**635**:1-26. DOI: 10.1017/S0022112009007551

[18] Kityk AV, Embs J, Mekhonoshin VV, Wagner C. Spatiotemporal characterization of interfacial Faraday waves by means of a light absorption technique. Physical Review E. 2005;**72**:036209. DOI: 10.1103/PhysRevE.72.036209

[19] Zhang W, Viñals J. Pattern formation in weakly damped Faraday waves. Journal of Fluid Mechanics. 1997;**336**:301-330. DOI: 10.1017/S0022112096004764

[20] Chen P, Viñals J. Amplitude equation and pattern selection in Faraday waves. Physical Review E. 1999;**60**:559-570. DOI: 10.1103/PhysRevE.60.559

[21] Skeldon AC, Guidoboni G. Pattern selection for Faraday waves in an incompressible viscous fluid. SIAM Journal on Applied Mathematics. 2007;**67**:1064-1100. DOI: 10.1137/050639223

[22] Silber M, Topaz C, Skeldon AC. Two-frequency forced Faraday waves: Weakly damped modes and pattern selection. Physica D. 2000;**143**:205-225. DOI: 10.1016/S0167-2789(00)00102-0

[23] Ballesta P, Lettingaa MP, Manneville S. Interplay between a hydrodynamic instability and a phase transition: The Faraday instability in dispersions of rodlike colloids. Soft Matter. 2011;**7**:11440-11446. DOI: 10.1039/c1sm06175e

[24] Epstein T, Deegan RD. Strip waves in vibrated shear-thickening wormlike micellar solutions. Physical Review E. 2010;**81**:066310. DOI: 10.1103/PhysRevE.81.066310

[25] Price AD, Schwartz DK. Fatty-acid monolayers at the nematic/water interface: Phases and liquid-crystal. Journal of Physical Chemistry B. 2007;**111**:1007-1015. DOI: 10.1021/jp066228b

[26] Clare BH, Abbott NL. Orientations of nematic liquid crystals on surfaces presenting controlled densities of peptides: Amplification of protein–peptide binding events. Langmuir. 2005;**21**:6451-6461. DOI: 10.1021/la050336s

[27] Popa-Nita V, Oswald P. Surface tension and capillary waves at the nematic-isotropic interface in ternary mixtures of liquid crystal, colloids, and impurities. Journal of Chemical Physics. 2007;**127**:104702. DOI: 10.1063/1.2772251

[28] Popa-Nita V, Sluckin TJ, Kralj S. Waves at the nematic-isotropic interface: Thermotropic nematogen–non-nematogen mixtures. Physical Review E. 2005;**71**:061706. DOI: 10.1103/PhysRevE.71.061706

[29] Popa-Nita V, Sluckin TJ. Surface modes at the nematic-isotropic interface. Physical Review E. 2002;**66**:041703. DOI: 0.1103/PhysRevE.66.041703

[30] Popa-Nita V, Oswald P. Waves at the nematic-isotropic interface: The role of surface tension anisotropy, curvature elasticity, and backflow effects. Physical Review E. 2003;**68**:061707. DOI: 10.1103/PhysRevE.68.061707

[31] Popa-Nita V, Oswald P, Sluckin TJ. Waves at the nematic-isotropic interface. Molecular Crystals and Liquid Crystals. 2005;**435**:215/[875]-232/[892]. DOI: 10.1080/15421400590955299

[32] Lishchuk SV. Waves at surfactant-laden liquid-liquid crystal interface. Physical Review E. 2007;**76**:011711. DOI: 10.1103/PhysRevE.76.011711

[33] Lishchuk SV. Scattering of light on surfactant-laden liquid-liquid crystal interface. Chemical Engineering Communications. 2009;**196**:854-871. DOI: 10.1080/00986440802668174

[34] Rey AD. Viscoelastic theory for nematic interfaces. Physical Review E. 2000;**61**:1540. DOI: 10.1103/PhysRevE.61.1540

[35] Rey AD. Mechanical model for anisotropic curved interfaces with applications to surfactant-laden liquid-liquid crystal interfaces. Langmuir. 2006;**22**:219-228. DOI: 10.1021/la051974d

[36] Rey AD. Polar fluid model of viscoelastic membranes and interfaces. Journal of Colloid and Interface Science. 2006;**304**:226-238. DOI: 10.1016/j.jcis.2006.08.027

[37] Schmid F, Germano G, Wolfsheimer S, Chilling T. Fluctuating interfaces in liquid crystals. Macromolecular Symposia. 2007;**252**:110-118. DOI: 10.1002/masy.200750611

[38] Germano G, Schmid F. Nematic-isotropic interfaces under shear: A molecular-dynamics simulation. Journal of Chemical Physics. 2005;**123**:214703. DOI: 10.1063/1.2131065

[39] Hernández-Contreras M. Faraday waves on nematic liquid crystals: Effect of Marangoni flow and thermal phase transition. Physical Review E. 2013;**88**:062311

[40] Hernández-Contreras M. Faraday waves in smectic A liquid crystal layers. Journal of Physics: Condensed Matter. 2010;**22**:035106. DOI: 10.1088/0953-8984/22/3/035106

[41] Lockwood NA, Abbott NL. SSelf-assembly of surfactants and phospholipids at interfaces between aqueous phases and thermotropic liquid crystals. Current Opinion in Colloid & Intface Science. 2005;**10**:111-120. DOI: 10.1016/j.cocis.2005.06.002

[42] Lockwood N.A, Gupta J.K, Abbot N.L. Self-assembly of amphiphiles, polymers and proteins at interfaces between thermotropic liquid crystals and aqueous phases. Surface Science Reports. 2008;**63**:255-293. DOI: 10.1016/j.surfrep.2008.02.002

[43] Crawford GP, Ondris-Crawford RJ, Doane JW, Žumer S. Systematic study of orientational wetting and anchoring at a liquid-crystal-surfactant interface. Physical Review E. 1996;**53**:3647. DOI: 10.1103/PhysRevE.53.3647

[44] Fazio VSU, Nannelli F, Komitov L. Sensitive methods for estimating the anchoring strength of nematic liquid crystals. Physical Review E. 2001;**63**:061712. DOI: 10.1103/PhysRevE.63.061712

[45] Bahr C. Surfactant-induced nematic wetting layer at a thermotropic liquid crystal/water interface. Physical Review E. 2006;**73**:030702. DOI: 10.1103/PhysRevE.73.030702

[46] Poulin P, Stark H, Lubensky TC, Weitz DA. Novel colloidal interactions in anisotropic fluids. Science. 1997;**275**:1770-1773. DOI: 10.1126/science.275.5307.1770

[47] Figuereido Neto A, SRA S, editors. The Physics of Lyotropic Liquid Crystals: Phase Transitions and Structural Properties. 1st ed. New York: Oxford University Press; 2005. p. 301. DOI: 10.0198525508

[48] Langevin D. Spectre des fluctuations thermiques à la surface libre d'un cristal liquide smectique A en présence d'un champ magnétique horizontal. Journal de Physique France. 1976;**37**:737-746. DOI: 10.1051/jphys:01976003706073700

[49] Buzza DMA, Jones JL, McLeish TCB, Richards RW. Theory of surface light scattering from a fluid–fluid interface with adsorbed polymeric surfactants. Journal of Chemical Physics. 1998;**109**:5008-5024. DOI: 10.1063/1.477114

[50] Crilly JF, Earnshaw JC. A light scattering study of thermotropic transitions of monoglyceride monolayers: Influence of molecular area fluctuations. Journal de Physique France. 1987;**48**:485-494. DOI: 10.1051/jphys:01987004803048500

[51] Langevin D. Analyse spectrale de la lumière diffusée par la surface libre d'un cristal liquide nématique. Mesure de la tension superficielle et des coefficients de viscosité. Journal de Physique France. 1972;**33**:249-256. DOI: 10.1051/jphys:01972003302-3024900

[52] Hernández-Contreras M, Kim MW, Pincus P. Surface hydrodynamics on a freely standing layer of a polymer solution. Physical Review E. 1999;**60**:4319-4329. DOI: 10.1103/PhysRevE.60.4319

[53] Langevin D. Light-scattering study of monolayer viscoelasticity. Journal of Colloid & Interface. Science. 1981;**80**:412-425. DOI: 10.1016/0021-9797(81)90200-9

[54] Langevin D, editor. Light Scattering by Liquid Surfaces and Complementary Techniques. 1st ed. New York: Marcel Dekker; 1992. p. 472. DOI: 10.0824786076

[55] Earnshaw JC. Light scattering studies of interfacial rheology. Journal of Dispersion Science and Technology. 1999;**20**:743-765. DOI: 10.1080/019322699908943818

[56] Giermanska-Khan J, Monroy F, Langevin D. Negative effective surface viscosities in insoluble fatty acid monolayers: Effect of phase transitions on dilational viscoelasticity. Physical Review E. 1999;**60**:7163-7173. DOI: 10.1103/PhysRevE.60.7163

[57] Huber P, Soprunyuk VP, Embs JP, Deutsch WCM. Faraday instability in a surface-frozen liquid. Physical Review Letters. 2005;**94**:184504. DOI: 10.1103/PhysRevLett.94.184504

[58] Langevin D, Bouchiat MA. Spectre des fluctuations thermiques a la surface libre d'un cristal liquide nématique. Journal de Physique France. 1972;**33**:101-111. DOI: 10.1051/jphys:01972003301010100

[59] Urbach W, Hervet H, Rondelez F. Thermal diffusivity measurements in nematic and smectic phases by forced Rayleigh light scattering. Molecular Crystals and Liquid Crystals. 1978;**46**:209-221. DOI: 10.1080/00268947808083724

[60] Dhont JKG, Briels WJ. Viscoelasticity of suspensions of long, rigid rods. Colloids and Surfaces A. 2003;**213**:131-156. DOI: 10.1016/S0927-7757(02)00508-3

[61] Périnet N, Falcón C, Chergui J, Juric D, Shin S. Hysteretic Faraday waves. Physical Review E. 2016;**93**:063114. DOI: 10.1103/PhysRevE.93.063114

[62] Singh S. Phase transitions in liquid crystals. Physics Reports. 2000;**324**:107-269. DOI: 10.1016/S0370-1573(99)00049-6

Chapter 5

Lyotropic Liquid Crystals Incorporated with Different Kinds of Carbon Nanomaterials or Biomolecules

Zhaohua Song, Yanzhao Yang and Xia Xin

Additional information is available at the end of the chapter

http://dx.doi.org/10.5772/intechopen.70392

Abstract

Liquid crystals (LCs) are considered as the "fourth state of matter," which can display properties between crystals and isotropic liquids. LCs can be classified into lyotropic liquid crystals (LLCs) and thermotropic liquid crystals (TLCs), among which LLCs are a kind of self-assemblies formed by amphiphile molecules in a given solvent within certain concentration ranges. The structures and properties of LLCs can be tuned by the incorporation of various kinds of additives, which represents an interesting and novel route for realizing functional composites. This review focuses on recent progress on LLCs-based materials assembled with diverse additives including carbon nanotubes, graphene, graphene oxide, and biomolecules. The thermal stability and mechanical strength of the host LLCs can be greatly improved after the guests are incorporated. In addition, new functions such as conductivity, photothermal effect, and bioactivity can be introduced by the incorporation of the guests, which significantly widens the applications of LLCs-based hybrids in nanotechnology, electrochemistry, drug delivery, and life science.

Keywords: liquid crystals, amphiphilic molecule, carbon nanotubes, graphene oxide, biomolecule

1. Introduction

Liquid crystals (LCs), including lyotropic liquid crystals (LLCs) and thermotropic liquid crystals (TLCs), are an intermediate state between isotropic liquid and ordered crystal [1]. They normally show the anisotropic physical properties because of the long orientational order of the molecular self-assemble aggregates. Thus, LCs are currently of great significance in nanotechnology to act as templates to devise, arrange, or even synthesize interesting composites due to the intrinsic self-assembly speciality [2–9]. Among them, LLCs were a special kind of self-assemblies of amphiphilic molecules within the scope of the long-range ordered arrangement, and their unique functions cannot be replaced by other materials. For example, the LLCs can be used as mediums for various organic and inorganic reactions that can be carried out in the hydrophilic or hydrophobic domains in confined spaces of those mesophases [10–13]. Moreover, the lamellar, hexagonal and cubic phases of LLCs can also be used as soft templates to control the structure and morphology of desired composites [14]. The Bi, PdS nanoparticles can also been synthesized by the LLCs [15]. A large variety of functional materials possessing mechanical or processing properties, biocompatibility, and so on, could be produced in this way. The LLCs have been the ideal self-organizing templates of carbon nanotubes, graphene oxide, biomolecules, and a large number of functional nanoparticles, because of their alignment regulated by the ordered matrix of LLCs [3, 16]. The preparation of LLCs-based hybrid materials has been proven to be an efficient approach to prepare ordered functional composites.

In the present chapter, a deep review on structure and property of LLCs-based materials is presented. Emphasis will be put on the hybrids composed of LLCs of amphiphilic molecules incorporated with different kinds of additives including carbon nanotubes, graphene oxide, and biomolecules.

2. Carbon nanotubes/LLC hybrid materials

Carbon nanotubes (CNTs) possess wide potential applications in physics, chemistry, and material and life sciences due to their distinct cylindrical π-conjugated structure and remarkable mechanical, electronic, optical, and thermal properties [17, 18]. The orienting and aligning of CNTs in CNTs-based composites is crucial for the advancement of their applications. Great efforts have been paid to disperse single-walled carbon nanotubes (SWNTs) into aqueous solutions by noncovalent method typically with surfactants, polymers, and polyelectrolytes [19, 20]. One of the efficient methods to develop aligned CNTs is to make use of the order and fluidity of LLCs [21]. The incorporation of CNTs in LLCs has achieved well-dispersed and uniformly-aligned CNTs, which are two key prerequisites during the application of CNTs.

Studies about the dispersion and alignment of CNTs in LLCs have been reported [4, 16, 22–24]. For example, Lagerwall et al. presented an approach of using self-organized lyotropic nematic LC to align and disperse SWNTs [21]. This work also pointed out the possibility of using other LLCs with higher orientational orders such as lamellar and hexagonal phases to align CNTs. However, the above LC composites must be presented in a test cell to avoid the fluiding of the materials. Okano et al. proposed an approach of using sulfonated polyaramide to prepare ordered SWCNTs/LLCs film. The fabrication process of this system is simple (LLCs in water) and low-cost, endowing the broad versatility performance of LLC polymer film with SWCNTs [25]. These characteristic properties also demonstrated that the LLC/SWNTs systems took an advantage over the TLC/SWCNTs systems.

Although using LLCs to align CNTs is a promising method to promote the application of CNTs, the direct incorporation of SWNTs into the LLCs with high viscosity is usually difficult.

A modified method for the introduction of the CNTs into LLCs is accompanied by rigorous sonication and heating, which is complicated and harmful for the LLC matrix. The heating process may also induce the desorption of the amphiphilic molecules from the surface of the CNTs, leading to the aggregation of the CNTs [26]. To overcome these problems, Xin and coworkers developed a nondestructive strategy to incorporate SWNTs into LLCs utilizing the spontaneous phase separation between a nonionic surfactant (n-dodecyl hexaoxyethylene glycol monoether, $C_{12}E_6$) and a hydrophilic polymer (poly(ethyiene glycol), (PEG)) [27]. After phase separation, an upper phase contains hexagonal LLCs incorporated by SWNTs together with a bottom isotropic phase consisting of PEG (**Figure 1A** and **B**). The type of the LLC phase (hexagonal or lamellar phase) could be regulated by varying the ratio of PEG and $C_{12}E_6$. The quality of SWNTs/LLC composites was characterized by polarized microscopy (POM) observations (**Figure 1C** and **D**) and small-angle X-ray scattering (SAXS) measurements (**Figure 1E** and **F**). The d-spacing of the upper hexagonal phase was improved with increasing amount of SWNTs, which was consistent with the results obtained from POM observations.

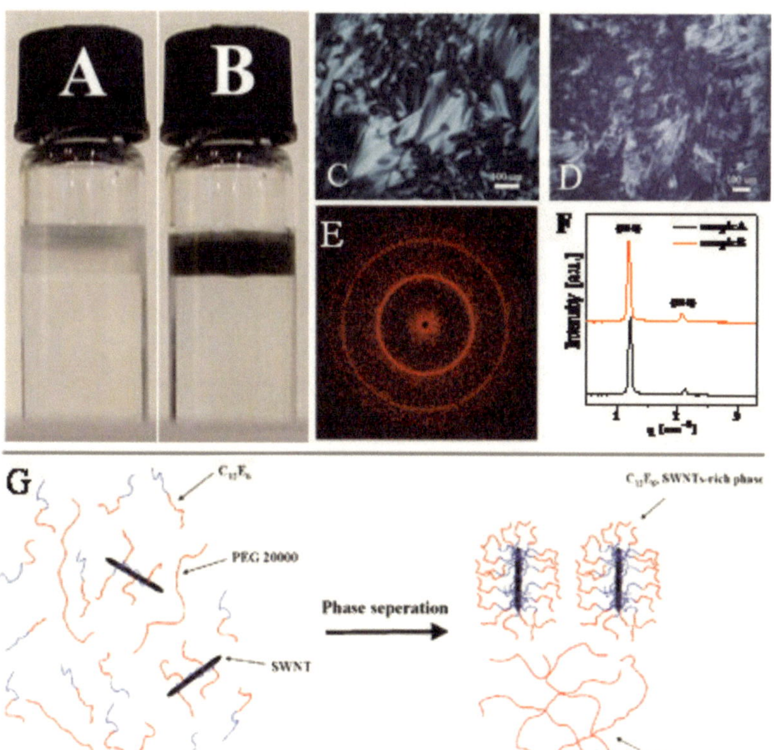

Figure 1. SWNTs embedded in the upper hexagonal phase formed by $C_{12}E_6$ (10 wt%) after phase separation induced by PEG 20,000 (20 wt%). The percent of SWNTs in the upper phase was calculated to be (A) 0 wt% and (B) 0.25 wt%. Corresponding polarized micrographs of the upper $C_{12}E_6$ hexagonal phases without (C) and with (D) incorporated SWNTs. (E) Small-angle diffraction rings of sample B and (F) SAXS results of SWNTs/LLC composites. (G) Schematic representation of the phase separation process in the four-component mixture of the surfactant $C_{12}E_6$, PEG 20,000, SWNTs, and water [27].

Next, the incorporation of CNTs into the LLC phase formed by n-dodecyl tetraethylene monoether ($C_{12}E_4$) through phase separation at the presence of PEG with different molecular weights has been systematically studied (**Figure 2**) [28]. The LLC/CNTs hybrid material maintained the lamellar organizations of the host LLCs according to POM observations and SAXS measurements. The increase of the d-spacing of the LLC/CNTs hybrid with increasing concentration of incorporated CNTs indicated that the CNTs have been successfully integrated within the layer of lamellar LLCs. UV-vis and Raman spectra further confirmed that CNTs have been incorporated into the LLC phase, which also revealed the alignment of CNTs in the LLC matrix. The mechanical strength of the hybrid material has also been improved after the introduction of CNTs.

Furthermore, the LLCs formed by ionic surfactants such as sodium dodecyl sulfate (SDS) and cetyltrimethylammonium bromide (CTAB) have also been selected as the host matrix for SWNTs incorporation. In these cases, polyelectrolytes such as poly(sodium styrenesulfonate) (PSS) or poly(diallyldimethylammonium chloride) (PDADMAC) instead of PEG were utilized to induce phase separation (**Figure 3**) [29]. The final concentration of the SWNTs in the upper LLC phase was increased a few times compared with that in the initial aqueous dispersion. Thus, it can be concluded that incorporation of SWNTs into LLC by phase separation method provides a practical way to achieve highly concentrated SWNTs aqueous

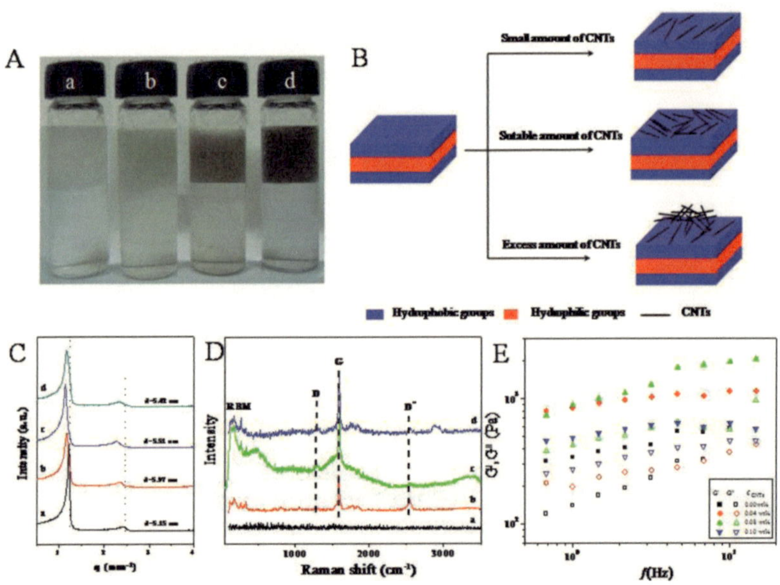

Figure 2. (A) CNTs embedded in the upper LLC phase formed by $C_{12}E_4$ (10 wt%) after phase separation induced by PEG 20,000 (20 wt%). (B) Schematic illustrations of the states of LLC/CNTs composites as a function of the concentration of incorporated CNTs in the upper LLC phase: (a) 0, (b) 0.04 wt%, (c) 0.08 wt% and (d) 0.10 wt%. (C) SAXS results of LLC/CNTs composites as a function of the concentration of incorporated CNTs in the upper LLC phase. (D) Raman spectra excited at 1064 nm of (a) pure $C_{12}E_4$ LLC, (b) the raw CNTs, (c) 0.05 wt% CNTs dispersed in 10 mL of 0.1 wt% $C_{12}E_4$ and (d) $C_{12}E_4$ LLC/CNTs composites. (E) Rheological results for LLC/CNTs composites with increasing concentration of CNTs [28].

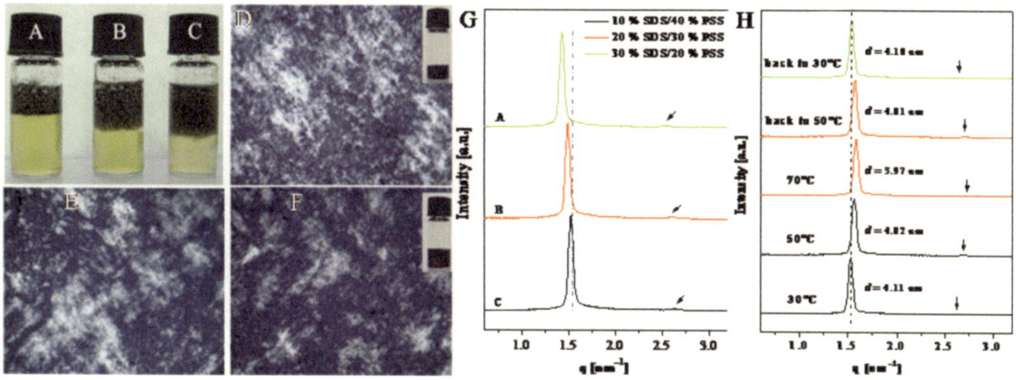

Figure 3. (A–C) Photographs of three typical samples of SWNTs/SDS/PSS/H_2O system after phase separation. (D–F) Corresponding polarized micrographs of the upper SWNTs/LLC hybrids. The weight percent of SDS and PSS to water are 10 wt%/40 wt% (D), 20 wt%/30 wt% (E), and 30 wt%/20 wt% (F). (G) SAXS results of the three typical samples of A–C. (H) SAXS results of one typical sample of SWNTs in CTAB/LLC as a function of temperature. The weight fractions of CTAB and PDADMAC to water are 20 and 15 wt%, respectively [29].

dispersion. Moreover, it is surprising to find the stability of SWNTs/LLC hybrids fabricated from ionic surfactants is much better than those prepared from nonionic surfactants, which indicates that the SWNTs/LLC hybrids prepared by the combination of ionic surfactant/polyelectrolyte may find potential applications at higher temperatures.

CNTs possess hollow lumens with diameters of a few to hundred nanometers, which are an ideal geometry for drug transport and delivery. However, CNTs are observed to exhibit weak infrared emissions. For diagnosis, the nanotubes must be functionalized with spectroscopically characteristic fluorescent dyes [30]. Xin and coworkers fabricated a luminescent CNTs-based hybrid material by anchoring lanthanide complexes (Eu or Tb) onto the surfaces of multi-walled CNTs (MWNTs) (**Figure 4A**). UV–vis measurements was used to demonstrate the successful coupling of Eu(III) complexes to MWCNTs (**Figure 4B**). Then, they incorporated the luminescent Eu-MWCNTs into the $C_{12}E_6$ LLC phase with a phase separation method using PEG [31]. The hexagonal phase of $C_{12}E_6$ was remained, and the Eu-MWNTs were ordered in the hexagonal liquid crystal phase (**Figure 4C**). The introduction of the Eu-MWNTs merely induced the swollen of the hexagonal lattice, and the luminescent property was retained (**Figure 4D**).

Cellulose nanocrystals (CNCs), obtained from hydrolysis of the cellulose, have attracted considerable attention due to their anisotropic properties and their self-assembly behavior [32]. Recently, Yuan and coworkers firstly fabricated the composite films with ordered oxidized CNTs (o-CNTs) using the lyotropic nematic liquid crystals (CNLCs), and the framework of the CNLCs can be retained in the final solid films (**Figure 5**) [33]. The randomly oriented o-CNTs in the aqueous dispersion were aligned orderly due to the confinement of the liquid crystal matrix of CNCs. The composite films were endowed with the anisotropic conductivity with the help of ordered arrangement of o-CNTs and the anisotropy of the CNLCs. Their work provides a new approach of fabricating LCs/CNTs material with application in sensors and photoelectronics.

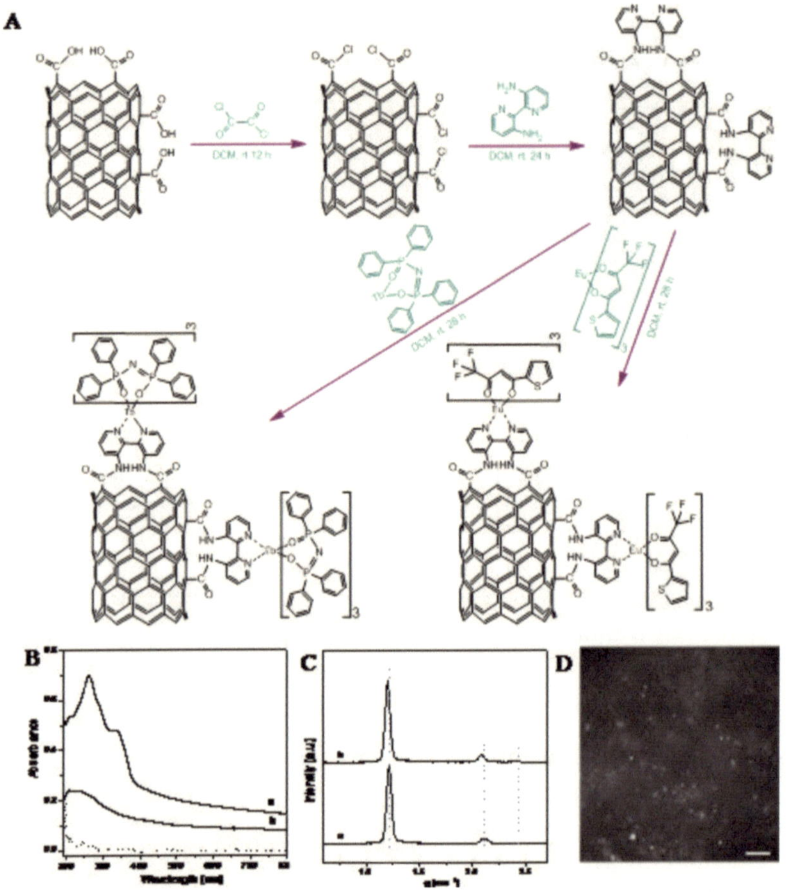

Figure 4. (A) Synthetic procedure for Eu(III) and Tb(III)-coupled MWNTs (Eu or Tb-MWNTs). (B) UV-vis spectra of Eu-MWNTs (a) and ligand-modified MWNTs (b) dispersed in 1 wt% SDS. (C) Typical SAXS results of the upper condensed phase formed by $C_{12}E_6$ at the presence of PEG without (a) and with (b) Eu-MWNTs incorporated. The dotted lines are guides for the eyes to highlight the peaks. (D) A typical image obtained under a fluorescence microscope for the upper $C_{12}E_6$ hexagonal phase incorporated with 0.1 wt% Eu-MWNTs [31].

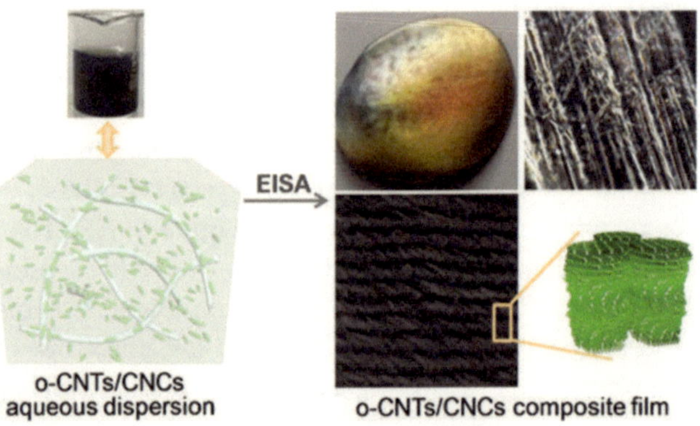

Figure 5. The composite films with ordered CNTs using the chiral nematic liquid crystals (CNLCs) prepared by the cellulose nanocrystals (CNCs) [33].

3. Graphene (or graphene oxide)/LLC hybrid materials

Apart from one-dimensional (1D) CNTs, the two-dimensional (2D) carbon-based nanomaterials, i.e., graphene nanosheets, have also attracted a great deal of attention. Graphene, a single-layer and 2D carbon lattice, is one of the most promising materials with great potential applications due to its unique mechanical, quantum, and electrical properties [34–36]. The properties of superior dispersibility, stability, and processability in water also promote the exploration of graphene assemblies [37]. For example, Behabtu et al. reported the formation of LC phase of graphene at high concentrations (~20–30 mg ml^{-1}) [38]. The graphite can be spontaneously exfoliated into single-layer graphene in chlorosulphonic acid without the need for surfactant stabilization and sonication. The LC phases are promising for functionalization, and for scalable manufacturing of nanocomposites, films, coatings, and high-performance fibers. The oxidized derivative of graphene, i.e., graphene oxide (GO), is a good hydrophilic and biocompatible material. Due to the presence of a variety of hydroxyl, epoxide, and carbonyl groups at their basal planes and edges, GO has been regarded as the most important substitute for graphene to form stable colloidal carbon-based composites in water and polar organic solvents [39–43]. Kim et al. prepared nematic LCs of exfoliated graphene oxide by a modified Hummer's method [44]. The ionic strength and pH are the influencing factors on the stability of the LCs. They also successfully tuned the macroscopic orientation of GO in LCs by applying a magnetic field (**Figure 6**). This method provided a viable route to engineering high-performance nanocomposites.

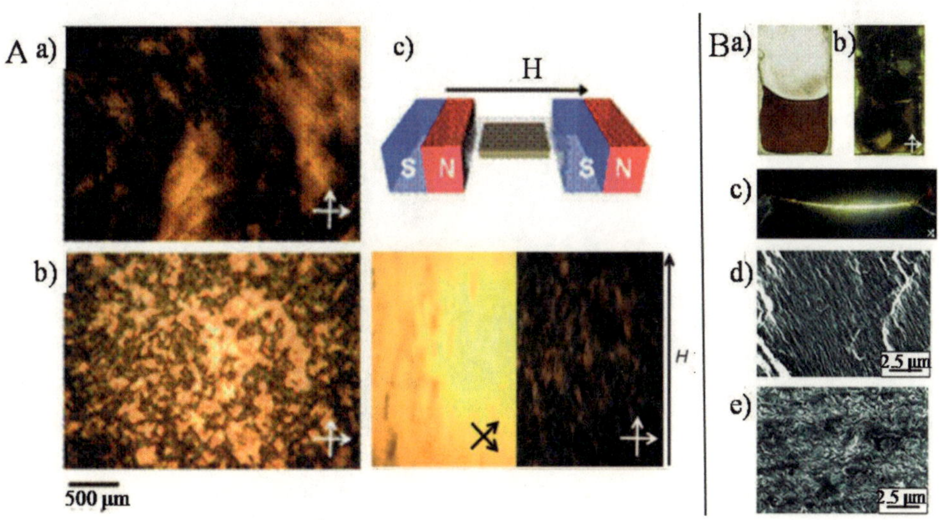

Figure 6. (A) Magnetic-field-induced alignment of graphene oxide liquid crystals. a) Shear-induced birefringent morphology formed after sample preparation. b) Nematic schlieren morphology formed about 3 h after sample preparation without any external field. c) Top: experimental scheme for magnetic field application; bottom: magnetic-field-induced highly aligned liquid-crystal texture. (B) Mechanical-deformation-induced alignment of PAA/grapheme oxide composites. Water/PAA/graphene oxide three-component liquidcrystal mixtures: a) without and b) with crossed polarizers. c) Handdrawn gel composite fiber. The strong optical birefringence was caused by homogeneously dispersed, uniaxially oriented graphene oxide platelets. d) Highly aligned graphene oxide morphology along the fiber axis. e) Randomly oriented graphene oxide morphology in the fiber cross section [44].

LLC phases containing graphene and GO sheets have received considerable attention because of the new dimension to soft self-assembly science. A series of LLCs containing graphene and GO have been fabricated employing novel amphiphilic molecules, which achieved functional materials with enhanced properties, self-assembly, and alignment of graphene and GO [45, 46]. Pasquali et al. explored the lyotropic phase behavior of giant GO flakes in water with an order of magnitude higher than other works in 2011 [47]. Wallace and coworkers produced the GO LLC phase in a wide range of organic solvents (ethanol, acetone, tetrahydrofuran, and some other organic solvents) fully using the ultralarge GO sheets (**Figure 7A**) [48]. The GO LLC phase could be used to take the exploitation of organizing and aligning SWNTs through the addition of LC GO to the SWNT dispersions (**Figure 7B**). The presence of fine interband transitions of the UV/vis-near-IR spectra of the LC GO and LC GO-SWNT demonstrated that SWNT sizes have been preserved in the composite formulation (**Figure 7C**).

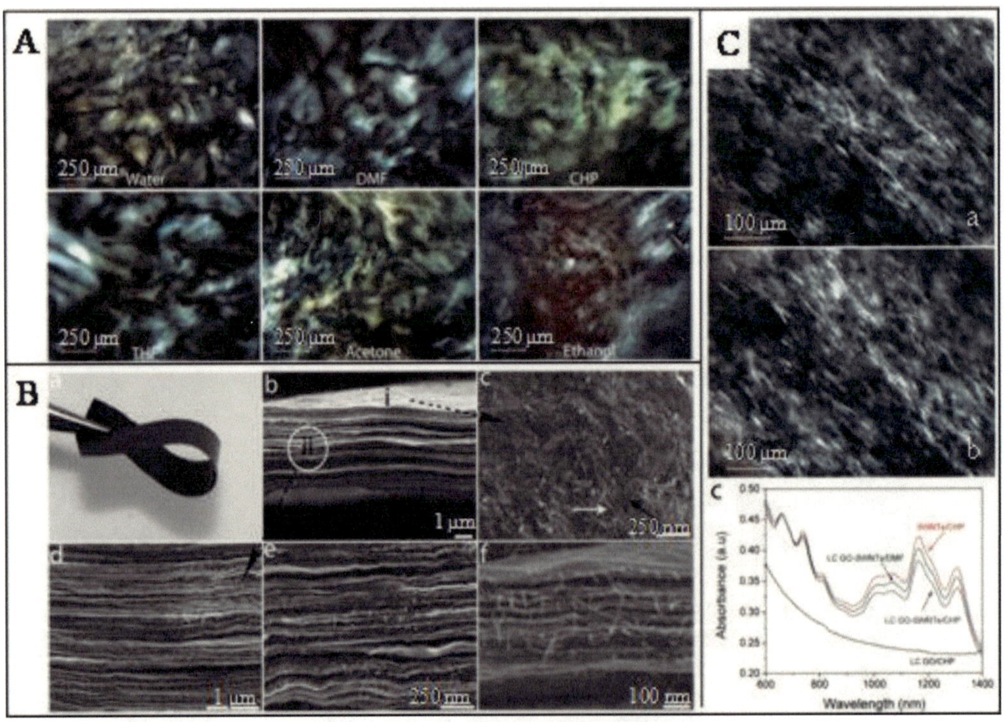

Figure 7. (A) Representative POM images of LC GO in various organic solvents at a concentration of 2.5 mg mL^{-1}. (B) (a) Representative photograph of a flexible free-standing paper of LC GO-SWNT made by cast drying method. (b) SEM image of the cross section of as-cast dried LC GO-SWNT paper. (c) SEM image of the surface of the layer-by-layer composite, which is marked as region (i) in (b). Some of the SWNTs are laid on the surface of the paper (white arrow), while others are placed between layers of GO sheets (black arrow). Transparency of the monolayer/few layers of GO sheets allows observing tube sites in different layers. (d-f) Cross section of composite paper at different magnifications (marked as (ii) in (b)) confirmed the self-oriented nature of the composite as well as maintaining SWNTs debundled after the fabrication of composite. (C) POM micrographs of LC GO-SWNTs/CHP (a), (b) LC GO-SWNTs/DMF. (c) UV-vis-near-IR spectra of SWNTs and LC GO dispersions before and after mixing together [48].

Xin and coworkers successfully incorporated graphene and GO into LLC matrix constructed by $C_{12}E_4$ (**Figure 8A and B**) [49]. Typical TEM images showed that graphene and GO were all fully exfoliated and well-dispersed in aqueous solutions, indicating that the carbon nanosheets can be efficiently stabilized by the steric repulsions created by $C_{12}E_4$ (**Figure 8C** and **D**). According to the polarized optical microscope (POM) and small-angle X-ray scattering (SAXS) results, all of the composites including graphene–$C_{12}E_4$ (**Figure 8E–G**) and graphene oxide–$C_{12}E_4$ (**Figure 8H** and **I**) are characteristic of lamellar structures. The strength of the composite could be enhanced by the addition of a small amount of well dispersed graphene and GO. A phase separation method was taken to demonstrate the difference between the interaction mechanisms of graphene and GO with $C_{12}E_4$ LLC because of the different nature of graphene and GO. The schematic illustration of graphene and GO incorporated into the $C_{12}E_4$ LLC phase is presented in **Figure 9**. The results indicated that the improved mechanical and electrical properties of the $C_{12}E_4$ LLC by graphene and GO facilitated the manipulation and processing of graphene and GO in nanotechnology, electrochemical and biochemical areas.

The graphene (or GO)/LLC matrix could be facilely regulated by changing its composition, for example, the type of amphiphilic molecules. Following the above work, Xin and coworkers successfully incorporated GO into a hybrid LLC matrix constructed by the mixture of $C_{12}E_4$ and 1-dodecyl-3-methylimidazolium bromide ionic liquid (C_{12}mimBr) [50]. The GO was well-dispersed in the $C_{12}E_4/C_{12}$mimBr hybrid LLC matrixes at room temperature according to POM observations and SAXS results, which indicated that GO/60 wt% $C_{12}E_4$/10 wt% C_{12}mimBr LLC composite is the lamellar phase and GO/10 wt% $C_{12}E_4$/60 wt% C_{12}mimBr LLC composite exhibits a hexagonal structure. GO could not only improve the mechanical properties, but also tune the phase state of hybrid LLC matrixes from lamellar to hexagonal state. The addition of large amounts of C_{12}mimBr greatly increased the thermal stability of GO/$C_{12}E_4/C_{12}$mimBr LLC

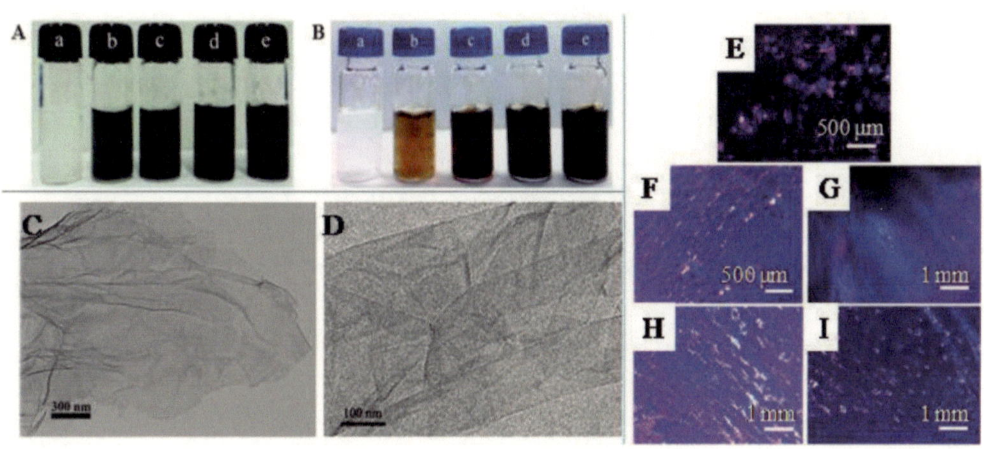

Figure 8. Graphene (A) and graphene oxide (B) incorporated into the LLC phase formed by 35 wt% $C_{12}E_4$. The weight percent of graphene and GO is 0 (a), 0.03 (b), 0.09 (c), 0.15 (d), 0.20 (e) mg mL-1 in (A) and 0 (a), 0.1 (b), 0.5 (c), 1.0 (d) 1.5 (e) mg mL-1 in (B), respectively. (C, D) Typical TEM images of grapheme and GO dispersed in water. POM images for graphene–$C_{12}E_4$ (E–G) and GO–$C_{12}E_4$ (H, I) LLC composites [49].

and there was no obvious structural change during the heating or cooling process (**Figure 10**). **Figure 11** gives the schematic illustration of the formation of GO/$C_{12}E_4$/C_{12}mimBr lamellar and hexagonal LLC composites.

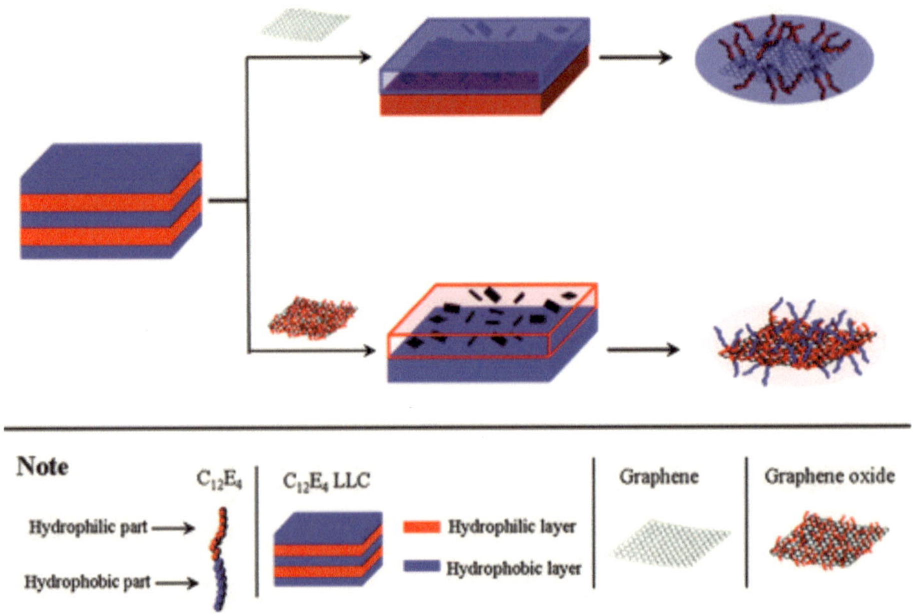

Figure 9. Schematic illustration of graphene and GO incorporated into the $C_{12}E_4$ LLC phase. Graphene incorporates into the hydrophobic layers of the $C_{12}E_4$ LLC phase while GO tends to stay in the hydrophilic layers [49].

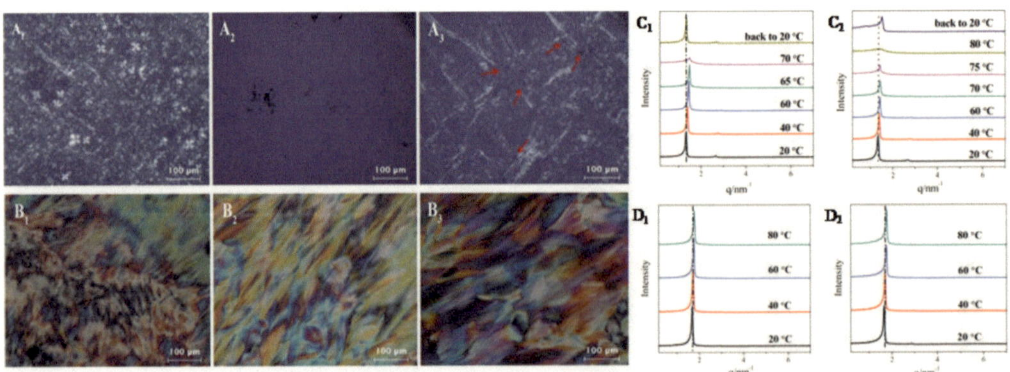

Figure 10. POM images of liquid crystal composites as the change of temperature, c_{GO} = 0.3 mg mL^{-1}, (A) 60 wt% $C_{12}E_4$/10 wt% C_{12}mimBr and (B) 10 wt% $C_{12}E_4$/60 wt% C_{12}mimBr. The temperature is (A_1 and B_1) 20°C, (A2) 80°C, (B2) 90°C and (A_3 and B_3) back to 20°C from high temperature. The SAXS results of LLC composites: 60 wt% $C_{12}E_4$/10 wt% C_{12}mimBr (C_1 and C_2); 10 wt% $C_{12}E_4$/60 wt% C_{12}mimBr (D_1 and D_2). The concentration of GO is 0 mg mL^{-1} (C_1 and D_1) and 0.3 mg mL^{-1} (C_2 and D_2) [50].

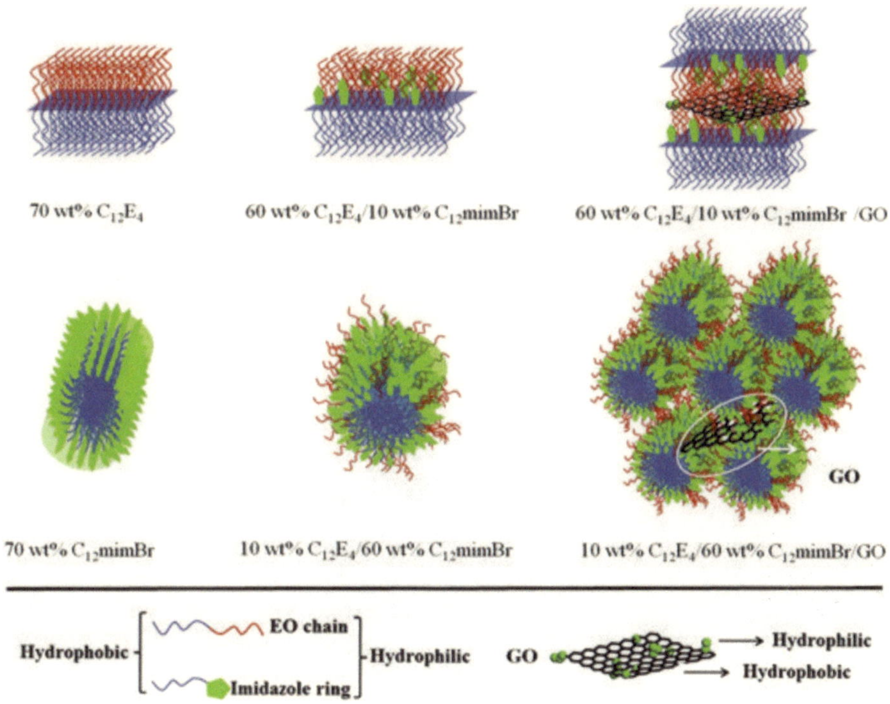

Figure 11. Schematic illustration of the formation of GO/$C_{12}E_4$/C_{12}mimBr lamellar and hexagonal LLC composites [50].

4. Manipulation of LLC behavior of amphiphilic molecules by biomolecules

Apart from varying the type of amphiphilic molecules, the structure and properties of LLC can also be tuned by changing the additives to the LLC composites. Much attention has been paid to study the effect of biomolecules to the amphiphilic molecules LLCs with new types of functions. The literatures have shown that LLCs with biomolecules are useful for biological sensing and NMR RDC analysis [51]. Clark et al. have reported the formation of LLC phases of double-stranded DNA and RNA oligomers in water [52, 53]. The formation of LLC by hydrated duplex DNA plays a crucial role in deciphering the structure of DNA and enables the alignment of the DNA chains.

Amino acids are the basic units of proteins and peptides, which give the proteins and peptides specific molecular structure features [54]. Oligomers of the β-amino acids (β-peptide), a second class of biomolecules, can aggregate into diverse nanostructures [55, 56]. Gellman et al. investigated the effect of β-peptide modifications on the propensity of these helical molecules to form LLC in water [57]. The side chain of β-peptides displayed an important role to well-defined nanostructures and rules for creating LLC phases, which can also endow

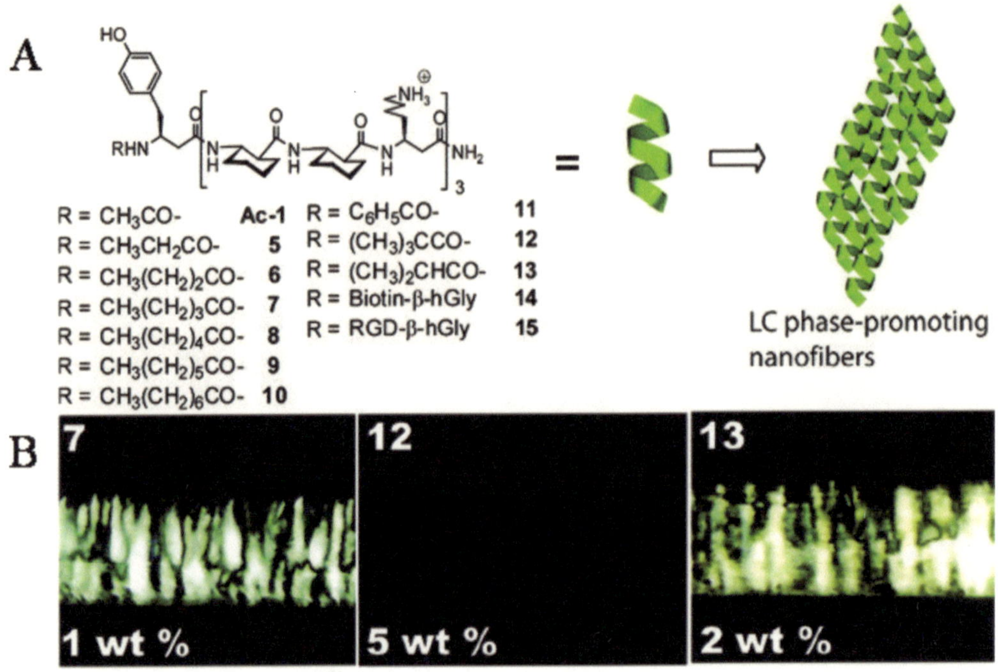

Figure 12. (A) Acylated β-peptides containing hydrocarbon acyl tails, Ac-1 and 5-13, and biological recognition groups, 14 and 15. Model for the multistate assembly of amphiphilic β-peptides, progressing of formation of nanofibers directly from monomeric β-peptides. (B) Optical micrographs of aqueous solutions of acylated β-peptides 7, 12, and 13 between crossed polarizers [57].

the LLCs useful properties (**Figure 12**). They also demonstrated that LLC formed by the modified β-peptides was useful as NMR alignment media to small organic molecules in aqueous solution and provided initial evidence for enantiodiscrimination [58].

The interactions between the surfactants and amino acids have also been investigated, which is not only of fundamental important in theoretics but also practical in industrial applications [59–62]. Except for traditional surfactants, long-chain alkyl ionic liquids are special kinds of amphiphilic surfactant molecules which can also form LLC. Xin and coworkers systematically studied the effects of alkaline amino acids L-Arginine (L-Arg) and L-Lysine (L-Lys) on the LLC behavior of C_{14}mimBr. C_{14}mimBr/L-Arg system remained the hexagonal phases and merely led to the variation of the mechanical strength (**Figure 13A–F**) [63]. L-Lys could induce a transition of the C_{14}mimBr LLC phase to worm-like micelles (WLMs). The balance among electrostatic interaction, H-bond interaction, and hydrophobic interaction between amino acids, C_{14}mimBr and water contributes to the phase transition. The schematic illustrations for phase transition introduced by the amino acids are shown in **Figure 13G**. These changes can contribute to a better understanding of the effect of the additives on the influence of the structure and macroscopic properties of LLCs.

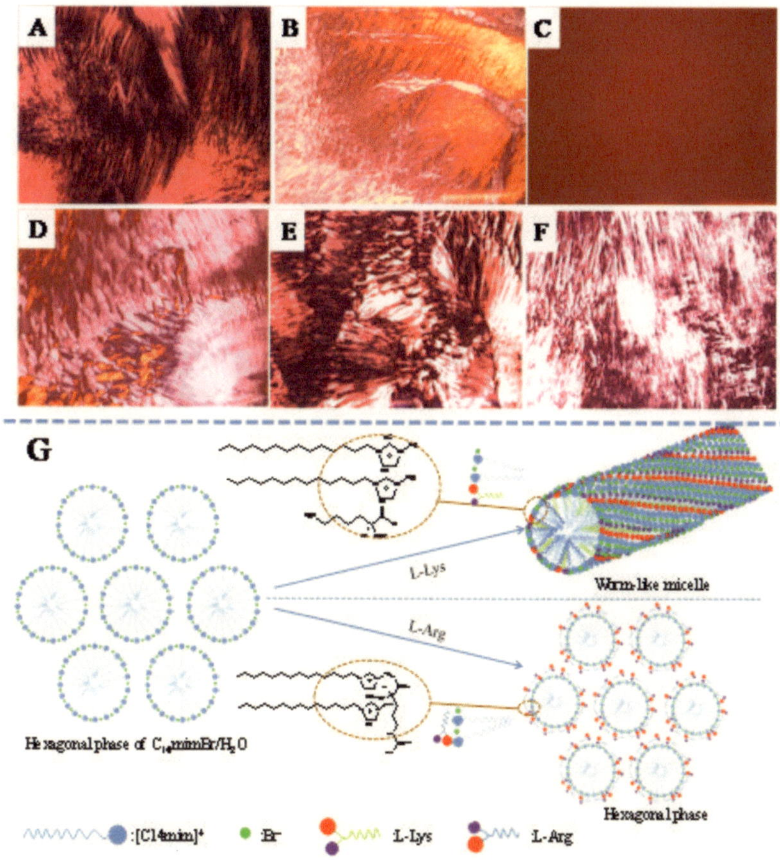

Figure 13. POM images (100×) for 38 wt% C_{14}mimBr–amino acids mixtures with different water contents at room temperature. The amount of the amino acids is: (A) 0; (B) 1.0 wt% L-Lys; (C) 5.0 wt% L-Lys; (D) 2.5 wt% L-Arg; (E) 5 wt% L-Arg; and (F) 7.5 wt% L-Arg. (G) Schematic illustrations of the phase transition between LLCs and WLMs by the L-Arg and L-Lys [63].

5. Conclusion and outlook

In summary, LCs, with their fluidity as well as long-range organization, represent an interesting and novel route for realizing functional composites. In the ordered LLC phases, the molecules tend to align along a common direction, forming orientationally ordered macroscopic domains which can provide a way to control the orientation of guest materials. Thus, different kinds of additives including SWNTs, MWNTs, graphene, GO, and biomolecules can be incorporated into LLCs, which will induce various different properties of LLCs. The mechanical, electric, physicochemical properties of the hybrid LLC materials will be improved largely after the incorporation, which will open the door for the applications of these interesting hybrid materials in nanotechnology, electrochemical and biochemical areas.

Acknowledgements

We gratefully acknowledge the financial support obtained from the National Natural Science Foundation of China (21476129, 21203109) and Young Scholars Program of Shandong University (2016WLJH20).

Author details

Zhaohua Song, Yanzhao Yang and Xia Xin*

*Address all correspondence to: xinx@sdu.edu.cn

National Engineering Technology Research Center for Colloidal Materials, Shandong University, Jinan, PR China

References

[1] Collings PJ, Hird M. Introduction to Liquid Crystals: Chemistry and Physics. CRC Press; 1997

[2] Hamley IW. Nanotechnology with soft materials. Angewandte Chemie International Edition. 2003;**42**:1692-1712. DOI: 10.1002/anie.200200546

[3] Lagerwall J, Scalia G, Haluska M, Dettlaff-Weglikowska U, Roth S, Giesselmann F. Nanotube alignment using lyotropic liquid crystals. Advance Materials. 2007;**19**:359-364. DOI: 10.1002/adma.200600889

[4] Dierking I, Scalia G, Morales P, LeClere D. Aligning and reorienting carbon nanotubes with nematic liquid crystals. Advance Materials. 2004;**16**:865-869. DOI: 10.1002/adma.200306196

[5] Kijima T, Yoshimura T, Uota M, Ikeda T, Fujikawa D, Mouri S, Uoyama S. Noble-metal nanotubes (Pt, Pd, Ag) from lyotropic mixed-surfactant liquid-crystal templates. Angewandte Chemie International Edition. 2004;**116**:230-234. DOI: 10.1002/ange.200352630

[6] Rabatic BM, Pralle MU, Tew GN, Stupp SI. Nanostructured semiconductors templated by cholesteryl-oligo(ethylene oxide) amphiphiles. Chemistry of Materials. 2003;**15**:1249-1255. DOI: 10.1021/cm020899e

[7] Monteiro MJ, Hall G, Gee S, Xie L. Protein transfer through polyacrylamide hydrogel membranes polymerized in lyotropic phases. Biomacromolecules. 2004;**5**:1637-1641. DOI: 10.1021/bm049789m

[8] Hulvat JF, Stupp SI. Anisotropic properties of conducting polymers prepared by liquid crystal templating. Advanced Materials. 2004;**16**:589-592. DOI: 10.1002/adma.200306263

[9] Kossyrev PA, Yin A, Cloutier SG, Cardimona DA, Huang DH, Alsing PM, Xu JM. Electric field tuning of plasmonic response of nanodot array in liquid crystal matrix. Nano Letters. 2005;**5**(10):1978-1981. DOI: 10.1021/nl0513535

[10] Attard GS, Glyde JC, Göltner CG. Liquid-crystalline phases as templates for the synthesis of mesoporous silica. Nature. 1995;**378**:366-368. DOI: 10.1038/378366a0

[11] Soberats B, Yoshio M, Ichikawa T, Taguchi S, Ohno H, Kato T. 3D anhydrous proton-transporting nanochannels formed by self-assembly of liquid crystals composed of a sulfobetaine and a sulfonic acid. Journal of American Chemical Society. 2013;**135**:15286-15289. DOI: 10.1021/ja407883b

[12] Olutas EB, Balci FM, Dag O. Strong acid-nonionic surfactant lyotropic liquid-crystalline mesophases as media for the synthesis of carbon quantum dots and highly proton conducting mesostructured silica thin films and monoliths. Langmuir 2015;**31**:10265-10271. DOI: 10.1021/acs.langmuir.5b02225

[13] Negrini R, Fong WK, Boyd BJ, Mezzenga R. pH-responsive lyotropic liquid crystals and their potential therapeutic role in cancer treatment. Chemical Communication. 2015;**51**:6671-6674. DOI: 10.1039/C4CC10274F

[14] Bryks W, Wette M, Velez N, Hsu SW, Tao AR. Supramolecular precursors for the synthesis of anisotropic Cu_2S nanocrystals. Journal of American Chemical Society. 2014;**136**:6175-6178. DOI: 10.1021/ja500786p

[15] Dellinger TM, Braun PV. Lyotropic liquid crystals as nanoreactors for nanoparticle synthesis. Chemical Materials. 2004;**16**:2201-2207. DOI: 10.1021/cm0349194

[16] Lynch MD, Patrick DL. Organizing carbon nanotubes with liquid crystals. Nano Letters. 2002;**2**:1197-1201. DOI: 10.1021/nl025694j

[17] Ajayan PM. Nanotubes from carbon. Chemical Review. 1999;**99**:1787-1800. DOI: 10.1021/cr970102g

[18] Niyogi S, Hamon MA, Hu H, Zhao B, Bhowmik P, Sen R, Itkis M, Haddon R. Chemistry of single-walled carbon nanotubes. Accounts of Chemical Research. 2002;**35**:1105-1113. DOI: 10.1021/ar010155r

[19] Badaire S, Zakri C, Maugey M, Derré A, Barisci JN, Wallace G, Poulin P. Liquid crystals of DNA-stabilized carbon nanotubes. Advanced Materials. 2005;**17**:1673-1676. DOI: 10.1002/adma.200401741

[20] Sinani VA, Gheith MK, Yaroslavov AA, Rakhnyanskaya AA, Sun K, Mamedov AA, Wicksted JP, Kotov NA. Aqueous dispersions of single-wall and multiwall carbon nanotubes with designed amphiphilic polycations. Journal of American Chemical Society. 2005;**127**:3463-3472. DOI: 10.1021/ja045670+

[21] Lagerwall J, Scalia G, Haluska M, Dettlaff-Weglikowska U, Roth S, Giesselmann F. Nanotube alignment using lyotropic liquid crystals. Advanced Materials. 2007;**19**:359-364. DOI: 10.1002/adma.200600889

[22] Jiang W, Yu B, Liu W, Hao J. Carbon nanotubes incorporated within lyotropic hexagonal liquid crystal formed in room-temperature ionic liquids. Langmuir. 2007;**23**:8549-8553. DOI: 10.1021/la700921w

[23] Weiss V, Thiruvengadathan R, Regev O. Preparation and characterization of a carbon nanotube–lyotropic liquid crystal composite. Langmuir. 2006;**22**:854-856. DOI: 10.1021/la052746m

[24] Xu W, Yin Q, Gao Y, Yu L. Solvent and substituent effects on the aggregation behavior of surface-active ionic liquids with aromatic counterions and the dispersion of carbon nanotubes in their hexagonal liquid crystalline phase. Langmuir. 2015;**31**:12644-12652. DOI: 10.1021/acs.langmuir.5b03586

[25] Okano K, Noguchi I, Yamashita T. Anisotropic carbon nanotube films fabricated from a lyotropic liquid-crystalline polymer. Macromolecules. 2010;**43**:5496-5499. DOI: 10.1021/ma101004s

[26] Zhang S, Kumar S. Carbon nanotubes as liquid crystals. Small. 2008;**4**:1270-1283. DOI: 10.1002/smll.200700082

[27] Xin X, Li H, Wieczorek S A, Szymborski T, Kalwarczyk E, Ziebacz N, Gorecka E, Pociecha D, Hołyst R. Incorporation of carbon nanotubes into a lyotropic liquid crystal by phase separation in the presence of a hydrophilic polymer. Langmuir. 2009;**26**:3562-3568. DOI: 10.1021/la902960a

[28] Wang L, Xin X, Guo K, Yang M, Ma X, Yuan J, Shen J, Yuan S. Ordered carbon nanotubes–n-dodecyl tetraethylene monoether liquid crystal composites through phase separation induced by poly (ethylene glycol). Physical Chemistry Chemical Physics. 2014;**16**:14771-14780. DOI: 10.1039/c4cp00622d

[29] Xin X, Li H, Kalwarczyk E, Kelm A, Fiałkowski M, Gorecka E, Pociecha D, Hołyst R. Single-walled carbon nanotube/lyotropic liquid crystal hybrid materials fabricated by a phase separation method in the presence of polyelectrolyte. Langmuir. 2010;**26**:8821-8828. DOI: 10.1021/la101032d

[30] Wang F, Dukovic G, Brus LE, Heinz TF. The optical resonances in carbon nanotubes arise from excitons. Science. 2005;**308**:838-841. DOI: 10.1126/science.1110265

[31] Xin X, Pietraszkiewicz M, Pietraszkiewicz O, Chernyayeva O, Kalwarczyk T, Gorecka E, Pociecha D, Li H, Hołyst R. Eu(III)-coupled luminescent multi-walled carbon nanotubes in surfactant solutions. Carbon. 2012;**50**:436-443. DOI:10.1016/j.carbon.2011.08.059

[32] Eyley S, Thielemans W. Surface modification of cellulose nanocrystals. Nanoscale. 2014;**6**:7764-7779. DOI: 10.1039/C4NR01756K

[33] Sun J, Zhang C, Yuan Z, Ji X, Fu Y, Li H, Qin M. Composite films with ordered carbon nanotubes and cellulose nanocrystals. Journal of Physical Chemistry C. 2017;**121**:8976-8981. DOI: 10.1021/acs.jpcc.7b01528

[34] Novoselov KS, Geim AK, Morozov SV, Jiang D, Katsnelson M, Grigorieva I, Dubonos S, Firsov A. Two-dimensional gas of massless Dirac fermions in graphene. Nature. 2005;**438**:197-200. DOI: 10.1038/nature04233

[35] Huang X, Yin Z, Wu S, Qi X, He Q, Zhang Q, Yan Q, Boey F, Zhang H. Graphene-based materials: Synthesis, characterization, properties, and applications. Small. 2011;**7**:1876-1902. DOI: 10.1002/smll.201002009

[36] Lee C, Wei X, Kysar J W, Hone J. Measurement of the elastic properties and intrinsic strength of monolayer graphene. Science. 2008;**321**:385-388. DOI: 10.1126/science.1157996

[37] Li Z, Liu Z, Sun H, Gao C. Superstructured assembly of nanocarbons: Fullerenes, nanotubes, and graphene. Chemical Review. 2015;**115**:7046-7117. DOI: 10.1021/acs.chemrev.5b00102

[38] Behabtu N, Lomeda JR, Green MJ, Higginbotham AL, Sinitskii A, Kosynkin DV. Tsentalovich D, Parra-Vasquez ANG, Schmidt J, Kesselman E. Spontaneous high-concentration dispersions and liquid crystals of graphene. Natural Nanotechnology. 2010;**5**:406-411. DOI: 10.1038/NNANO.2010.86

[39] Stankovich S, Dikin DA, Piner RD, Kohlhaas KA, Kleinhammes A, Jia Y, Wu Y, Nguyen ST, Ruoff RS. Synthesis of graphene-based nanosheets via chemical reduction of exfoliated graphite oxide. Carbon. 2007;**45**:1558-1565. DOI: 10.1016/j.carbon.2007.02.034

[40] Zhang C, Ren L, Wang X, Liu T. Graphene oxide-assisted dispersion of pristine multiwalled carbon nanotubes in aqueous media. Journal of Physical Chemistry C. 2010;**114**:11435-11440. DOI: 10.1021/jp103745g

[41] Zhu Y, Murali S, Cai W, Li X, Suk JW, Potts JR, Ruoff RS. Graphene and graphene oxide: Synthesis, properties, and applications. Advanced Materials. 2010;**22**:3906-3924. DOI: 10.1002/adma.201001068

[42] Xu Y, Hong W, Bai H, Li C, Shi G. Strong and ductile poly (vinyl alcohol)/graphene oxide composite films with a layered structure. Carbon. 2009;**47**:3538-3543. DOI: 10.1016/j.carbon.2009.08.022

[43] Chen D, Feng H, Li J. Graphene oxide: Preparation, functionalization, and electrochemical applications. Chemical Review. 2012;**112**:6027-6053. DOI: 10.1021/cr300115g

[44] Kim JE, Han TH, Lee SH, Kim JY, Ahn CW, Yun JM, Kim SO. Graphene oxide liquid crystals. Angewandte Chemie International Edition. 2011;**50**:3043-3047. DOI: 10.1002/anie.201004692

[45] Aboutalebi SH, Gudarzi MM, Zheng QB, Kim JK. Spontaneous formation of liquid crystals in ultralarge graphene oxide dispersions. Advanced Functional Materials. 2011;**21**:2978-2988. DOI: 10.1002/adfm.201100448

[46] Xu Z, Gao C. Graphene chiral liquid crystals and macroscopic assembled fibres. Natural Communication. 2011;**2**:571. DOI: 10.1038/ncomms1583

[47] Dan B, Behabtu N, Martinez A, Evans JS, Kosynkin DV, Tour JM, Pasquali M, Smalyukh II. Liquid crystals of aqueous, giant graphene oxide flakes. Soft Matter. 2011;**7**:11154-11159. DOI: 10.1039/C1SM06418E

[48] Jalili R, Aboutalebi SH, Esrafilzadeh D, Konstantinov K, Moulton SE, Razal JM, Wallace GG. Organic solvent-based graphene oxide liquid crystals: A facile route toward the next generation of self-assembled layer-by-layer multifunctional 3D architectures. Acs Nano. 2013;**7**:3981-3990. DOI: 10.1021/nn305906z

[49] Wang L, Xin X, Yang M, Ma X, Feng Z, Chen R, Shen J, Yuan S. Comparative study of n-dodecyl tetraethylene monoether lyotropic liquid crystals incorporated with graphene and graphene oxide. Physical Chemistry Chemical Physics. 2014;**16**:20932-20940. DOI: 10.1039/c4cp02634a

[50] Wang L, Xin X, Yang M, Shen J, Yuan S. Incorporation of graphene oxide into C 12 E 4/C 12 mimBr hybrid lyotropic liquid crystal and its thermo-sensitive properties. RSC Advances. 2015;**5**:68404-68412. DOI: 10.1039/c5ra13497h

[51] Van Nelson JA, Kim SR, Abbott NL. Amplification of specific binding events between biological species using lyotropic liquid crystals. Langmuir. 2002;**18**:5031-5035. DOI: 10.1021/la0118715

[52] Nakata M, Zanchetta G, Chapman BD, Jones CD, Cross JO, Pindak R, Bellini T, Clark NA. End-to-end stacking and liquid crystal condensation of 6–to 20–base pair DNA duplexes. Science. 2007;**318**:1276-1279. DOI: 10.1126/science.1143826

[53] Pan J, Han J, Borchers C H, Konermann L. Electron capture dissociation of electrosprayed protein ions for spatially resolved hydrogen exchange measurements. Journal of American Chemical Society. 2008;**130**:11574-11575. DOI: 10.1021/ja802871c

[54] He F, Xu G, Pang J, Ao M, Han T, Gong H. Effect of amino acids on aggregation behaviors of sodium deoxycholate at air/water surface: Surface tension and oscillating bubble studies. Langmuir. 2010;**27**:538-545. DOI: 10.1021/la103478c

[55] Guichard G, Huc I. Synthetic foldamers. Chemical Communication. 2011;**47**:5933-5941. DOI: 10.1039/C1CC11137J

[56] Molski MA, Goodman JL, Craig CJ, Meng H, Kumar K, Schepartz A. β-peptide bundles with fluorous cores. Journal of American Chemical Society. 2010;**132**:3658-3659. DOI: 10.1021/ja910903c

[57] Pomerantz WC, Yuwono VM, Drake R, Hartgerink JD, Abbott NL, Gellman SH. Lyotropic liquid crystals formed from ACHC-rich β-peptides. Journal of American Chemical Society. 2011;**133**:13604-13613. DOI: 10.1021/ja204874h

[58] Thiele CM, Pomerantz WC, Abbott NL, Gellman SH. Lyotropic liquid crystalline phases from helical β-peptides as alignment media. Chemical Communication. 2011;**47**:502-504. DOI: 10.1039/C0CC02123G

[59] Sun X, Xin X, Tang N, Guo L, Wang L, Xu G. Manipulation of the gel behavior of biological surfactant sodium deoxycholate by amino acids. Journal of Physical Chemistry B 2014;**118**:824-832. DOI: 10.1021/jp409626s

[60] Zhang Y, Xin X, Shen J, Tang W, Ren Y, Wang L. Biodegradable, multiple stimuli-responsive sodium deoxycholate–amino acids–NaCl mixed systems for dye delivery. RSC Advances. 2014;**4**:62262-62271. DOI: 10.1039/C4RA13353F

[61] Geng F, Zheng L, Liu J, Yu L, Tung C. Interactions between a surface active imidazolium ionic liquid and BSA. Colloid and Polymer Science. 2009;**287**:1253-1259. DOI:10.1007/s00396-009-2085-1

[62] Ruiz C C, Hierrezuelo J, Molina-Bolívar J. Effect of glycine on the surface activity and micellar properties of N-decanoyl-N-methylglucamide. Colloid and Polymer Science. 2008;**286**:1281-1259. DOI: 10.1007/s00396-008-1893-z

[63] Song Z, Xin X, Shen J, Jiao J, Xia C, Wang S, Yang Y. Manipulation of lyotropic liquid crystal behavior of ionic liquid-type imidazolium surfactant by amino acids. Colloids and Surfaces A. 2017;**518**:7-14. DOI: 10.1016/j.colsurfa.2017.01.004

Chapter 6

Electrical and Thermal Tuning of Band Structure and Defect Modes in Multilayer Photonic Crystals

Carlos G. Avendaño, Daniel Martínez and
Ismael Molina

Additional information is available at the end of the chapter

http://dx.doi.org/10.5772/intechopen.70473

Abstract

We describe the main results previously studied concerning the thermal and electrical tuning of photonic band gap structures and the temperature-dependent defect modes in multilayer photonic liquid crystals using nematic liquid crystal slabs in a twisted configuration. In addition to this, we present new results regarding the electrical control of defect modes in such multilayer structures. In order to achieve this goal, we establish and solve numerically the equation governing the twisted nematic configurations under the action of the external electric field by assuming arbitrary anchoring conditions at the boundaries. After this, we write Maxwell's equations in a 4×4 matrix representation and, by using the matrix transfer technique, we obtain the transmittance and reflectance for incident circularly polarized waves.

Keywords: photonic band bap, electrical and thermal tuning, nematic liquid crystal, multilayer structure, defect mode

1. Introduction

Photonic crystals (PCs) are artificial structures with spatially periodic dielectric permittivity whose interesting optical properties have attracted the attention of the scientific community since the seminal works made by Yablonovitch [1] and John [2]. The most attractive attribute of these periodic structures is the existence of photonic band gaps (PBGs) in which the propagation of electromagnetic waves is prohibited for a specific wavelength range. In one-dimensional PCs, this phenomenon is usually called Bragg reflection. Liquid crystals (LCs) are anisotropic intermediate phases between the solid and liquid states of matter that possess positional and orientational order just like those of the solid crystals, and they can flow as a conventional liquid [3]. For many decades, LCs have been used as optoelectronic substances on account of easy tunability of their properties under the stimuli of external agents as temperature, pressure and electromagnetic fields. This fact suggests the conception of new artificial structures by making a convenient combination of LCs with PCs, whose most prominent feature is the externally controlled PBG. First studies reported on this subject in three- and two-dimensional structures are attributed to Busch and John [4] and Leonard et al. [5], respectively. In [4], it was demonstrated the tunability of the PBG under the action of an external electric field meanwhile the temperature tuning when a nematic LC is infiltrated into the void regions of solid PCs is showed in [5].

Multilayer photonic liquid crystals (MPLCs) consisting of LCs alternated by transparent isotropic dielectric films have been previously studied. In Ref. [6], Ha et al. demonstrated experimentally simultaneous red, green and blue reflections (multiple PBGs) using the single-pitched polymeric cholesteric LC films. Later, Molina et al. [7] investigated the strong dependence of electric field on the PBG for incident waves of left- and right-circular polarization at arbitrary incidence angles using nematic liquid crystal (NLC) slabs in a twisted configuration. In a recent paper, Avendaño and Reyes [8] studied the optical band structure for reflectance and transmittance considering that the dielectric matrix of a similar one-dimensional photonic structure to that studied in Ref. [7] depends on temperature and wavelength. Twisted nematic LCs, where the molecular orientation exhibits a 90° twist, have proven technological advantages to control light flow. They have been used to switch effectively the pass of polarized light in nematic displays by means of a normally applied low-frequency electric field.

Surface anchoring plays an essential role in the science and technology of LCs. The structure of LCs in the bulk is different than that near the interface, and the boundary conditions established from this interface structure influence the behaviour of the LCs in the bulk. There are two cases of surface anchoring of particular interest. First, a strong anchoring case in which the molecules near the surface adopt a rigidly fixed orientation, and the anchoring energies are very large. Second, a weak anchoring case where the surface strengths are not strong enough to impose a well-defined molecular orientation at the interface, and the expression for the anchoring energy is some finite function that depends on the LC properties at the surface, the surface properties and the external fields (e.g., electric and magnetic fields) and temperature [9, 10]. Anchoring effects on the electrically controlled PBG in MPLCs were previously investigated by Avendaño [11]. They considered a generalization of the model studied in [7] for which arbitrary anchoring of the nematic at the boundaries is taken into account. They also found the nematic configuration versus the anchoring forces and the PBG under the action of a strong enough external field parallel to the periodicity axis, which is able to modify the configuration of the nematic-twisted LC in the whole material including at the boundaries of each nematic slab. Later, Avendaño and Martínez [12] theoretically exhibited that this system is able to produce an omnidirectional PBG that can be electrically controlled for circularly polarized incident waves. An omnidirectional PBG requires that there be no states in the given frequency range for propagation in any direction in the material for both polarizations, which implies the total reflectivity for all incident angles.

Resonant transmittance peaks in the PBG can be induced in PCs when defects are introduced in the periodic lattice. In this case, standing waves with a huge energy density are localized in the proximity of the defects (defect modes). Ozaki et al. [13] developed the first tunable PC/LC hybrid structure by using a planar aligned NLC as a defect layer sandwiched between two one-dimensional periodical multilayers (dielectric materials with high- and low-refractive-index layers stacked alternatively) and demonstrated the electrical tuning of the defect modes. Thermal tunability of one-dimensional PC/LC cells was demonstrated by Arkhipkin et al. [14]. Electrical-dependent defect mode in PC/LC hybrid structures using a twisted nematic LC as defect layer was studied by Lin et al. [15] and Timofeev et al. [16]. Thermal tuning of defect modes in MPLCs using twisted nematic LC was recently shown in Ref. [8].

In this chapter, we describe the main results previously studied concerning the thermal and electrical tuning of PBG and the temperature-dependent defect modes in MPLCs using NLC slabs in a twisted configuration. In addition to this, we present new results regarding the electrical control of defect modes in MPLCs. In order to achieve this goal, we establish and solve numerically the equation governing the twisted nematic configurations under the action of the external electric field by assuming arbitrary anchoring conditions at the boundaries. After this, we write Maxwell's equations in a 4×4 matrix representation and, by using the matrix transfer technique, we obtain the transmittance and reflectance for incident circularly polarized waves.

2. Nematic-twisted configuration

As said above, we are focused on a 1D structure consisting in N NLC slabs in a twisted configuration alternated by N transparent isotropic dielectric films as it is illustrated in **Figure 1 (a)** and **(b)**. For each of the NLC cells, the nematic is sandwiched between two dielectric layers in such a way that its director is aligned parallel in both frontiers. A twist is then imposed on the NLC by rotating an angle $2\varphi_t$, one of the dielectric layers about its own normal direction. Because of the possibility of molecular reorientation under the influence of external stimuli, such as electromagnetic fields, the director $\boldsymbol{n}$ takes the general form

$$\boldsymbol{n} \equiv \boldsymbol{n}[\alpha(z), \varphi(z)] = [\cos\alpha(z)\cos\varphi(z),\ \cos\alpha(z)\sin\varphi(z),\ \sin\alpha(z)], \tag{1}$$

with $\alpha(z)$ and $\varphi(z)$, the polar (zenithal) and azimuthal angles made by $\boldsymbol{n}$ with the xy plane and the x-axis, respectively. For the present physical system and assuming small distortions in the nematic [17], the expression that describes the elastic energy density of the nematic is

$$f_{el} = \frac{1}{2}K_1(\nabla\cdot\boldsymbol{n})^2 + \frac{1}{2}K_2(\boldsymbol{n}\cdot\nabla\times\boldsymbol{n})^2 + \frac{1}{2}K_3(\boldsymbol{n}\times\nabla\times\boldsymbol{n})^2, \tag{2}$$

where the positive elastic moduli K_1, K_2 and K_3 refer to splay, twist and bend bulk deformations, respectively. At this point, it is important to mention that it has experimentally found that when the nematic temperature is increasing to its transition temperature T_{NI} (where the NLC becomes isotropic), a reduction of the values of the elastic moduli is induced [3, 17]. On the other hand, if we take into account the presence of an external electric field $\boldsymbol{E}$, the interaction of this field with the LC is described through the energy density $f_{em} = -\frac{1}{2}\mathrm{Re}\{\boldsymbol{E}\cdot\boldsymbol{D}^*\}$,

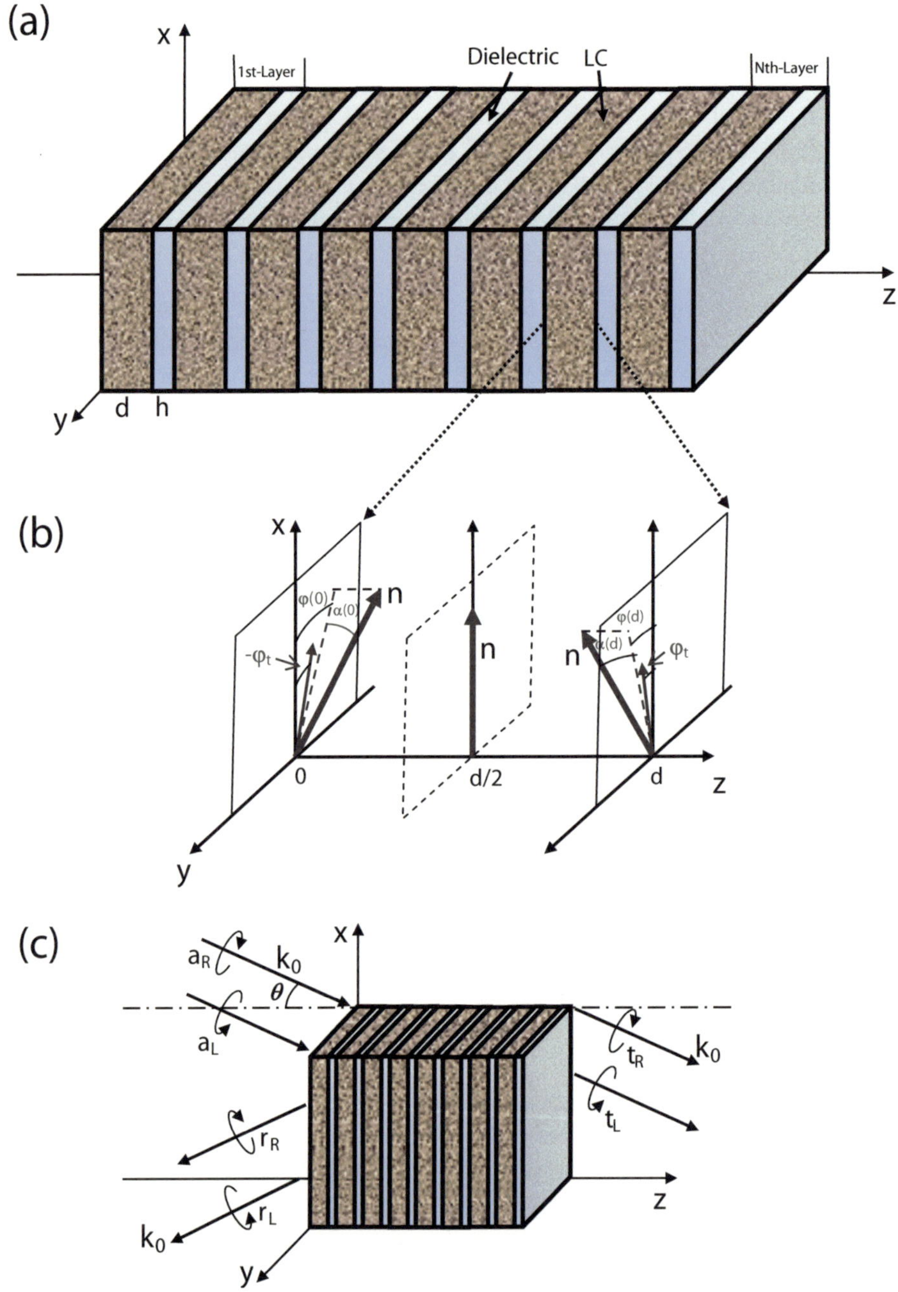

Figure 1. (a) Schematic of a MPLC consisting of N nematic LC slabs in a twisted configuration alternated by N transparent homogeneous isotropic dielectric films with thicknesses d and h, respectively. (b) Schematic of the polar α and azimuthal φ angles made by the director $\boldsymbol{n}$ with the xy-plane and the x-axis, respectively, at the boundaries of each of nematic LC slabs; the twist angle is given by ϕ_t and, at the middle of the slab, $\alpha=\varphi=0$. (c) An obliquely incident electromagnetic field with wave vector $\mathbf{k}_0$ impinges on the structure in the xz-plane and it makes an angle θ with respect to the z-axis. Here, a_L and a_R represent the amplitudes of left- and right-circularly polarized components of incident wave, respectively, and r_L, r_R, t_L and t_R correspond to those of the reflected and transmitted waves.

where we have assumed that the nematic follows the constitutive relation $\boldsymbol{D}=\varepsilon_0\boldsymbol{\varepsilon}\cdot\boldsymbol{E}$ characterized by the uniaxial dielectric tensor

$$\boldsymbol{\varepsilon} = \varepsilon_{\perp}\delta_{ij} + \varepsilon_a \boldsymbol{nn}, \tag{3}$$

with $\varepsilon_a=\varepsilon_{\parallel}-\varepsilon_{\perp}$ the dielectric anisotropy of the medium and ε_0 the permittivity of free space. Here, δ_{ij} is the Kronecker delta, $\varepsilon_{\perp}$ and $\varepsilon_{\parallel}$ denote the relative dielectric permittivity perpendicular and parallel to the nematic axis, respectively, and they are related to the ordinary n_o and extraordinary n_e refractive indices by $\varepsilon_{\perp} = n_o^2$ and $\varepsilon_{\parallel} = n_e^2$.

The study of confined nematic liquid crystals is strongly influenced by the physical properties of the boundary walls [18]. From the macroscopic-geometrical and microscopic interactions between the molecules of such surfaces and of the nematic, the alignment of the director $\boldsymbol{n}$ on the boundary surfaces, known as *anchoring*, can be completely determined. Once the anchoring conditions are established, the orientation of the NLC molecules at the substrate surface determined the director in the bulk.

Several methods and techniques for surface alignment have been developed [19, 20]. In the case of rubbed polymer films [21, 22], it has been observed that NLC molecules are *strongly anchored* at the surface, and the alignment is parallel to the grooves produced by the rubbing process. Also, the orientation of NLC molecules at the surfaces is preserved even if an external field (electric or magnetic) is applied and removed.

On the other hand, photoalignment [23] and nanostructuring polymer surfaces [24] are contact-free methods where it is induced a surface ordering that causes an anchoring of controllable strength, which corresponds to a *weak anchoring*. For this anchoring condition, alignment of the NLC molecules before and after the application of external fields is different.

Anchoring energy can be expressed in terms of the surface anchoring coefficients which are related to the interaction strength between the NLC and the wall substrate for the deviation of the easy axis along the correspondent directions. It is experimentally found that these coefficients are temperature dependent [18] and their values for specific NLCs can be obtained by using the dynamic light scattering [25]. Thus, if we write the director in terms of $\alpha(z)$ and $\varphi(z)$, as in expression (1), the anchoring energy of each NLC slab can be expressed in terms of the surface anchoring coefficients W_α and W_φ [26] as follows:

$$\begin{aligned} g^L &= W_\alpha \sin^2\alpha_m^L + W_\varphi \cos^2\alpha_m^L \sin^2\left(\varphi_m^L + \varphi_t\right) \\ g^R &= W_\alpha \sin^2\alpha_m^R + W_\varphi \cos^2\alpha_m^R \sin^2\left(\varphi_m^R - \varphi_t\right), \end{aligned} \tag{4}$$

which is an extension of the Rapini-Papoular model [9] and where $\alpha_m^L\left(\alpha_m^R\right)$ and $\varphi_m^L\left(\varphi_m^R\right)$ are the polar and azimuthal angles at the left (right) boundary of each NLC, respectively, and φ_t the twist angle. These anchoring coefficients are measured in energy per area units.

Thus, strong anchoring conditions are achieved when the anchoring coefficients are sufficiently large and can be modelled by considering that $W_\alpha\to\infty$ and $W_\varphi\to\infty$. In contrast, for weak anchoring conditions, it is taken that $W_\alpha\to 0$ and $W_\varphi\to 0$. Another criterion to establish whether the anchoring is strong or weak is based on the *extrapolation length* [27].

Thus, the equations governing the equilibrium configuration of the system are obtained by considering specific anchoring conditions and by minimizing the total free energy

$$F = \int_V (f_{el} + f_{em}) dV + \frac{1}{2}\int_{S_0} g^L dS + \frac{1}{2}\int_{S_d} g^R dS, \tag{5}$$

that can be achieved by considering strong or weak anchoring conditions.

2.1. Strong anchoring

Let us assume that the structure shown in **Figure 1(a)** is subjected to a DC electric field $\boldsymbol{E}_{dc}$=(0, 0, E_{dc}) parallel to z-axis, and we consider that the orientation of the director at the surfaces of each nematic cell are fixed and given by α=0°, $\varphi\left[z_m^L = (m-1)(d+h)\right] = -\varphi_t$ and $\varphi\left[z_m^R = (m-1)(d+h)+d\right] = \varphi_t$ for m=1,2,3, …,N. Here, z_m^L and z_m^R represent the positions of the left and right boundaries of the N nematic layers, respectively. Under these circumstances and by using a standard variational calculus procedure, the minimum free-energy condition δF=0 together with the restriction $\delta\boldsymbol{n}$=0 at the surface of each slab generate the equations [12]

$$0 = f(\alpha)\frac{d^2\alpha}{dz^2} + \frac{1}{2}\frac{df(\alpha)}{d\alpha}\left(\frac{d\alpha}{dz}\right)^2 - \frac{1}{2}\frac{dg(\alpha)}{d\alpha}\left(\frac{d\varphi}{dz}\right)^2 + \frac{1}{2}\left(\frac{\sigma}{d}\right)^2 \sin 2\alpha, \tag{6}$$

$$0 = g(\alpha)\frac{d^2\varphi}{dz^2} + \frac{dg(\alpha)}{d\alpha}\frac{d\alpha}{dz}\frac{d\varphi}{dz}, \tag{7}$$

where we have defined the dimensionless parameter $\sigma^2 = \varepsilon_0\varepsilon_a E_{dc}^2/\left(K_1/d^2\right)$ which represents the ratio between the electric and elastic energies. The functions $f(\alpha)$ and $g(\alpha)$ are defined as

$$\begin{aligned} f(\alpha) &= \cos^2\alpha + \frac{K_3}{K_1}\sin^2\alpha \\ g(\alpha) &= \left(\frac{K_2}{K_1}\cos^2\alpha + \frac{K_3}{K_1}\sin^2\alpha\right)\cos^2\alpha. \end{aligned} \tag{8}$$

In absence of the dc electric field, the polar angle $\alpha(z)$=0° for any value of z and, the solution of Eqs. (6) and (7) is simply

$$\varphi_m(z) = \frac{2\varphi_t}{d}(z - (m-1)(d+h)) - \varphi_t, \tag{9}$$

where $\varphi_m(z)$ represents the configuration of the mth layer in the region $(m-1)(d+h) \leq z \leq (m-1)(d+h)+d$ and the nematic director (1) is reduced to

$$\boldsymbol{n} \equiv \boldsymbol{n}[\varphi(z)] = [\cos\varphi(z), \sin\varphi(z), 0]. \tag{10}$$

2.2. Weak anchoring

In this case, we consider a free-end-point variation for which the director orientation is affected by the existence of finite anchoring coefficients [12]. This minimization procedure leads to the same set of coupled equations given by (6) and (7) subjected to boundary conditions at each layer:

$$\left.\frac{d\alpha}{dz}\right|_{z=z_m^L} = \frac{1}{d\,\Gamma_\alpha}\left.\frac{\left(1-\Gamma\sin^2(\varphi+\varphi_t)\right)\sin 2\alpha}{f(\alpha)}\right|_{z=z_m^L}, \tag{11}$$

$$\left.\frac{d\alpha}{dz}\right|_{z=z_m^R} = -\frac{1}{d\,\Gamma_\alpha}\left.\frac{\left(1-\Gamma\sin^2(\varphi-\varphi_t)\right)\sin 2\alpha}{f(\alpha)}\right|_{z=z_m^R}, \tag{12}$$

$$\left.\frac{d\varphi}{dz}\right|_{z=z_m^L} = \frac{\Gamma}{d\,\Gamma_\alpha}\left.\frac{\sin 2(\varphi+\varphi_t)\cos^2\alpha}{g(\alpha)}\right|_{z=z_m^L}, \tag{13}$$

$$\left.\frac{d\varphi}{dz}\right|_{z=z_m^R} = -\frac{\Gamma}{d\,\Gamma_\alpha}\left.\frac{\sin 2(\varphi-\varphi_t)\cos^2\alpha}{g(\alpha)}\right|_{z=z_m^R}, \tag{14}$$

With $\Gamma_\alpha=1/\gamma_a, \Gamma=\gamma_\varphi/\gamma_a$, $\gamma_a=W_\alpha d/K_1$ and $\gamma_\varphi=W_\varphi d/K_1$.

3. Electromagnetic propagation in a layered medium

The interaction between electromagnetic fields and matter is governed by Maxwell's equations and their corresponding constitutive equations. Optical propagation in layered media can be studied by conveniently writing Maxwell's equations in a 4×4 matrix representation. In this matrix representation, the boundary conditions of waves impinging on material can be imposed in such a way that the transfer and scattering matrix formalism to obtain the transmittances and reflectances can be used in a natural way [28, 29].

3.1. 4 × 4 matrix representation

In systems where boundary conditions cannot be avoided, Maxwell's equations require the continuity of tangential components of electric **E** and magnetic **H** fields at the boundaries. In studying the optical properties of dielectric layers which are confined between parallel walls, it is useful to write the set of Maxwell's equations in a representation where only appears, at the same time, the transversal components of **E** and **H** (two components for **E** and two components for **H**). This formalism is frequently referred to as Marcuvitz-Schwinger representation [30]. If we consider that the optical properties of a multilayer structure depends only on spatial variable z, we define the time-harmonic transversal four-vector

$$\mathbf{\Psi}(x,y,z) = \boldsymbol{\psi}(z)e^{ik_x x - i\omega t} = \begin{pmatrix} e_x(z) \\ e_y(z) \\ h_x(z) \\ h_y(z) \end{pmatrix} e^{ik_x x}e^{-i\omega t}, \tag{15}$$

with ω the angular frequency of the propagating wave and k_x the transversal component of the wave vector. Maxwell's equations, inside a nonmagnetic medium, can be written in the following matrix form:

$$\frac{\partial\psi(z)}{\partial z} = i\boldsymbol{A}(z).\psi(z), \tag{16}$$

for which the 4×4 matrix $\boldsymbol{A}(z)$ is given by

$$\mathbf{A}(z) = \begin{pmatrix} -\frac{k_x\varepsilon_{zx}}{k_0\varepsilon_{zz}} & -\frac{k_x\varepsilon_{zy}}{k_0\varepsilon_{zz}} & 0 & 1-\frac{k_x^2}{k_0^2\varepsilon_{zz}} \\ 0 & 0 & -1 & 0 \\ -\varepsilon_{yx}+\frac{\varepsilon_{yz}\varepsilon_{zx}}{\varepsilon_{zz}} & \frac{k_x^2}{k_0^2}-\varepsilon_{yy}+\frac{\varepsilon_{yz}\varepsilon_{zy}}{\varepsilon_{zz}} & 0 & \frac{k_x\varepsilon_{yz}}{k_0\varepsilon_{zz}} \\ \varepsilon_{xx}-\frac{\varepsilon_{xz}\varepsilon_{zx}}{\varepsilon_{zz}} & \varepsilon_{xy}-\frac{\varepsilon_{xz}\varepsilon_{zy}}{\varepsilon_{zz}} & 0 & -\frac{k_x\varepsilon_{xz}}{k_0\varepsilon_{zz}} \end{pmatrix} \tag{17}$$

where ε_{ij} $(i,j=x,y,z)$ represents the elements of dielectric matrix in the structure, $k_0=2\pi/\lambda=\omega/c$ is the wavenumber in free space, λ is the wavelength and c denotes the speed of light in vacuum. Also the fields $\mathbf{e}(z)=(e_x(z), e_y(z), e_z(z))$ and $\mathbf{h}(z)=(h_x(z), h_y(z), h_z(z))$ are related to the electric $\mathbf{E}(z)$ and magnetic $\mathbf{H}(z)$ fields by the following expressions $\mathbf{e}(z) \equiv Z_0^{-1/2}\mathbf{E}(z)$ and $\mathbf{h}(z) \equiv Z_0^{1/2}\mathbf{H}(z)$, with $Z_0 = \sqrt{\mu_0/\epsilon_0}$ the impedance in vacuum, ε_0 and μ_0 the permittivity and permeability of free space, respectively.

For a homogeneous and isotropic dielectric medium, the matrix $\boldsymbol{\varepsilon}(z)$ is diagonal and independent of the position, whereas for a nematic slab, $\boldsymbol{\varepsilon}(z)$ depends on the local orientation of the principal axis of the liquid crystal molecules characterized by expression (3).

3.2. Boundary condition

Let us consider a multilayer structure where each of the layers is confined between two planes, and the whole structure is surrounded by air. An electromagnetic wave impinging from the left side of the multilayer structure will propagate through the sample, and it will be transmitted and reflected outside the medium (see **Figure 1 (c)**).

The general solution of the differential equation (16) for electromagnetic waves propagating in homogeneous media is the superposition of four plane waves: two left-going and two right-going waves. With this in mind, we state the procedure to find the amplitudes of the transmitted and reflected waves in terms of incident ones (at plane z=0). This implies the definition of the following quantities [31]:

(i) The propagation matrix $\boldsymbol{U}(0,z)$ that is implicitly defined by the equations

$$\psi(z) = \boldsymbol{U}(0,z).\psi(0), \qquad \boldsymbol{U}(0,0) = \mathbf{1}, \tag{18}$$

where $\mathbf{1}$ is the identity matrix and $\boldsymbol{U}(0,z)$ satisfies the same propagation equation (16) found for ψ:

$$\partial_z\boldsymbol{U}(0,z) = i\boldsymbol{A}(z).\boldsymbol{U}(0,z). \tag{19}$$

This propagation matrix gives the right-side field amplitudes of the multilayer structure as function of the left-side ones.

(ii) For a specific value z=z_0, the transfer matrix is defined as $\boldsymbol{U}(0, z_0)$.

(iii) The scattering matrix $\boldsymbol{S}$ giving the output field as function of the incident one. The matrix $\boldsymbol{S}$ is defined through the relation $\boldsymbol{\alpha}_{out}$=$\boldsymbol{S}\cdot\boldsymbol{\alpha}_{in}$, where $\boldsymbol{\alpha}_{in}$ and $\boldsymbol{\alpha}_{out}$ are the amplitudes of the in-going and out-going waves.

To find out S, the field must expressed, in any one of the external media, as a superposition of planes waves by setting:

$$\psi = \boldsymbol{T}\cdot\boldsymbol{\alpha}; \qquad \boldsymbol{U}_{\boldsymbol{\alpha}}(0, z_0) = \boldsymbol{T}^{-1}\cdot\boldsymbol{U}(0, z_0)\cdot\boldsymbol{T}, \tag{20}$$

where $\boldsymbol{\alpha} = \left(a_1^+, a_2^+, a_1^-, a_2^-\right)^T$.

The relation ψ=$\boldsymbol{T}\cdot\boldsymbol{\alpha}$ can be interpreted as a basis change in the four dimensional space of the *state vectors* ψ. The columns of $\boldsymbol{T}$ are the ψ vectors representing the four plane waves generated by the incident waves in the two external media (assumed as identical). The elements of vector $\boldsymbol{\alpha}$ are the amplitudes of the four plane wave. The choice of the new basis could be different depending on the particular problem. By setting

$$\boldsymbol{U}(\boldsymbol{\alpha}) = \begin{pmatrix} \boldsymbol{U}_{ff} & \boldsymbol{U}_{bf} \\ \boldsymbol{U}_{fb} & \boldsymbol{U}_{bb} \end{pmatrix}, \tag{21}$$

the scattering matrix writes:

$$\boldsymbol{S} = \begin{pmatrix} \boldsymbol{U}_{ff} - \boldsymbol{U}_{bf}\boldsymbol{U}_{bb}^{-1}\boldsymbol{U}_{fb} & \boldsymbol{U}_{bf}\boldsymbol{U}_{bb}^{-1} \\ -\boldsymbol{U}_{bb}^{-1}\boldsymbol{U}_{fb} & \boldsymbol{U}_{bb}^{-1} \end{pmatrix}. \tag{22}$$

where the symbols + and f (– and b) mean forward (backward) propagating waves.

We point out that the methods of transfer and scattering matrices are very useful in studying the plane wave transmission and reflection from surfaces of multilayer structures.

Differential equation (16) can be formally integrated over a certain distance z_0 of the medium

$$\psi(z_0) = e^{i\int_0^{z_0} A(z')dz'}\cdot\psi(0), \tag{23}$$

and by straight comparison of Eqs. (18) and (23), the transfer matrix $\boldsymbol{U}(0, z_0)$ is defined as:

$$\boldsymbol{U}(0, z_0) = e^{i\int_0^{z_0} A(z')dz'}, \tag{24}$$

where plane waves are incident and reflected in the half-space $z < 0$, and plane waves are transmitted on the half-space $z > z_0$.

It can be seen immediately that the problem of finding $\boldsymbol{U}(0,z_0)$ is reduced to find a method to integrate expression (24) on the whole multilayer structure. Because of the non-homogeneity of the medium proposed here, we consider it as broken up into many very thin parallel layers, each of them with homogeneous anisotropic optical parameters [32]. In this way, $\boldsymbol{U}(0,z_0)$ is obtained by multiplying iteratively the matrix for each sublayer from $z = 0$ to $z = z_0$.

3.3. Transmission and reflection by multilayer structures

As said above, the general solution of the differential equation (16) for electromagnetic waves propagating in homogeneous media is the superposition of forward and backward propagating waves. The obliquely incident and reflected electromagnetic fields in free half-space $z \leq 0$ (**Figure 1 (c)**), for an arbitrary polarization state which are solutions of equation (16), can be expressed as:

$$\begin{pmatrix} \mathbf{e} \\ \mathbf{h} \end{pmatrix}_{inc} = \begin{pmatrix} [a_L(i\mathbf{u} - \mathbf{v}_+) - a_R(i\mathbf{u} + \mathbf{v}_+)] \exp(ik_{0z}z) \\ -i[a_L(i\mathbf{u} - \mathbf{v}_+) + a_R(i\mathbf{u} + \mathbf{v}_+)] \exp(ik_{0z}z) \end{pmatrix} \tag{25}$$

and

$$\begin{pmatrix} \mathbf{e} \\ \mathbf{h} \end{pmatrix}_{ref} = \begin{pmatrix} -[r_L(i\mathbf{u} - \mathbf{v}_-) - r_R(i\mathbf{u} + \mathbf{v}_-)] \exp(-ik_{0z}z) \\ i[r_L(i\mathbf{u} - \mathbf{v}_-) + r_R(i\mathbf{u} + \mathbf{v}_-)] \exp(-ik_{0z}z) \end{pmatrix}, \tag{26}$$

where $\mathbf{k}_0=(k_{0x}, k_{0y}, k_{0z})=k_0(\sin\theta, 0, \cos\theta)$ is the wave vector of the incident wave making an angle θ with respect to the z-axis, a_L and a_R represent the amplitudes of left- and right-circularly polarized (LCP and RCP) components of incident wave, respectively, and r_L and r_R correspond to those of the reflected wave (see **Figure 1(c)**). The unit vectors **u** and **v** are defined as

$$\mathbf{u} = \frac{\mathbf{u}_y}{\sqrt{2}}, \qquad \mathbf{v}_\pm = \frac{\mp \cos\theta \mathbf{u}_x + \sin\theta \mathbf{u}_z}{\sqrt{2}}, \tag{27}$$

with $\mathbf{u_x}, \mathbf{u_y}, \mathbf{u_z}$ the triad of Cartesian unit vectors. In the region $z \geq z_0$, the transmitted electromagnetic field is

$$\begin{pmatrix} \mathbf{e} \\ \mathbf{h} \end{pmatrix}_{tr} = \begin{pmatrix} [t_L(i\mathbf{u} - \mathbf{v}_+) - t_R(i\mathbf{u} + \mathbf{v}_+)] \exp(ik_{0z}(z - N(d+h) - h)) \\ -i[t_L(i\mathbf{u} - \mathbf{v}_+) + t_R(i\mathbf{u} + \mathbf{v}_+)] \exp(ik_{0z}(z - N(d+h) - h)) \end{pmatrix}, \tag{28}$$

where t_L and t_R are the amplitudes of LCP and RCP components, respectively, of transmitted wave. As the tangential components of **e** and **h** must be continuous across the planes $z=0$ and $z=z_0$, the boundary values $\psi(0)$ and $\psi(z_0)$ can be fixed as:

$$\psi(0) = \frac{\boldsymbol{P}}{\sqrt{2}} \cdot \begin{pmatrix} a_R \\ a_L \\ r_R \\ r_L \end{pmatrix} \quad \text{and} \quad \psi(z_0) = \frac{\boldsymbol{P}}{\sqrt{2}} \cdot \begin{pmatrix} t_R \\ t_L \\ 0 \\ 0 \end{pmatrix}, \tag{29}$$

with

$$\mathbf{P}=\begin{pmatrix}\cos\theta & \cos\theta & \cos\theta & \cos\theta\\ -i & i & i & -i\\ i\cos\theta & -i\cos\theta & i\cos\theta & -i\cos\theta\\ 1 & 1 & -1 & -1\end{pmatrix}. \tag{30}$$

By using Eqs. (23), (24) and (29), the problem of reflection-transmission can be established as follows

$$\begin{pmatrix}t_R\\ t_L\\ 0\\ 0\end{pmatrix}=\boldsymbol{M}\cdot\begin{pmatrix}a_R\\ a_L\\ r_R\\ r_L\end{pmatrix}, \tag{31}$$

where $\boldsymbol{M}=\boldsymbol{P}^{-1}\cdot\boldsymbol{U}(0,z_0)\cdot\boldsymbol{P}$ and $\boldsymbol{U}(0,z_0)$ are defined in (24). Notice that the matrix equation (31) gives a set of coupled equations relating the amplitudes a_L, a_R, r_L and r_R (from $z \le 0$) to the transmitted amplitudes t_L and t_R (for $z \ge z_0$).

The scattering matrix $\boldsymbol{S}$ relates the amplitudes t_L, t_R, r_L and r_R to the known incident amplitudes a_L and a_R. This relation can be expressed in terms of matrix $\boldsymbol{M}$ as [33]

$$\begin{pmatrix}t_R\\ t_L\\ r_R\\ r_L\end{pmatrix}=\boldsymbol{S}\cdot\begin{pmatrix}a_R\\ a_L\end{pmatrix}. \tag{32}$$

where

$$\boldsymbol{S}=\begin{pmatrix}t_{RR} & t_{RL}\\ t_{LR} & t_{LL}\\ r_{RR} & r_{RL}\\ r_{LR} & r_{LL}\end{pmatrix}=(\boldsymbol{Q}_1-\boldsymbol{M}\boldsymbol{Q}_2)^{-1}(\boldsymbol{M}\boldsymbol{Q}_1-\boldsymbol{Q}_2) \tag{33}$$

and

$$\boldsymbol{Q}_1=\begin{pmatrix}1&0&0&0\\0&1&0&0\\0&0&0&0\\0&0&0&0\end{pmatrix},\quad \boldsymbol{Q}_2=\begin{pmatrix}0&0&0&0\\0&0&0&0\\0&0&1&0\\0&0&0&1\end{pmatrix}. \tag{34}$$

Co-polarized coefficients have both subscripts identical meanwhile cross-polarized coefficients have different subscripts. The square of the amplitudes of t and r is the corresponding transmittance and reflectance, respectively; thus, $T_{RR}=|t_{RR}|^2$ is the co-polarized transmittance corresponding to the transmission coefficient t_{RR}, $T_{RL}=|t_{RL}|^2$ is the cross-polarized transmittance corresponding to the transmission coefficient t_{RL}, and so forth. In the absence of dissipation of energy inside the sample, the principle of conservation of energy must be satisfied from which we have that

$$T_{RR} + T_{LR} + R_{RR} + R_{LR} = 1 \qquad \text{and} \qquad T_{RL} + T_{LL} + R_{RL} + R_{LL} = 1. \tag{35}$$

Before ending this section, we mention that an alternative way to find the transmission and reflection coefficients is using the expressions given by (21) and (22). Also, the system of equations (31) can be solved numerically to find the scattering matrix.

4. Numerical results and discussion

In previous sections, we have presented in detail a general mathematical formalism to determine the reflectances and transmittances by multilayer structures. In this section, we apply this formalism to MPLCs using NLC slabs in a twisted configuration considering that circularly polarized light impinges on the structure in order to analyse the optical spectra and their dependence on external agents. In particular, we describe the main results previously studied concerning the thermal and electrical tuning of optical spectra and the temperature-dependent defect modes. In addition to this, we present new results regarding the electrical control of defect modes.

4.1. Electrical tuning of band structure and defect mode

In this section, we present the influence of the electric field on the optical band structure and defect mode by considering arbitrary anchoring conditions at the boundaries. To this aim, the equilibrium configuration of each NLC layer as a function of σ is obtained by solving the second order differential equations (6) and (7) for $\alpha(z)$ and $\varphi(z)$ subjected to the conditions expressed in Eqs. (11)–(14). Then, this configuration is substituted into Eq. (23) in order to obtain the transfer matrix $\boldsymbol{M}$ as function of σ for circularly polarized incident waves.

Numerical calculations were performed by considering a NLC phase 5CB for which K_1=0.62 × $10^{-11}N$, K_2=0.39 × $10^{-11}N$, K_3=0.82 × $10^{-11}N$ [17] and refractive indices at optical frequencies $n_o = \sqrt{\varepsilon_\perp} = 1.53$ and $n_e = \sqrt{\varepsilon_\parallel} = 1.717$. The twist angle is taken $2\varphi_t$=90°, and the homogeneous isotropic dielectric medium is zinc sulphide (ZnS) with refractive index n_d=2.35. The MPLC consists of N=11 NLC layers alternating with N=11 dielectric slabs with the same thickness. Finally, we report our results parameterizing all the spatial variables by the NLC thickness d. In this way, the dimensionless thickness of each NLC cell is $h'=d/d=1$, whereas for each ZnS slab is $h'=h/d=1$, and so forth.

Due to the competition between orientation produced by influence of the external electric field and by surface anchoring effects, we expect a deformation in the NLC only above a certain critical value σ_c. This critical electric field is expected to be maximum for the case of strong anchoring conditions, whereas for the weak anchoring case, σ_c will decrease as the surface forces get smaller [7, 11].

4.1.1. Strong anchoring conditions

For strong anchoring conditions, the orientation of the nematic molecules at the walls of each NLC is specified in Section 2.1. The curves for $\alpha(z)$ and $\varphi(z)$ are shown in **Figure 2(a)** and **(b)**, respectively, as function of dimensionless variable $w = z/d$ above the critical value $\sigma_c = 3.26$. As it can be noticed in **Figure 2(a)**, an increment in the electric field involves the augmentation in the polar angle α. Owing to the influence of the external field, the nematic molecules tend to be aligned parallel to it (z-axis). As expected, for $\sigma < \sigma_c$, $\alpha=0°$ for all values of w, which means that the director in this case is perpendicular to the z-axis. In **Figure 2(b)**, we observe that for $\sigma < \sigma_c$, the curves for azimuthal angle φ are reduced to straight lines with slope equal to $2\varphi_t=90°$, that corresponds to the configuration of a pure twisted NLC. Above the critical value, the strong anchoring condition is really dominant on the parameter φ as the electric field increases. Indeed, most of molecules tend to spread far from xz plane.

Figure 3 exhibits the co-polarized and cross-polarized transmittances and reflectances for LCP and RCP waves impinging normally on the structure as function of the dimensionless parameter d/λ for continuous values of the electric field above the critical value σ_c (it is worth to mention that below this value, the not-shown curves are very similar to that of $\sigma = \sigma_c$). Note the strong influence of σ on the transmission and reflection spectra in enhancing and extinguishing bands. Indeed, **Figure 3** clearly shows that for $\sigma = \sigma_c$, the curves for transmittances exhibit several stop bands of different widths in the plotted interval and, as σ increases, each stop band gets wider for co-polarized transmittances T_{RR} and T_{LL}. Also, cross-polarized transmittances T_{LR} and T_{RL} are totally absent for high enough values of the electric field. On the other hand, at the critical value, co-polarized reflectances R_{RR} and R_{LL} exhibit narrow-reflection bands with relatively high amplitudes and cross-polarized reflectances R_{LR} and R_{RL} show one-dominant high-reflection band. In this case, co-polarized reflectances reduce their band amplitudes practically to zero, and reflection bands of cross-polarized reflectances are highly enhanced for larger values of σ. These optical properties allow us to use this MPLC as an electrically shiftable universal rejection filter for incident RCP and LCP waves where, by

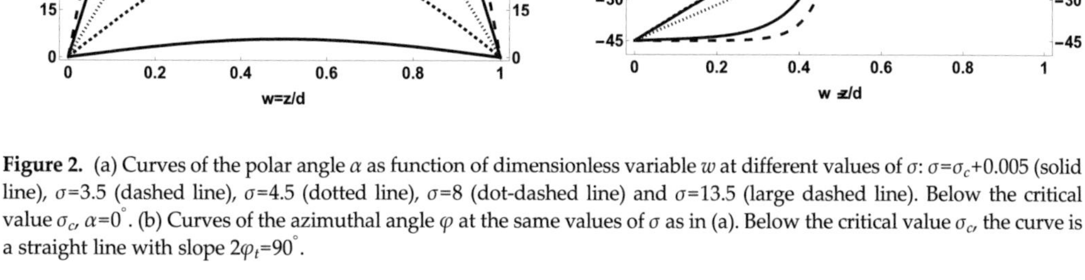

Figure 2. (a) Curves of the polar angle α as function of dimensionless variable w at different values of σ: $\sigma=\sigma_c+0.005$ (solid line), $\sigma=3.5$ (dashed line), $\sigma=4.5$ (dotted line), $\sigma=8$ (dot-dashed line) and $\sigma=13.5$ (large dashed line). Below the critical value σ_c, $\alpha=0^{\circ}$. (b) Curves of the azimuthal angle φ at the same values of σ as in (a). Below the critical value σ_c, the curve is a straight line with slope $2\varphi_t=90^{\circ}$.

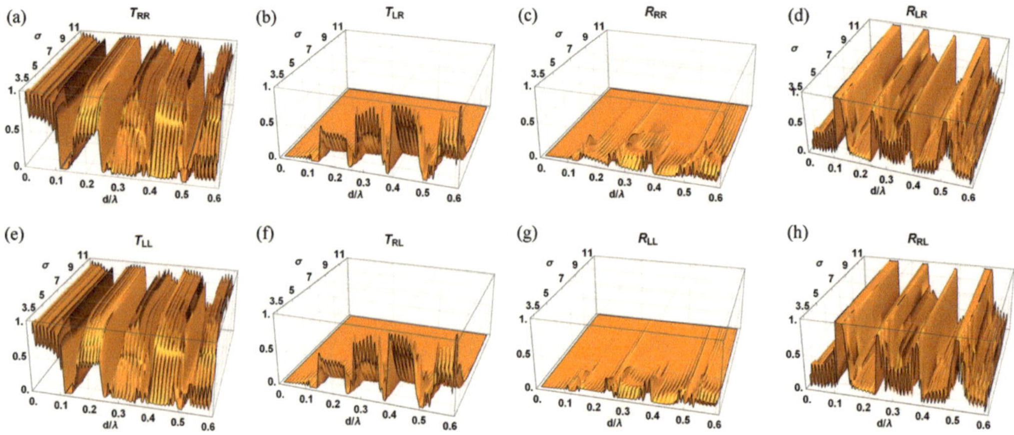

Figure 3. (a)–(h) Plots of co-polarized and cross-polarized reflectances and transmittances for LCP and RCP waves impinging normally on a MPLC as function of the dimensionless parameter d/λ and continuous values of σ within the interval $\sigma_c < \sigma < 13$.

increasing the electric field, one can highly enhance the cross-polarized reflection bands and supress the co-polarized ones.

In [7], it is shown that for a fixed value of σ the band structure of the reflectances and transmittances are shifted towards smaller wavelength regions as the incident angle θ increases. This behaviour results from the fact that for plane electromagnetic waves propagating obliquely with respect to the layer interfaces, only the normal component of the wave vector is involved in the photonic band formation. Hence, as the incident angle augments, the relative position of the bands is moved towards smaller wavelengths.

If one of the layers possesses a different size compared with the remaining ones, this layer can act as a defect, and an optical defect mode can be induced. Here, we specifically consider that the middle NLC-ZnS stack of the MPLC has a different size compared with the remaining ones. We choose specific values d_d=2d' and h_d=2h', where d_d and h_d are the dimensionless thicknesses of the NLC and ZnS defect layers, respectively. **Figure 4(a)** and **(b)** displays the defect mode induced in the photonic band of the co-polarized transmittance T_{RR} and cross-polarized reflectance R_{LR}, respectively, by LCP waves impinging normally on the MPLC. We notice that as the parameter σ increases two important facts occur: (i) two defect modes with small amplitude are induced within the first stop band (see **Figure 3**) which gradually merge into only one; the position of the defect mode possessing the largest wavelength moves toward regions of smaller wavelengths, keeping fixed the position of the other one and (ii) the amplitude of the defect modes gets larger. Physically, the origin of the defect mode is the phase change due to the variation in the optical path length caused by the defective medium. Once the defect mode is created at specific position, it can be controlled by inducing reorientation in the nematic molecules by means of an external electric field [7]. Indeed, since the refractive index of the LC depends on the angle β between the wave vector **k** of the electromagnetic wave in the LC and the local orientation of the director ***n***, the refractive index (and the optical path length) can be changed by varying β. At normal incidence and $\sigma < \sigma_c$, $\beta = 90°$ for all positions z.

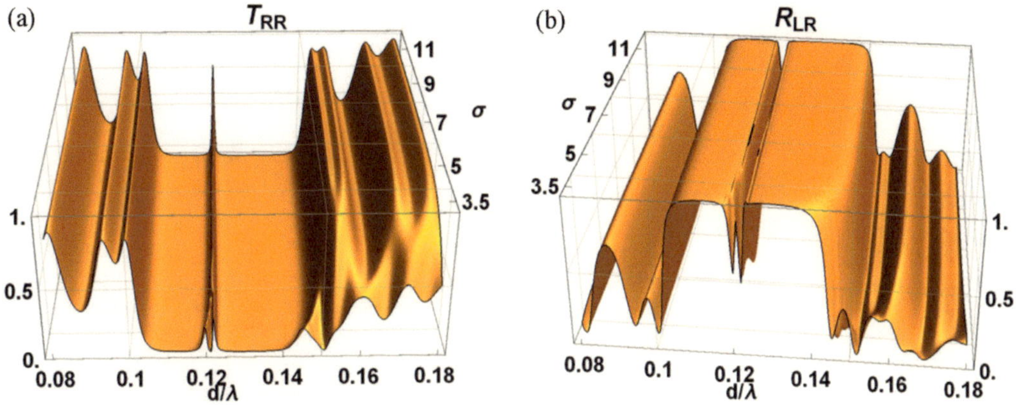

Figure 4. (a) At normal incidence, plots of co-polarized transmittance T_{RR} for LCP incident waves as function of the dimensionless parameter d/λ and continuous values of σ. (b) At normal incidence, cross-polarized reflectance R_{LR} for LCP incident waves at the same values of σ as in (a).

Nevertheless, as σ increases, most of the molecules tend to be aligned parallel to z-axis (see **Figure 2(a)**) and $\beta \rightarrow 0°$. These results show that the amplitude of defect mode and its position can be tuned by a DC electric field.

4.1.2. Weak anchoring conditions

It is experimentally found that for a LC phase 5CB, the polar anchoring γ_α is of the order of 10^1, and this value is one or two orders stronger than the azimuthal anchoring γ_φ [34]. Under these considerations, the values of the dimensionless anchoring parameters are taken as $\Gamma=\Gamma_\alpha=0.1$.

The curves for $\alpha(z)$ and $\varphi(z)$ are shown in **Figure 5(a)** and **(b)**, respectively, as function of dimensionless variable $w=z/d$ above the critical value $\sigma_c=2.86$. In **Figure 5(a)**, we can notice that, as σ augments, the values of α increase, getting a maximum at the middle of the cell. Because of the influence of external electric field, the polar angle at both borders enlarges by increasing σ highlighting the fact that even at the borders, the field is able to distort the configuration. **Figure 5 (b)** shows two interesting phenomena: (i) for $\sigma < \sigma_c$, the curves are reduced to straight lines with slope equal to $2\varphi_{0c}$, where φ_{0c} represents the azimuthal angle adopted by the MPLC at the walls of each NLC cell for values of electric field below the critical field; (ii) above the critical value, most of the molecules tend to acquire an angle $\varphi_t=-45°$ for $0< w < 0.5$ and $\varphi_t = 45°$ for $0.5 < w<1$.

Figure 6 shows the co-polarized and cross-polarized transmittances and reflectances for LCP and RCP waves impinging normally on the structure as function of the dimensionless parameter d/λ for continuous values of the electric field above the critical value σ_c (below this value, the not-shown curves are very similar to those corresponding to $\sigma=\sigma_c$). Although, the optical properties shown in **Figure 6** are qualitatively similar to those of **Figure 3** where strong anchoring conditions were considered, we notice that in the case of weak anchoring conditions, the behaviour of transmittances and reflectances in **Figure 6** is enhanced in comparison with **Figure 3**. Because of the strong influence of the electric field on the molecular orientation

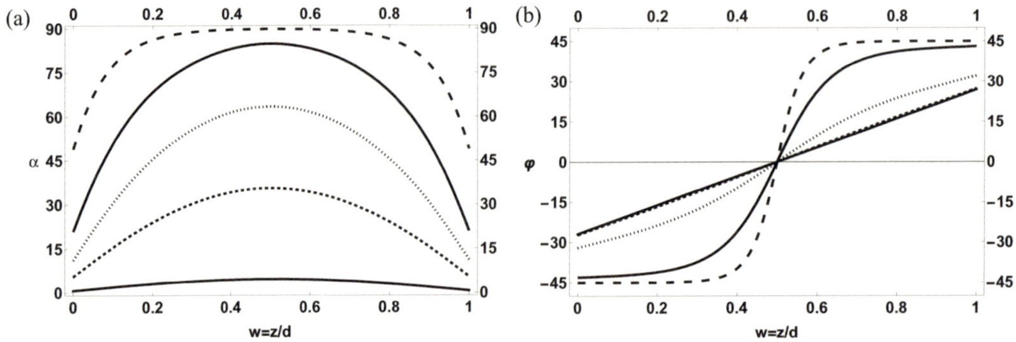

Figure 5. (a) Curves of the polar angle α as function of dimensionless variable w at different values of σ: $\sigma=\sigma_c+0.005$ (solid line), $\sigma=3.5$ (dashed line), $\sigma=4.5$ (dotted line), $\sigma=8$ (dot-dashed line) and $\sigma=13.5$ (large dashed line). (b) Curves of the azimuthal angle φ at the same values of σ as in (a).

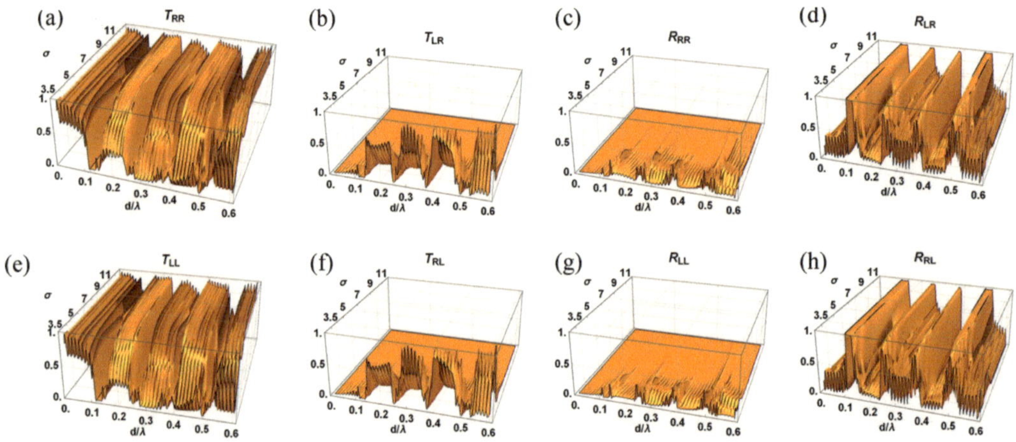

Figure 6. (a)–(h) Plots of co-polarized and cross-polarized reflectances and transmittances for LCP and RCP waves impinging normally on a MPLC as function of the dimensionless parameter d/λ and continuous values of σ within the interval $\sigma_c<\sigma<13$. Here, we consider weak anchoring conditions at the walls of each NLC.

for all values of z (including the walls of each cell), the alignment of most of the nematic molecules parallel to z-axis occurs at smaller values of electric field unlike for strong anchoring. Hence, the phenomenon of extinguishing and enhancing bands is present at smaller values of σ.

Now, we induce a defect mode in the photonic band structure by generating a defect in the MPLC in the same way as explained in Section 4.1.1. **Figure 7(a)** and **(b)** displays the defect mode induced in the photonic band of the co-polarized transmittance T_{RR} and cross-polarized reflectance R_{LR}, respectively, for LCP waves impinging normally on the MPLC. Similar to the case of strong anchoring conditions, we can observe that when the parameter σ augments the amplitude of the defect modes gets larger, and the position of the defect mode possessing the largest wavelength moves toward regions of smaller wavelengths, while the position of the other defect mode remains fixed. These facts are enhanced in comparison to those of strong anchoring assumptions because of the strong influence of the electric field on the molecular

http://dx.doi.org/10.5772/intechopen.70473

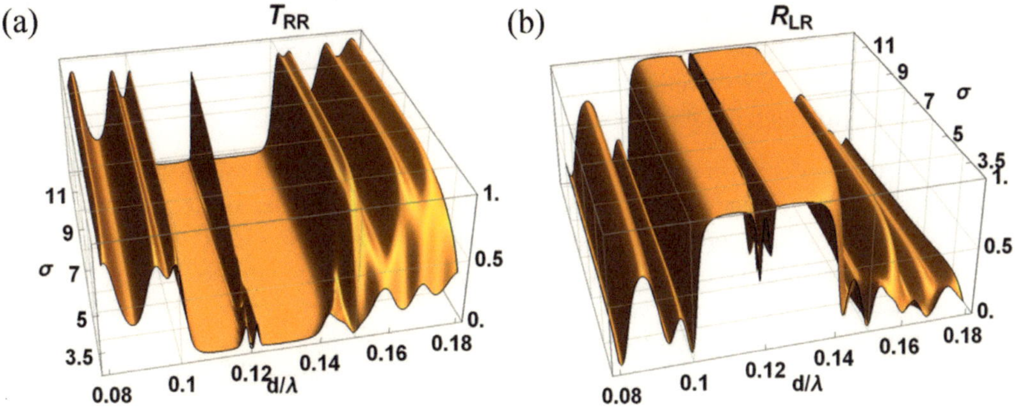

Figure 7. (a) At normal incidence, plots of co-polarized transmittance T_{RR} for LCP incident waves as function of the dimensionless parameter d/λ and continuous values of σ. (b) At normal incidence, cross-polarized reflectance R_{LR} for LCP incident waves at the same values of σ as in (a). Here, we consider weak anchoring conditions at the walls of each NLC.

orientation for all values of z, including the walls of each NLC slab in the MPLC. This implies that the defect-mode amplitude gets larger for smaller values of σ in comparison to that of the strong anchoring case.

4.2. Temperature-dependent band structure and defect mode

Here, we assume that the orientation of the director at the surfaces of each nematic cell is strongly anchored at the boundaries. In order to obtain the band structure, we apply the same mathematical procedure as depicted in Section 4.1, but in this case, we have to take into account that the director $\boldsymbol{n}$ is given by expressions (9) and (10) and the elements of dielectric tensor $\varepsilon(z)$ depend on the wavelength and temperature [8]. By considering E7 LC mixture slabs and ZnS dielectric layers, it is found that in the interval of temperatures [15°C,50°C] and for RCP waves impinging normally, the position of the photonic bands of the co-polarized transmittance T_{RR} and cross-polarized reflectance R_{LR} (analogous to those of **Figure 3** with σ=0) can be shifted from regions of small wavelengths toward regions of higher wavelengths by increasing the thickness d of the NLC layers. In addition to this, bandwidth increases for thicker layers, and new narrower transmission bands are created in regions of smaller wavelengths. In summary, the position, the width, and the number of bands augment as the thickness d is increased. Physically, when the magnitude of d gets larger, the optical path lengths increase, and hence, the wavelength zones of destructive or constructive interference are shifted towards higher wavelength regions. For constant thickness and temperature, it is observed that as the incident angle θ augments, the photonic bands undergo a shift towards smaller wavelengths and their widths get narrower. As said previously, this behaviour results from the fact that for plane electromagnetic waves propagating obliquely with respect to the layer interfaces, only the normal component of the wave vector is involved in the photonic band formation. Thus, as the incident angle augments the relative position of the bands moves towards smaller wavelengths, and the overall band is always closed up. On the other hand, for constant thickness and a fixed incident angle, as the temperature augments, the photonic bands move towards the short-wavelength region. Physically, since the average refractive index of the liquid crystal decreases as the temperature gets larger, the optical path length diminishes, and thus, the wavelength regions where the waves are able to undergo constructive or destructive interference shift towards smaller wavelengths zones.

In a similar way as demonstrated above, a defect mode can be induced by considering that the middle layer of the homogeneous and isotropic slabs (ZnS) has a different size compared with the remaining ones. If we consider normal incident RCP waves, for h_d=1.991$h^{'}$ and temperature values in the interval [15°C, 50°C], the position of the defect mode induced in the photonic band of the co-polarized transmittance T_{RR} and cross-polarized reflectance R_{LR} (analogous to those of **Figure 4** with σ=0) shifts from larger wavelengths toward smaller ones as the temperature gets increasing. Because the origin of the defect mode is the phase change due to the variation in the optical path length caused by the defective medium, the defect wavelength can be shifted towards smaller wavelength regions as the temperature is increased by taking into account that the average refractive index of the NLC decreases as temperature increases.

5. Conclusion

We presented a series of results concerning the thermal and electrical tuning of photonic band gaps and defect modes in multilayer photonic liquid crystals consisting of liquid crystal layers alternated by transparent isotropic dielectric films using nematic liquid crystal slabs in a twisted configuration. We exhibited that the position and width of the band gaps can be electrically and thermally controlled. When one of the homogeneous and isotropic slabs has a different size compared with the remaining ones, a defect mode is induced in the band structure whose wavelength can be tuned. Tuning of the transmission and reflection bands and the defect mode investigated here could be useful in the implementation of tunable optical filters and waveguides.

Author details

Carlos G. Avendaño*, Daniel Martínez and Ismael Molina

*Address all correspondence to: caravelo2000@gmail.com

Autonomous University of Mexico City, Mexico City, Mexico

References

[1] Yablonovitch E. Inhibited spontaneous emission in solid-state physics and electronics. Physical Review Letters. 1987;**58**:2059-2062. DOI: 10.1103/PhysRevLett.58.2059

[2] John S. Strong localization of photons in certain disordered dielectric superlattices. Physical Review Letters. 1987;**58**:2486-2489. DOI: 10.1103/PhysRevLett.58.2486

[3] de Gennes PG, Prost J. The Physics of Liquid Crystals. 2nd ed. Oxford: Oxford Science Publications; 1993. 597 p. ISBN: 978-0198517856

[4] Busch K, John S. Liquid-crystal photonic-band-gap materials: The tunable electromagnetic vacuum. Physical Review Letters. 1999;**83**:967-970. DOI: 10.1103/PhysRevLett.83.967

[5] Leonard SW, van Driel H, Toader O, John S, Busch K, Birner A, Gosele U. Tunable two-dimensional photonic crystals using liquid crystal infiltration. Physical Review B. 2000;**61**:R2389-R2392. DOI: 10.1103/Phys.RevB.61.R2389

[6] Ha NY, Ohtsuka Y, Jeong S, Nishimura S, Suzaki G, Takanishi Y, Ishikawa K, Takezoe H. Fabrication of a simultaneous red-green-blue reflector using single-pitched cholesteric liquid crystals. Nature Materials. 2008;**7**:43-47. DOI: 10.1038/nmat2045

[7] Molina I, Reyes JA, Avendaño CG. Electrically controlled optical bandgap in a twisted photonic liquid crystal. Journal of Applied Physics. 2011;**109**:113510-113516. DOI: 10.1063/1.3575154

[8] Avendaño C, Reyes A. Temperature-dependent optical band structure and defect mode in a one-dimensional photonic liquid crystal. Liquid Crystals. 2017;1-12. DOI: 10.1080/02678292.2017

[9] Rapini A, Papoular M. Distorsion d'une lamelle nématique sous champ magnétique conditions d'ancrage aux parois. Journal de Physique, Colloque. 1969;**30**:C4-54-C4-56. DOI: 10.1051/jphyscol:1969413

[10] Sonin AA. The Surface Physics of Liquid Crystals. Amsterdam: Gordon & Breach; 1995. ISBN: 978-2881249952

[11] Avendaño CG, Molina I, Reyes JA. Anchoring effects on the electrically controlled optical band gap in twisted photonic liquid crystals. Liquid Crystals. 2013;**40**:172-184. DOI: 10.1080/02678292.2012.735706

[12] Avendaño CG, Martinez D. Tunable omni-directional mirror based on one-dimensional photonic structure using twisted nematic liquid crystal: The anchoring effects. Applied Optics. 2014;**53**:4683-4690. DOI: 10.1364/AO.53.004683

[13] Ozaki R, Matsui T, Ozaki M, Yoshino K. Electro-tunable defect mode in one-dimensional periodic structure containing nematic liquid crystal as a defect layer. Japanese Journal of Applied Physics. 2002;**41**:L1482-L1484. DOI: 10.1143/JJAP.41.L1482

[14] Arkhipkin VG, Gunyakov VA, Myslivets SA, Gerasimov VP, Zyryanov VY, Vetrov SY, Shabanov VF. One-dimensional photonic crystals with a planar oriented nematic layer: Temperature and angular dependence of the spectra of defect modes. JETP. 2008;**106**:388-398. DOI: 10.1134/S1063776108020179

[15] Lin Y-T, Chang W-Y, Wu C-Y, Zyryanov VY, Lee W. Optical properties of one-dimensional photonic crystal with a twisted-nematic defect layer. Optics Express. 2010;**18**:26959-26964. DOI: 10.1364/OE.18.026959

[16] Timofeev IV, Lin Y-T, Gunyakov VA, Myslivets SA, Arkhipkin VG, Vetrov SY, Lee W, Zyryanov VY. Voltage-induce defect mode coupling in a one-dimensional photonic crystal with a twisted-nematic defect layer. Physical Review E. 2012;**85**:011705. DOI: 10.1103/PhysRevE.85.011705

[17] Stewart IW. The Static and Dynamic Continuum Theory of Liquid Crystals: A Mathematical Introduction. 1st ed. New York: Taylor & Francis Group; 2004. 351 p. ISBN: 978-0748408962

[18] Rasing T, Muševič I. Surfaces and Interfaces of Liquid Crystals. 1st ed. New York: Springer; 2004. p. 298. DOI: 10.1007/978-3-662-10157-5

[19] Castellano JA. Surface anchoring of liquid crystal molecules on various substrates. Molecular Crystals and Liquid Crystals. 1983;**94**:33-41. DOI: 10.1080/00268948308084245

[20] Takatoh K, Hasegawa M, Koden M, Itoh N, Hasegawa R, Sakamoto M. Alignment Technologies and Applications of Liquid Crystals Devices. 1st ed. New York: Taylor & Francis Inc.; 2005. p. 320. DOI: 10.1201/9781420023015

[21] Rüetschi M, Grütter P, Fünfschilling J, Güntherodt H. Creation of liquid crystal waveguides with scanning force microscopy. Science. 1994;**265**:512-514. DOI: 10.1126/science.265.5171.512

[22] Pidduck AJ, Haslam SD, Bryan-Brown GP, Bannister R, Kitely ID. Control of liquid crystal alignment by polyimide surface modification using atomic force microscopy. Applied Physics Letters. 1997;**71**:2907. DOI: 10.1063/1.120212

[23] Gibbons WM, Shannon PJ, Sun S, Swtlin BJ. Surface-mediated alignment of nematic liquid crystals with polarized laser light. Nature. 1991;**351**:49-50. DOI: 10.1038/351049a0

[24] Chigrinov VG, Kozenkov VM, Kwok H-S. Photoalignement of Liquid Crystalline Materials: Physics and Applications. West Sussex: John Wiley & Sons; 2008. p. 248. DOI: 10.1002/9780470751800

[25] Vilfan M, Mertelj A, Copic M. Dynamic light scattering measurements of azimuthal and zenithal anchoring of nematic liquid crystals. Physical Review E. 2002;**65**:041712-1-041712-7. DOI: 10.1103/PhysRevE.65.041712

[26] Zhao W, Wu C, Iwamoto M. Weak boundary anchoring, twisted nematic effect and homeotropic to twisted-planar transition. Physical Review E. 2002;**65**:031709. DOI: 10.1103/PhysRevE.65.031709

[27] Baek S-I, Kim S-J, Kim J-H. Measurement of anchoring coefficient of homeotropically aligned nematic liquid crystal using a polarizing optical microscope in reflective mode. AIP Advances. 2015;**5**:097170. DOI: 10.1063/1.4931950

[28] Chuang SL. Physics of Photonic Devices. 2nd ed. New Jersey: John Wiley & Sons; 2009. p. 840

[29] Hecht E, Zajac A. Óptica. 3rd ed. Madrid: Addison Wesley Iberoamericana; 2010. 722 p. ISBN: 84-7829-025-7

[30] Marcuvitz N, Schwinger J. On the representation of electric and magnetic field produced by currents and discontinuities in waveguides. Journal of Applied Physics. 1951;**22**:806-819. DOI: 10.1063/1.1700052

[31] Altman C, Suchy K. Reciprocity, Spatial Mapping and Time Reversal in Electromagnetics. 2nd ed. Heidelberg: Springer; 2011. DOI: 10.1007/978-94-007-1530-1

[32] Berreman DW, Scheffer TJ. Bragg reflection of light from single-domain cholesteric liquid-crystal films. Physical Reviews Letters. 1970;**25**:577-581. DOI: 10.1103/PhysRevLett. 25.902.4

[33] Avendaño CG, Ponti S, Reyes JA, Oldano C. Multiplet structure of the defect modes in 1D helical photonic crystals with twist defects. Journal of Physics A: Mathematical and General. 2005;**38**:8821-8840. DOI: 10.1088/0305-4470/38/41/001

[34] Nastishin YA, Polak RD, Shiyanovskii SV, Bodnar VH, Lavrentovich OD. Nematic polar anchoring strength measured by electric field techniques. Journal of Applied Physics. 1999;**86**:4199-4213. DOI: 10.1063/1.371347

Chapter 7

Nonlinear Optical Phenomena in Smectic A Liquid Crystals

Boris I. Lembrikov, David Ianetz and Yossef Ben Ezra

Additional information is available at the end of the chapter

http://dx.doi.org/10.5772/intechopen.70997

Abstract

Liquid crystals (LC) are the materials characterized by extremely high optical nonlinearity. Their physical properties such as temperature, molecular orientation, density, and electronic structure can be easily perturbed by an applied optical field. In particular, in smectic A LC (SALC), there is a specific mechanism of the cubic optical nonlinearity determined by the smectic layer normal displacement. The physical processes related to this mechanism are characterized by a comparatively large cubic susceptibility, short time response, strong dependence on the optical wave polarization and propagation direction, resonant spectral form, low scattering losses as compared to other LC phases, and weak temperature dependence in the region far from the phase transition. We investigated theoretically the nonlinear optical phenomena caused by this type of the cubic nonlinearity in SALC. It has been shown that the light self-focusing, self-trapping, Brillouin-like stimulated light scattering (SLS), and four-wave mixing (FWM) related to the smectic layer normal displacement are strongly manifested in SALC. We obtained the exact analytical solutions in some cases and made the numerical evaluations of the basic parameters such as the optical beam width and SLS gain.

Keywords: smectic liquid crystals, second sound, nonlinear optics, cubic nonlinearity, stimulated scattering of light, four-wave mixing, surface plasmon polariton

1. Introduction

Liquid crystals (LC) are characterized by the physical properties intermediate between ordinary isotropic fluids and solids [1]. LC flow like liquids but also exhibit some properties of crystals [1, 2]. The various phases in which such materials can exist are called mesophases [1, 2]. The LC molecules are large, anisotropic, and complex [2]. Dielectric constants, elastic constants, viscosities, absorption spectra, transition temperatures, anisotropies, and optical nonlinearities of LC are determined by the structure of these molecules [1, 2]. There exist three different types of LC: lyotropic, polymeric, and thermotropic [1, 2]. Lyotropic LC are obtained when an appropriate concentration of a material is dissolved in a solvent [2]. They can demonstrate a one-, two-, or three-dimensional positional order [2]. Liquid crystalline polymers are built up by the joining together the rigid mesogenic monomers [2]. Thermotropic LC exhibit different mesophases depending on temperature [1, 2]. Typically, they consist of organic molecules elongated in one direction and represented as rigid rods [2]. There are two types of LC sample orientation with respect to the boundary: (i) a homeotropic orientation when the long molecular axes are perpendicular to the boundary and (ii) a planar orientation when the long molecular axes are parallel to the boundary [1].

In this work, we consider only thermotropic LC, which are divided into three groups according to their symmetry: nematic LC (NLC), cholesteric LC (CLC), and smectic LC (SLC) [1, 2]. NLC are characterized by some long-range order in the direction of the molecular long axes, while the centers of gravity of the molecules do not have any long range order [1, 2]. The general direction of the molecules is defined by a unit vector function $\vec{n}\ (x, y, z, t)$; $\left|\vec{n}\ (x, y, z, t)\right| = 1$ called director [1, 2]. NLC molecules are centrosymmetric such that the $\vec{n}$ and $-\vec{n}$ directions are equivalent; NLC are optically uniaxial media with a comparatively large birefringence of about 0.2 [1, 2]. LC consisting of chiral molecules yield CLC phase with the helical structure [1, 2]. The molecule centers of gravity in CLC do not have a long range order like in NLC, while the direction of the molecular orientation rotates in space about the helical axis Z with a period of about 300 nm [1, 2]. The smectic LC (SLC) are characterized by the positional long range order in the direction of the elongated molecular axis and exhibit a layer structure [1, 2]. The layer thickness $d \approx 2nm$ is approximately equal to the length of the constituent molecule [1, 2]. SLC can be considered as natural nanostructures. Inside a layer the molecules form a two-dimensional liquid [1, 2]. The layers can easily move one along another because the elastic constant $B \approx 10^6 - 10^7 Jm^{-3}$ related to the layer compression is two orders of magnitude less than the elastic constant related to the bulk compression [1]. There exist different phases of SLC: (i) smectic A LC (SALC) where the molecule long axes are perpendicular to the layer plane; (ii) smectic B LC with the in-layer hexagonal ordering of the molecules; (iii) smectic C LC where the molecules are tilted with respect to the layers; (iv) smectic C^* LC consisting of the chiral molecules and possessing the spontaneous polarization; (v) different exotic smectic mesophases [1]. In this work, we consider only SALC. The SALC layered structure can be described by the one-dimensional mass density wave characterized by the complex order parameter. The modulus of this order parameter describes the mass density and its phase is related to smectic layer displacement $u(x, y, z, t)$ along the direction perpendicular to the layer plane [1]. SALC is an optically uniaxial medium [2].

LC are highly nonlinear optical materials due to their complex physical structures, and their temperature, molecular orientation, mass density, electronic structure can be easily perturbed by an external optical field [2–4]. Almost all known nonlinear optical phenomena have been observed in LC in time scale range from picoseconds to hours, involving laser powers from 10^6 Watt to 10^{-9} Watt, in different configurations such as bulk media, optical waveguides, optical resonators and cavities, and spatial light modulators [3]. For instance, a typical LC slab optical waveguide is a thin film of LC with a thickness of about $1\mu m$ sandwiched between two glass slides of lower refracted index than LC [2]. Stimulated light scattering (SLS), self-phase modulation (SPM), self-focusing, spatial soliton formation, optical wave mixing, harmonic generation, optical phase conjugation, and other nonlinear optical effects in LC have been investigated [3]. NLC is the most useful and widely studied type of LC [2–4]. However, the practical integrated electro-optical applications of NLC are limited by their large losses of about 20 dB/cm and relatively slow responses [2]. The scattering losses in SALC are much lower, and they can be useful in nonlinear optical applications [2]. Recently, the LC applications in plasmonics attracted a wide interest due to the combination of the surface plasmon polaritons (SPP) strong electric fields and the unique electro-optical properties of LC [4].

We investigated theoretically the nonlinear optical phenomena in SALC related to the specific mechanism of the cubic nonlinearity, which is determined by the smectic layer normal displacement $u(x, y, z, t)$ in the electric field of optical waves and SPP [5–13]. This mechanism combining the properties of the orientational and electrostrictive nonlinearities [2] occurs without the mass density change, strongly depends on the optical wave polarization and propagation direction, and has a resonant form of the frequency dependence. It is characterized by a comparatively short response time similar to acousto-optic processes [2, 14].

The theoretical analysis of the nonlinear optical phenomena in SALC related to the layer displacement was based on the simultaneous solution of the Maxwell equations for the optical waves propagating in SALC and the equation of motion for the SALC layers in the electric field of these waves. We used the slowly varying amplitude approximation (SVAA) [14]. We investigated the following nonlinear optical effects in SALC based on the layer displacement nonlinearity: self-focusing and self-trapping, SLS, and four-wave mixing (FWM) [5–10]. We applied the developed theory of the nonlinear optical phenomena in SALC to the SPP interactions in SALC [11–13]. The SPP stimulated scattering in SALC and the metal/insulator/metal (MIM) plasmonic waveguide with the SALC core are theoretically studied [11–13]. The detailed calculations and complicated explicit analytical expressions can be found in Refs [5–13]. In this chapter, we describe the general approach to the theoretical analysis of the nonlinear optical phenomena in SALC and present the main results.

The chapter is constructed as follows. The equation of motion for the smectic layer normal displacement $u(x, y, z, t)$ in the electric field is derived in Section 2. The self-focusing and self-trapping of the optical wave in SALC are considered in Section 3. The SLS in SALC is investigated in Section 4. The FWM in SALC is analyzed in Section 5. The SPP interaction in SALC is discussed in Section 6. The conclusions are presented in Section 7.

2. The smectic layer equation of motion

The structure of the homeotropically oriented SALC in an external electric field $\vec{E}(x, y, z, t)$ is presented in **Figure 1**.

The hydrodynamics of SALC is described by the following system of Eq. [1]

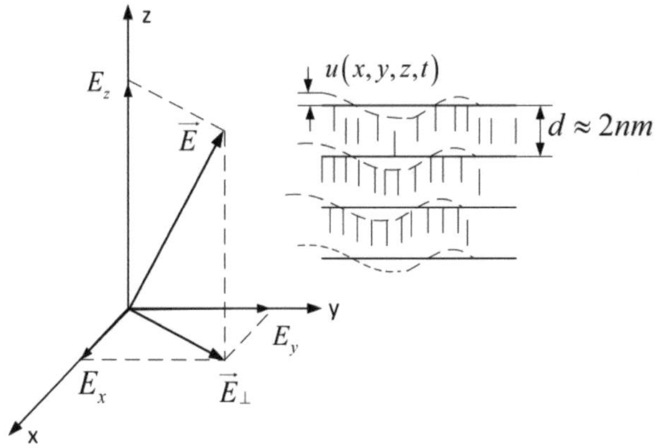

Figure 1. Homeotropically oriented SALC in an external electric field $\vec{E}\,(x, y, z, t)$.

$$\operatorname{div} \vec{v} = 0 \tag{1}$$

$$\rho \frac{\partial v_i}{\partial t} = -\frac{\partial \Pi}{\partial x_i} + \Lambda_i + \frac{\partial \sigma'_{ik}}{\partial x_k} \tag{2}$$

$$\Lambda_i = -\frac{\delta F}{\delta u_i} \tag{3}$$

$$\sigma'_{ik} = \alpha_0 \delta_{ik} A_{ll} + \alpha_1 \delta_{iz} A_{zz} + \alpha_4 A_{ik} + \alpha_{56}(\delta_{iz} A_{zk} + \delta_{kz} A_{zi}) + \alpha_7 \delta_{iz} \delta_{kz} A_{ll} \tag{4}$$

$$A_{ik} = \frac{1}{2}\left(\frac{\partial v_i}{\partial x_k} + \frac{\partial v_k}{\partial x_i}\right) \tag{5}$$

$$v_z = \frac{\partial u}{\partial t} \tag{6}$$

Here, $\vec{v}$ is the hydrodynamic velocity, ρ is the mass density, Π is the pressure, $\vec{\Lambda}$ is the generalized force density, σ'_{ik} is the viscous stress tensor, α_i are the viscosity Leslie coefficients, $\delta_{ik} = 1, i = k; \delta_{ik} = 0, i \neq k$, and F is the free energy density of SALC. Typically, SALC is supposed to be incompressible liquid according to Eq. (1) [1]. For this reason, we assume that the pressure $\Pi = 0$ and the SALC-free energy density F do not depend on the bulk compression [1]. The normal layer displacement $u(x, y, z, t)$ by definition has only one component along the Z axis. In such a case, the generalized force density $\vec{\Lambda}$ has only the Z component according to Eq. (3): $\vec{\Lambda} = (0, 0, \Lambda_z)$. Eq. (6) is specific for SALC since it determines the condition of the smectic layer continuity [1]. The SALC free energy density F in the presence of the external electric field $\vec{E}\,(x, y, z, t)$ has the form [1]

$$F = \frac{1}{2} B \left(\frac{\partial u}{\partial z}\right)^2 + \frac{1}{2} K \left(\frac{\partial^2 u}{\partial x^2} + \frac{\partial^2 u}{\partial y^2}\right)^2 - \frac{1}{2} \varepsilon_0 \varepsilon_{ik} E_i E_k \tag{7}$$

where K is the Frank elastic constant associated with the SALC purely orientational energy, ε_0 is the free space permittivity, and ε_{ik} is the SALC permittivity tensor including the nonlinear terms related to the smectic layer strains. It is given by [1]

$$\begin{aligned}&\varepsilon_{xx} = \varepsilon_{yy} = \varepsilon_\perp + a_\perp \frac{\partial u}{\partial z}; \varepsilon_{zz} = \varepsilon_\parallel + a_\parallel \frac{\partial u}{\partial z};\\&\varepsilon_{xz} = \varepsilon_{zx} = -\varepsilon_a \frac{\partial u}{\partial x}; \varepsilon_{yz} = \varepsilon_{zy} = -\varepsilon_a \frac{\partial u}{\partial y}; \varepsilon_a = \varepsilon_\parallel - \varepsilon_\perp\end{aligned} \tag{8}$$

Here, $\varepsilon_\perp, \varepsilon_\parallel$ are the diagonal components of the uniaxial SALC permittivity tensor perpendicular and parallel to the optical axis Z, respectively, and $a_\perp \sim 1; a_\parallel \sim 1$ are the phenomenological dimensionless coefficients. For the smectic layer displacement $u(x, y, z, t)$ depending on z, the purely orientational second term in the free energy density F (7) can be neglected. Indeed, for the typical values of B and $K \sim 10^{-11} N$ [1], the following inequality is valid: $Kk_{S\perp}^2 \ll B$ where $k_{S\perp}$ is the in-plane component of the smectic layer displacement wave vector. The contribution of the first term containing the normal layer strain is dominant. We consider the smectic layer normal displacement with $k_{Sz} \neq 0$. Taking into account the assumptions mentioned above and combining Eqs. (1)–(9), we obtain the equation of motion for the smectic layer normal displacement $u(x, y, z, t)$ [5, 6, 10]

$$\begin{aligned}&-\rho\nabla^2 \frac{\partial^2 u}{\partial t^2} + \left[\alpha_1 \nabla_\perp^2 \frac{\partial^2}{\partial z^2} + \frac{1}{2}(\alpha_4 + \alpha_{56})\nabla^2\nabla^2\right]\frac{\partial u}{\partial t} + B\nabla_\perp^2 \frac{\partial^2 u}{\partial z^2}\\&= \frac{\varepsilon_0}{2}\nabla_\perp^2\left[\frac{\partial}{\partial z}\left(a_\perp\left(E_x^2 + E_y^2\right) + a_\parallel E_z^2\right) - 2\varepsilon_a\left(\frac{\partial}{\partial x}(E_x E_z) + \frac{\partial}{\partial y}(E_y E_z)\right)\right].\end{aligned} \tag{9}$$

.Here, $\nabla_\perp^2 = \partial^2 u/\partial x^2 + \partial^2 u/\partial y^2$. If the external electric field is absent and the viscosity terms responsible for the decay of the smectic layer displacement are neglected, Eq. (10) coincides with the equation of the so-called second sound (SS) [1]

$$\rho\nabla^2 \frac{\partial^2 u}{\partial t^2} = B\nabla_\perp^2 \frac{\partial^2 u}{\partial z^2}. \tag{10}$$

Generally, for an arbitrary direction of the wave vector $\vec{k}_S$ in SALC, there exist two practically uncoupled acoustic modes: (i) the ordinary longitudinal sound wave caused by the mass density oscillations; (ii) SS wave caused by the smectic layer oscillations [1]. The SS propagation may be considered separately from ordinary sound since B is much less than the elastic constant of the mass density oscillations [1]. The SS dispersion relation corresponding to Eq. (11) has the form [1]

$$\Omega_S = s_0 \frac{k_{S\perp} k_{Sz}}{k_S}; s_0 = \sqrt{\frac{B}{\rho}} \tag{11}$$

Here, $\Omega_S, \vec{k}_S, s_0$ are SS frequency, wave vector and phase velocity, respectively. It is seen from Eq. (11) that SS is neither longitudinal, nor purely transverse, and it vanishes for the wave vectors $\vec{k}_S$ perpendicular or parallel to the layer plane. SS represents the oscillations of the

SALC complex order parameter phase [1]. If we take into account the viscosity terms in Eq. (10), then we can obtain the SS relaxation time τ_S given by

$$\tau_S = 2\rho\left[\alpha_1\frac{\left(k_{Sx}^2 + k_{Sy}^2\right)k_{Sz}^2}{k_S^2} + \frac{1}{2}(\alpha_4 + \alpha_{56})k_S^2\right]^{-1} \tag{12}$$

SS has been observed experimentally [15–17].

3. Self-focusing and self-trapping of optical beams in SALC

We first consider the self-action effects of the optical waves propagating in an anisotropic inhomogeneous nonlinear medium. The light beam propagation through a nonlinear medium is accompanied by the intensity-dependent phase shift on the wavefront of the beam [2]. Self-focusing of light results from the wavefront distortion inflicted on the beam by itself while propagating in a nonlinear medium [14]. In such a case, the field-induced refractive change Δn has the form $\Delta n = n_2|E|^2$ where $n_2 = const$ [14]. A light beam with a finite cross section also diffracts [14]. At a certain optical power level, the beam self-focusing and diffraction can be balanced in such a way that the beam propagates in the nonlinear medium with a plane wavefront and a constant transverse intensity profile [1]. This phenomenon is called self-trapping of an optical beam [1]. The optical wave propagation in a nonlinear medium is described by the following wave equation for the electric field $\vec{E}\,(x, y, z, t)$ [14]

$$curlcurl\,\vec{E} + \mu_0\frac{\partial^2\vec{D}\,L}{\partial t^2} = -\mu_0\frac{\partial^2\vec{D}\,NL}{\partial t^2} \tag{13}$$

Here, μ_0 is the free space permeability, $\vec{D}^L$ and $\vec{D}^{NL}$ are the linear and nonlinear parts of the electric induction. In SALC as a uniaxial medium two waves with the same frequency ω can propagate: an ordinary wave with the wave vector $\vec{k}_o$ and an extraordinary one with the wave vector $\vec{k}_e$ [2, 18]. Taking into account the SALC symmetry, we can choose the xz plane as a propagation plane. Then, the ordinary wave is polarized along the Y axis, and its electric field is given by

$$E_{oy} = A_o \exp[i(k_{ox}x + k_{oz}z - \omega t)] + c.c. \tag{14}$$

The extraordinary wave is polarized in the XZ plane having a component along the optical axis Z. The electric field of the extraordinary wave has the form

$$\vec{E}_e = \vec{e}_e A_e \exp[i(k_{ex}x + k_{ez}z - \omega t)] + c.c. \tag{15}$$

Here, $\vec{e}_e = \vec{a}_x e_{ex} + \vec{a}_z e_{ez}$ is the polarization unit vector of the extraordinary wave, $\vec{a}_{x,z}$ are the unit vectors of the X, Z axes, and c.c. stands for complex conjugate. The propagation direction and polarization of the ordinary and extraordinary waves in SALC are shown in **Figure 2**.

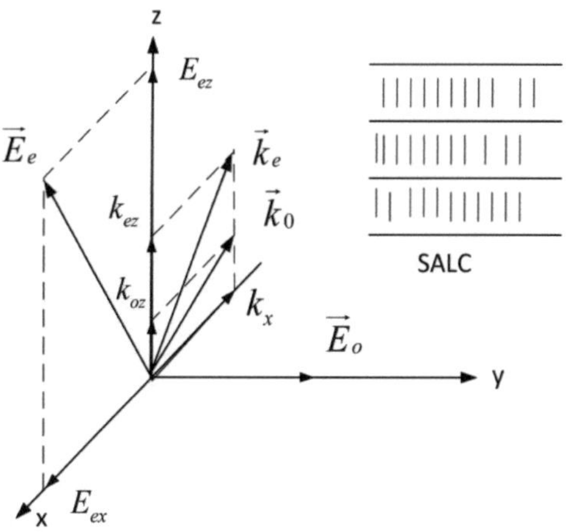

Figure 2. Propagation direction and polarization of the ordinary wave $\vec{E}_o$ and extraordinary wave $\vec{E}_e$ in SALC.

The corresponding linear electric induction vectors $\vec{D}_o^L, \vec{D}_e^L$ are given by [7, 18]

$$D_{oy}^L = \varepsilon_0\varepsilon_\perp E_y; \vec{D}_e^L = \varepsilon_0\left(\vec{a}_x\varepsilon_\perp e_{ex} + \vec{a}_z\varepsilon_\| e_{ez}\right)A_e\exp[i(k_{ex}x + k_{ez}z - \omega t)] + c.c. \quad (16)$$

In the linear approximation, substituting Eqs. (14)–(16) into the wave Eq. (13) we obtain the dispersion relations for the ordinary and extraordinary waves, respectively [7, 18]

$$k_o^2 = \varepsilon_\perp\frac{\omega^2}{c^2}; \frac{k_{ex}^2}{\varepsilon_\|} + \frac{k_{ez}^2}{\varepsilon_\perp} = \frac{\omega^2}{c^2} \quad (17)$$

It should be noted that the ordinary and extraordinary beams in the uniaxial medium propagate in different directions and the vectors $\vec{E}_e$ and $\vec{D}_e^L$ are not parallel [18]. The extraordinary wave propagates in the direction of the beam vector $\vec{s} \perp \vec{E}_e$, which is determined by the angle $\theta_e = \arctan\{(\varepsilon_\perp/\varepsilon_\|)\tan\theta_1\}$ with respect to the Z axis [18]. Here, θ_1 is the angle between $\vec{k}_e$ and the Z axis.

We consider separately the self-focusing and self-trapping of the ordinary and extraordinary beams [7]. We start with the analysis of the slab-shaped ordinary beam with the dimension in the Y direction much greater than in the incidence XZ plane. In such a case, the dependence on the coordinate y may be neglected [7, 9]. Substituting expression (14) into the equation of motion (9), we obtain [7, 9]

$$\frac{\partial u}{\partial z} = \frac{\varepsilon_0 a_\perp}{B}|A_o|^2; D_o^{NL} = \varepsilon_0 a_\perp\frac{\partial u}{\partial z}E_o \quad (18)$$

Expression (18) shows that the nonlinearity related to the smectic layer normal strain is the Kerr-type nonlinearity [14]. We introduce now the coordinates (x', z') parallel and normal to the ordinary beam propagation direction, respectively [7, 9]

$$x' = x\sin\theta_o + z\cos\theta_o; z' = -x\cos\theta_o + z\sin\theta_o \tag{19}$$

Here, θ_o is the angle between $\vec{k}_o$ and the Z axis. We use the SVAA for the ordinary beam amplitude A_o [14]

$$\left|\frac{\partial^2 A_o}{\partial x'^2}\right| \ll \left|k_o \frac{\partial A_o}{\partial x'}\right| \sim \left|\frac{\partial^2 A_o}{\partial z'^2}\right| \tag{20}$$

We are interested in the spatially localized solutions with the following boundary conditions [7, 9]

$$\lim_{z'\to\infty} |A_o(z')| = 0; \left.\frac{\partial|A_o(z')|}{\partial z'}\right|_{z'=0} = 0; |A_o(z'=0)| = |A_o|_{\max} \tag{21}$$

Then, substituting expressions (14), (18), (19) and the first ones of Eq. (16), (17) into Eq. (13) and taking into account the SVAA conditions (20), we obtain the truncated equation for the SVA $A_o(x', z')$, which has the form [7]

$$i\frac{\partial A_o}{\partial x'} + \frac{1}{2k_o}\frac{\partial^2 A_o}{\partial z'^2} + \frac{\omega^2}{c^2}\frac{\varepsilon_0 a_\perp^2}{2Bk_o}|A_o|^2 A_o = 0 \tag{22}$$

Eq. (22) is the nonlinear Schrodinger equation (NSE) [19]. The coefficient of the last term in the left-hand side (LHS) of Eq. (22) is positive definite $\omega^2\varepsilon_0^2 a_\perp^2/(4c^2Bk_o) > 0$, which corresponds to the stationary two-dimensional self-focusing of the light beam. The solution of Eq. (22) with the boundary conditions (21) has the form [7]

$$A_o(x', z') = |A_o|_{\max}\exp\left(i\frac{\varepsilon_0 a_\perp^2|A_o|_{\max}^2}{4B\varepsilon_\perp}k_o x'\right)\left[\cosh\left(\frac{\sqrt{\varepsilon_0}a_\perp|A_o|_{\max}}{\sqrt{2B\varepsilon_\perp}}k_o z'\right)\right]^{-1} \tag{23}$$

The self-trapped beam (23) is the so-called spatial soliton with the width $w_o = \sqrt{2B\varepsilon_\perp}\left(\sqrt{\varepsilon_0}a_\perp |A_o|_{\max}k_o\right)^{-1}$ [7].

The self-trapped ordinary beam normalized intensity spatial distribution is shown in **Figure 3**.

The self-trapping of the extraordinary beam (15) can be realized only when the anisotropy angle $(\theta_1 - \theta_e)$ is small enough: $\tan(\theta_1 - \theta_e) \ll (k_e w_o)^{-1}$ [7]. For the typical values of $\varepsilon_\perp, \varepsilon_\parallel$ [2], the following condition is valid: $0 \le \tan(\theta_1 - \theta_e) \le 0.12$, and the self-trapping condition for the extraordinary beam can be satisfied [7]. Then, using the procedure described above for the ordinary beam, we obtain the spatial soliton of the extraordinary beam. It has the form [7, 10]

$$A_e = |A_e|_{\max}\exp\left[i\frac{\varepsilon_0 h_e^2|A_e|_{\max}^2\omega^2}{4Bl_{e\parallel}\left(1+\frac{\varepsilon_a}{\varepsilon_\perp}e_{ez}\sin\theta_e\right)c^2\sin^2\theta_e}x''\right]\left[\cosh\left(\frac{z''}{w_e}\right)\right]^{-1} \tag{24}$$

Here, $x'' = x\sin\theta_e + z\cos\theta_e; z'' = -x\cos\theta_e + z\sin\theta_e$ are the coordinates parallel and perpendicular to the beam vector, respectively,

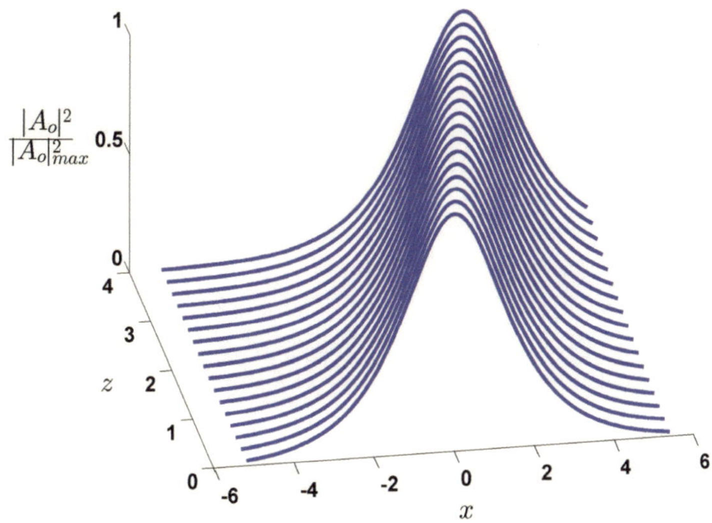

Figure 3. The self-trapped ordinary beam normalized intensity for the maximum amplitude $|A_o|_{\max} = 10^5 V/m$ and $\theta_o = \pi/6$.

$h_e = \left(a_\perp e_{ex}^2 + a_\| e_{ez}^2\right)\sin\theta_e + 2\varepsilon_a e_{ex} e_{ez}\cos\theta_e$, $l_{e\|} = k_{e\|}\left(1 + (\varepsilon_a/\varepsilon_\perp)\sin\theta_e\right)^{-1}$, and $k_{e\|}$ is the wave vector component parallel to the beam vector. The width w_e of the extraordinary beam spatial soliton is given by [7, 10]

$$w_e = \frac{\sqrt{2B\left(1 + \frac{\varepsilon_a}{\varepsilon_\perp} e_{ez}\sin\theta_e\right)(\sin\theta_e)c}}{\sqrt{\varepsilon_0 h_e}|A_e|_{\max}\omega} \tag{25}$$

For the typical values of the SALC parameters [1, 2], the optical beam electric field $|A_{o,e}|_{\max} \sim 10^5 V/m$, $\omega \sim 10^{15} s^{-1}$ and small angle θ_e the spatial soliton width is $w_{o,e} \sim 10^{-4} m$ [7]. SALC samples with a thickness of $10^{-4} m$ have been demonstrated experimentally [2, 17].

The optical wave self-trapping can occur also at the interface between the linear medium in the region $z < 0$ with the permittivity ε_s and the SALC cladding $(z > 0)$. For the light wave $E_y = A(z)\exp i(k_o x - \omega t)$ propagating along the interface parallel to the X axis, the self-trapped solution represents a bright surface wave with the amplitude $A(z)$ given by [7]

$$A(z) = A_{\max}\left[\cosh\left(\frac{z - z_0}{w_o}\right)\right]^{-1}; A_{\max} = A(z_0) \tag{26}$$

The cubic susceptibility of SALC $\chi_{SALC}^{(3)}$ related to the smectic layer compression is larger than $\chi^{(3)}$ related to the Kerr nonlinearity in organic liquids [14], but it is much less than the giant orientational nonlinearity (GON) in NLC [2]. However, the optical beam intensity in SALC may be much greater than in NLC, which are extremely sensitive to the strong optical fields [2]. In such cases, the approach based on the purely orientational mechanism of the optical nonlinearity is invalid.

4. Stimulated light scattering (SLS) in SALC

SLS is a process of parametric coupling between light waves and the material excitations of the medium [14]. We consider the SLS in SALC related to the smectic layer normal displacement and SS excited by the interfering optical waves [5, 6, 8–10]. We have taken into account the combined effect of SALC layered structure and anisotropy. It should be noted that SS propagates in SALC without the change of the mass density in such a way that the SS wave and the ordinary sound wave are decoupled [1].

In general case when the coupled optical waves have arbitrary polarizations and propagation directions, each optical wave in SALC ($z > 0$) splits into the extraordinary and ordinary ones with the same frequency and different wave vectors due to the strong anisotropy of SALC [6, 10, 18]. The polarizations of these waves are shown in **Figure 4**. The XZ plane is chosen to be the propagation plane of the waves $\vec{E}_1^{o,e}$. In such a case, the extraordinary wave $\vec{E}_1^{e}$ is polarized in the XZ plane, while the ordinary wave $\vec{E}_1^{o}$ is parallel to the Y axis [18]. The ordinary wave $\vec{E}_2^{o}$ is polarized in the XY plane perpendicular to the optical Z axis, and the extraordinary wave $\vec{E}_2^{e}$ possesses a three-dimensional polarization vector $\vec{e}_2^{\,e}$ [18]. The wave vectors $\vec{k}_{1,2}^{o}$ and $\vec{k}_1^{e}$ of these waves satisfy the dispersion relations (17) while the three-dimensional wave vector $\vec{k}_2^{e}$ satisfies the dispersion relation $\left(\left(k_{2x}^{e}\right)^2 + \left(k_{2y}^{e}\right)^2\right)\varepsilon_{\parallel}^{-1} + \left(k_{2z}^{e}\right)^2\varepsilon_{\perp}^{-1} = (\omega_2/c)^2$ [18]. The fundamental ordinary and extraordinary waves have the form, respectively

$$\begin{aligned}\vec{E}_1^{o,e} &= \vec{e}_1^{\,o,e}\left\{A_1^{o,e}(z)\exp i\left[\left(\vec{k}_1^{o,e}\cdot\vec{r}\right) - \omega_1 t\right] + c.c.\right\}\\ \vec{E}_2^{o,e} &= \vec{e}_2^{\,o,e}\left\{A_2^{o,e}(z)\exp i\left[\left(\vec{k}_2^{o,e}\cdot\vec{r}\right) - \omega_2 t\right] + c.c.\right\}\end{aligned} \tag{27}$$

Here, $\omega_1 > \omega_2$ and $\Delta\omega = \omega_1 - \omega_2 \ll \omega_1$. Each pair of the waves (27) has the same frequency, and for this reason, we define the nonlinear mixing of these waves as partially frequency degenerate FWM [6]. We assume that the complex amplitudes $A_{1,2}^{o,e}(z) = \left|A_{1,2}^{o,e}(z)\right|\exp i\gamma_{1,2}^{o,e}(z)$ are slowly varying along the optical axis Z: $\left|\partial^2 A_{1,2}^{o,e}/\partial z^2\right| \ll \left|k_{1,2z}^{o,e}\partial A_{1,2}^{o,e}/\partial z\right|$. As a result, the nonlinear two-wave mixing analyzed in Ref. [5] transforms into a partially degenerate FWM [6, 10]. We substitute the waves (27) into equation of motion (9). The interfering optical waves (27) with close frequencies $\omega_{1,2}$ create a dynamic grating of the smectic layer normal displacement $u(x, y, z, t)$ consisting of four propagating harmonics with the same frequency and different wave vectors. It has the form [6]

$$u(x, y, z, t) = \frac{i\varepsilon_0}{\rho}\sum_{j=1}^{4}\frac{\left(\Delta k_{j\perp}\right)^2 h_j M_j}{\left(\Delta k_j\right)^2 G_j\left(\Delta\omega, \Delta\vec{k}_j\right)}\exp i\left[\left(\Delta\vec{k}_j\cdot\vec{r}\right) - \Delta\omega t\right] + c.c. \tag{28}$$

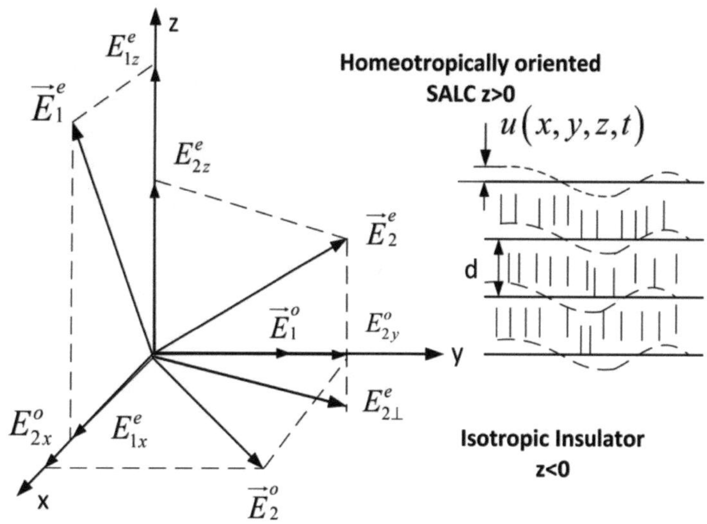

Figure 4. The polarizations of the fundamental ordinary waves $\vec{E}^{o}_{1,2}$ and extraordinary waves $\vec{E}^{e}_{1,2}$ in SALC (z > 0).

Here, $\Delta\vec{k}_1 = \vec{k}^e_1 - \vec{k}^o_2; \Delta\vec{k}_2 = \vec{k}^e_1 - \vec{k}^e_2; \Delta\vec{k}_3 = \vec{k}^o_1 - \vec{k}^o_2; \Delta\vec{k}_4 = \vec{k}^o_1 - \vec{k}^e_2;$

$h_1 = a_\perp \Delta k_{1z} e^e_{1x} e^o_{2x} - \varepsilon_a \left[\Delta k_{1x} e^e_{1z} e^o_{2x} + \Delta k_{1y} e^e_{1z} e^o_{2y}\right],$

$h_2 = a_\perp \Delta k_{2z} e^e_{1x} e^e_{2x} + a_\parallel \Delta k_{2z} e^e_{1z} e^e_{2z} - \varepsilon_a \left[\Delta k_{2x}\left(e^e_{1x} e^e_{2z} + e^e_{1z} e^e_{2x}\right) + \Delta k_{2y} e^e_{1z} e^e_{2y}\right]; h_3 = a_\perp \Delta k_{3z} e^o_{2y};$

$h_4 = a_\perp \Delta k_{4z} e^e_{2y} - \varepsilon_a \Delta k_{4y} e^e_{2z}; M_1 = A^e_1 (A^o_2)^*; M_2 = A^e_1 (A^e_2)^* \; M_3 = A^o_1 (A^o_2)^*, \; M_4 = A^o_1 (A^e_2)^*$, and

$$G_j\left(\Delta\omega, \Delta\vec{k}_j\right) = (\Delta\omega)^2 + i\Delta\omega\Gamma_j - \Omega_j^2$$
$$\Gamma_j = \frac{1}{\rho}\left[\alpha_1 \frac{(\Delta k_{j\perp})^2 (\Delta k_{jz})^2}{(\Delta k_j)^2} + \frac{1}{2}(\alpha_4 + \alpha_{56})(\Delta k_j)^2\right]; \Omega_j^2 = s_0^2 \frac{(\Delta k_{j\perp})^2 (\Delta k_{jz})^2}{(\Delta k_j)^2} \tag{29}$$

The parametric amplification of the fundamental optical waves $\vec{E}^{o,e}_2$ with the lower frequency ω_2 by the other pair of optical waves $\vec{E}^{o,e}_1$ with the higher frequency ω_1 occurs in SALC due to the SLS on the light-induced smectic layer dynamic grating (28) [6, 10]. It is actually the Stokes SLS [14]. The fundamental optical waves also create Stokes and anti-Stokes small harmonics with the combination frequencies and wave vectors. The analysis of SLS in SALC is based on the simultaneous solution of the smectic layer equation of motion (9), the wave Eq. (13) for ordinary waves (14) and extraordinary waves (15) with the permittivity tensor (8). The nonlinear part of the permittivity tensor (8) ε^N_{ik} in the three-dimensional case can be written as follows [6]

$$\varepsilon^N_{ik} = \widehat{N}_{ik} u(x,y,z,t); \widehat{N}_{xx} = \widehat{N}_{yy} = a_\perp \frac{\partial}{\partial z}; \widehat{N}_{xy} = \widehat{N}_{yx} = 0;$$
$$\widehat{N}_{xz} = \widehat{N}_{zx} = -\varepsilon_a \frac{\partial}{\partial x}; \widehat{N}_{yz} = \widehat{N}_{zy} = -\varepsilon_a \frac{\partial}{\partial y}; \widehat{N}_{zz} = a_\parallel \frac{\partial}{\partial z} \tag{30}$$

Combining Eqs. (27)–(30), we obtain the nonlinear part of the electric induction, or the nonlinear polarization $D_i^{NL} = \varepsilon_0 \varepsilon_{ik}^N E_k$ [6]. This nonlinear polarization consists of two types of terms: (i) four harmonics, which are phase-matched with fundamental waves (27); (ii) all other terms with the combination frequencies and wave vectors, which give rise to the small scattered Stokes and anti-Stokes harmonics similar to the Brillouin scattering [6, 10, 14]. The combination of the anisotropy and nonlinearity also results in the creation of the small additional components of the waves $\vec{E}_1^{o,e}$ and $\vec{E}_2^{o}$ polarized in the XZ plane and XY plane, respectively [6].

We start with the analysis of the parametric coupling among the waves (27). Substituting expressions (27)–(30) and the phase-matched part of $\vec{D}^{NL}$ into wave Eq. (13), taking into account SVAA for the complex amplitudes $A_{1,2}^{o,e}(z)$, and separating the real and imaginary parts, we obtain the truncated equations for the magnitudes $\left|A_{1,2}^{o,e}(z)\right|$ and phases $\gamma_{1,2}^{o,e}(z)$ of these SVA [6]

$$2l_1^{o,e}\frac{\partial\left|A_1^{o,e}\right|}{\partial z} = -\left(\frac{\omega_1}{c}\right)^2 \times \frac{\varepsilon_0 \Delta\omega}{\rho}\left\{\frac{h_{3,2}^2(\Delta k_{3,2\perp})^2\Gamma_{3,2}}{|G_{3,2}|^2(\Delta k_{3,2})^2}\left|A_2^{o,e}\right|^2 + \frac{h_{4,1}^2(\Delta k_{4,1\perp})^2\Gamma_{4,1}}{|G_{4,1}|^2(\Delta k_{4,1})^2}\left|A_2^{e,o}\right|^2\right\}\left|A_1^{o,e}\right| \tag{31}$$

$$2l_2^{o,e}\frac{\partial\left|A_2^{o,e}\right|}{\partial z} = \left(\frac{\omega_2}{c}\right)^2 \times \frac{\varepsilon_0 \Delta\omega}{\rho}\left\{\frac{h_{3,2}^2(\Delta k_{3,2\perp})^2\Gamma_{3,2}}{|G_{3,2}|^2(\Delta k_{3,2})^2}\left|A_1^{o,e}\right|^2 + \frac{h_{1,4}^2(\Delta k_{1,4\perp})^2\Gamma_{1,4}}{|G_{1,4}|^2(\Delta k_{1,4})^2}\left|A_1^{e,o}\right|^2\right\}\left|A_2^{o,e}\right| \tag{32}$$

$$2l_1^{o,e}\frac{\partial\gamma_1^{o,e}}{\partial z} = -\left(\frac{\omega_1}{c}\right)^2\frac{\varepsilon_0}{\rho} \times \left\{\frac{h_{3,2}^2(\Delta k_{3,2\perp})^2\left[(\Delta\omega)^2 - \Omega_{3,2}^2\right]}{|G_{3,2}|^2(\Delta k_{3,2})^2}\left|A_2^{o,e}\right|^2 + \frac{h_{4,1}^2(\Delta k_{4,1\perp})^2\left[(\Delta\omega)^2 - \Omega_{4,1}^2\right]}{|G_{4,1}|^2(\Delta k_{4,1})^2}\left|A_2^{e,o}\right|^2\right\} \tag{33}$$

$$2l_2^{o,e}\frac{\partial\gamma_2^{o,e}}{\partial z} = -\left(\frac{\omega_2}{c}\right)^2\frac{\varepsilon_0}{\rho} \times \left\{\frac{h_{3,2}^2(\Delta k_{3,2\perp})^2\left[(\Delta\omega)^2 - \Omega_{3,2}^2\right]}{|G_{3,2}|^2(\Delta k_{3,2})^2}\left|A_1^{o,e}\right|^2 + \frac{h_{1,4}^2(\Delta k_{1,4\perp})^2\left[(\Delta\omega)^2 - \Omega_{1,4}^2\right]}{|G_{1,4}|^2(\Delta k_{1,4})^2}\left|A_1^{e,o}\right|^2\right\} \tag{34}$$

Here, $l_{1,2}^o = k_{1,2z}^o; l_{1,2}^e = k_{1,2z}^e\left[1 - e_{1,2z}^e\left(\vec{k}_{1,2}^e \cdot \vec{e}_{1,2}^e\right)\left(k_{1,2z}^e\right)^{-1}\right]$.

Eqs. (31) and (32) describe the parametric energy exchange between the fundamental waves, Eqs. (33) and (34) describe the cross-phase modulation (XPM) of these waves [6]. Combining Eqs. (31) and (32), we obtain the Manley-Rowe relation, which expresses the conservation of the total photon number [14]. In our case, it has the form [6]

$$\left(\frac{\omega_1}{c}\right)^{-2}\left[l_1^o|A_1^o|^2 + l_1^e|A_1^e|^2\right] + \left(\frac{\omega_2}{c}\right)^{-2}\left[l_2^o|A_2^o|^2 + l_2^e|A_2^e|^2\right] = const = I_0 \tag{35}$$

The solution of the system of Eqs. (31)–(34) can be written in the integral form [6, 10]

$$w_1^{o,e} = w_1^{o,e}(0)\exp\left\{-\int_0^z \left(\beta_{3,2}w_2^{o,e} + \beta_{4,1}w_2^{e,o}\right)dz'\right\} \tag{36}$$

$$w_2^{o,e} = w_2^{o,e}(0)\exp\left\{\int_0^z \left(\beta_{3,2}w_1^{o,e} + \beta_{1,4}w_1^{e,o}\right)dz'\right\} \tag{37}$$

$$\gamma_1^{o,e} - \gamma_1^{o,e}(0) = -\frac{1}{2}\int_0^z \left(\delta_{3,2}w_2^{o,e} + \delta_{4,1}w_2^{e,o}\right)dz' \tag{38}$$

$$\gamma_2^{o,e} - \gamma_2^{o,e}(0) = -\frac{1}{2}\int_0^z \left(\delta_{3,2}w_1^{o,e} + \delta_{1,4}w_1^{e,o}\right)dz' \tag{39}$$

Here, the dimensionless variables are given by [6]:

$$w_{1,2}^{o,e} = \frac{1}{I_0}\left(\frac{\omega_{1,2}}{c}\right)^{-2} l_{1,2}^{o,e}\left|A_{1,2}^{o,e}\right|^2; w_1^o + w_1^e + w_2^o + w_2^e = 1 \tag{40}$$

$$\beta_j = C_j\Gamma_j\Delta\omega, \delta_j = C_j\left[(\Delta\omega)^2 - \Omega_j^2\right], j = 1, 2, 3, 4,$$

$$C_j = \left(\frac{\omega_1\omega_2}{c^2}\right)^2 \frac{\varepsilon_0 I_0 \left(\Delta k_{j\perp}\right)^2}{\rho|G_j|^2 d_j\left(\Delta k_j\right)^2}; d_1 = l_1^e l_2^o, d_2 = l_1^e l_2^e, d_3 = l_1^o l_2^o, d_4 = l_1^o l_2^e \tag{41}$$

Comparison of Eq. (36), (37), and (40) shows that for $z \to \infty$ $w_1^{o,e} \to 0$ and $w_2^o + w_2^e \to 1$. Hence, the pumping waves with the larger frequency ω_1 are depleted, the signal waves with smaller frequency $\omega_2 < \omega_1$ are amplified with the saturation at the sufficiently large distances, and the system is stable. The gain terms β_j reach their maximal values close to the SS resonance condition $\Delta\omega \approx \Omega_j$, which can be satisfied for $\Delta\omega \sim \omega_1 s_0/c \ll \omega_1$ [6]. In such a case, $\beta_j \gg \delta_j$, the parametric amplification process is dominant while XPM can be neglected.

In general case, the exact analytical solution of Eqs. (31)–(34) is hardly possible. However, the explicit expressions for the coupled wave SVA have been obtained when both waves are propagating in the same XZ plane [5, 6]. For instance, assume that the pumping extraordinary wave with the frequency $\omega_1 > \omega_2$ is mainly polarized in the XZ plane, the signal ordinary wave with the frequency ω_2 is mainly polarized along the Y axis, and the intensities of the components with other polarizations are small in such a way that $w_1^e \gg w_1^o; w_2^o \gg w_2^e$. Then, in the first approximation, the normalized intensities w_1^e, w_2^o of the main components have the form [6, 10]

$$w_1^e = \frac{1}{2}J_1\left[1 - \tanh\left(\eta - \eta_0\right)\right]; w_2^o = \frac{1}{2}J_1\left[1 + \tanh\left(\eta - \eta_0\right)\right] \tag{42}$$

Here, $w_1^e + w_2^o = J_1 = const = w_1^e(0) + w_2^o(0)$, $\eta = \beta_1 J_1 z/2$. It is seen from Eq. (42) that for $\eta \to \infty$ $w_1^e \to 0; w_2^o \to J_1$, and the crossing point $z_0 = \left(\beta_1 J_1\right)^{-1}\ln\left(w_1^e(0)/w_2^o(0)\right)$ exists only for $w_1^e(0)/w_2^o(0) > 1$. The coordinate dependence of the normalized intensities w_1^e, w_2^o is presented in **Figure 5**. The numerical estimations show that for the typical values of SALC parameters [1–3] in the resonant case the coupling constant per unit optical intensity $\beta_1/P_{opt} \sim (0.01\ -10)cm/MW$ [6]. For the optical intensity $P_{opt} \sim 10^6 - 10^7 Wcm^{-2}$, the SLS gain $\beta_{1\max} \sim 10^2 cm^{-1}$, which is at least an order of magnitude larger than the gain at Brillouin SLS in isotropic organic liquids [14]. Such optical intensities are feasible [20, 21].

The explicit expressions of the small component intensities w_1^o and w_2^e can be obtained in the second approximation. They have the form [6]

$$\begin{aligned} w_1^o &= w_1^o(0)\left\{\frac{\exp(-\eta)\cosh\left(\eta_0\right)}{\cosh\left(\eta-\eta_0\right)}\right\}^{\beta_3/\beta_1} \\ w_2^e &= w_2^e(0)\left\{\frac{\exp(\eta)\cosh\left(\eta_0\right)}{\cosh\left(\eta-\eta_0\right)}\right\}^{\beta_2/\beta_1} \end{aligned} \tag{43}$$

It is easy to see from Eq. (43) that for $\eta \to \infty$ $w_1^o \to 0$ and $w_2^e \to w_2^e(0)\left[1 + w_1^e(0)/w_2^o(0)\right]^{\beta_2/\beta_1}$ $= const.$

The evaluation of the phases $\gamma_{1,2}^{o,e}$ shows that the pumping wave phases $\gamma_1^{o,e}$ rapidly increase that results in the fast oscillations of the depleted amplitudes $A_1^{o,e}(z)$ [6]. The phases $\gamma_2^{o,e}$ of the signal waves tend to the constant values at sufficiently large η [6].

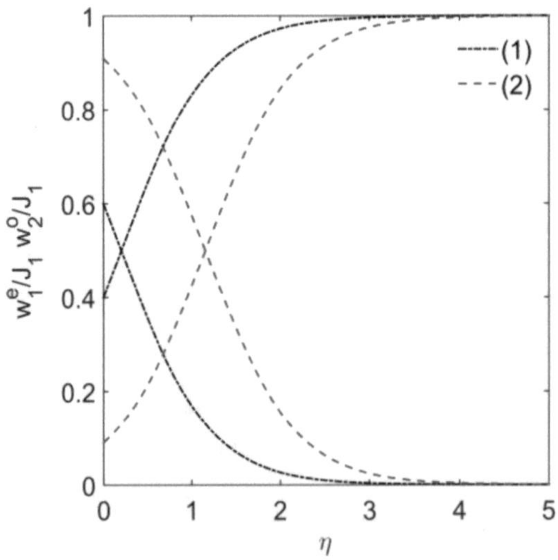

Figure 5. The dependence of the normalized pumping and signal intensities $w_1^e/J_1; w_2^o/J_1$ on the dimensionless coordinate η for the pumping-to-signal ratio $w_1^e(0)/w_2^o(0) = 1.5; 5$ (curves 1 and 2, respectively).

The Brillouin-like SLS also results in the generation of six Stokes small harmonics with the frequency $(\omega_2 - \Delta\omega)$ and combination wave vectors, six anti-Stokes small harmonics with the frequency $(\omega_1 + \Delta\omega)$ and combination wave vectors, and eight small harmonics with the fundamental frequencies $\omega_{1,2}$ and combination wave vectors [6].

5. The nondegenerate FWM in SALC

Consider now the nondegenerate FWM in SALC [8–10]. Assume that four coupled fundamental optical waves have different close frequencies ω_n such that $\Delta\omega_{mn} = \omega_m - \omega_n \sim s_0\omega_n/c \ll \omega_n$. For the sake of definiteness, we suppose that $\omega_1 < \omega_2 < \omega_3 < \omega_4$. These fundamental waves have the form [8–10]

$$\vec{E}_m = \vec{e}_m\left\{A_m(z)\exp i\left[\left(\vec{k}_m\cdot\vec{r}\right) - \omega t\right] + c.c.\right\}, m = 1, \ldots, 4 \tag{44}$$

The interfering waves (44) create a dynamic grating of the smectic layer normal displacement of the type (28), but this time each harmonic has a different frequency $\Delta\omega_{mn} = \omega_m - \omega_n$, $m, n = 1, \ldots, 4$. We discuss two cases: (i) all waves (44) are polarized in the directions perpendicular to the propagation plane and propagate as ordinary waves; (ii) all waves (44) are polarized in the propagation plane and behave as extraordinary waves [8–10]. Using the SVAA and the theory developed in the previous section, we obtain the truncated equations for the slowly varying magnitudes $|A_m(z)|$ and phases $\gamma_m(z)$ similar to Eqs. (31)–(34). The analysis of these equations shows that the wave with the lowest frequency ω_1 is amplified up to the saturation level determined by the integral of motion I_0 similar to the one from Eq. (35) [8–10]

$$I_0 = \sum_{m=1}^{4} l_m\left(\frac{\omega_m}{c}\right)^{-2}|A_m|^2 = const \tag{45}$$

Here, the factors l_m are defined above for the ordinary or extraordinary wave, respectively. Three other waves with the higher frequencies $\omega_{2,3,4}$ undergo the depletion like the pumping waves [8–10]. The depletion of the waves $\vec{E}_{2,3,4}$ is accompanied by the XPM with the rapidly increasing phases while the phase of the signal wave $\vec{E}_1$ saturates at large distances. The analytical solution of the type (42) and (43) has been obtained for the case when the pumping wave $\vec{E}_4$ and the signal wave $\vec{E}_1$ are much stronger than the idler waves $\vec{E}_{2,3}$ with the intermediate frequencies $\omega_{2,3}$ [8–10].

In the special case when some ordinary optical waves (44) have perpendicular polarizations vectors $\vec{e}_{m\perp}\perp\vec{e}_{n\perp}$ the polarization-decoupled FWM is possible [8–10]. Such waves do not excite the dynamic grating since the corresponding coupling constants $h^o_{mn} = a_\perp\Delta k_{mnz}\left(\vec{e}_{m\perp}\cdot\vec{e}_{n\perp}\right)$ vanish [8–10]. In the case of the extraordinary wave mixing, the polarization-decoupled FWM is impossible because of the SALC anisotropy. If the electric field of the signal ordinary wave $\vec{E}_1$ is perpendicular to the fields of all other waves than this wave propagates though SALC without any change of its SVA: $A_1^o = const$. If $\vec{E}_1^o \perp \vec{E}_{2,3}^o$ and $\vec{E}_1^o \| \vec{E}_4^o$, then FWM is divided in two separate two-wave mixing processes between the waves $\vec{E}_{1,4}^o$ and the waves $\vec{E}_{2,3}^o$ with the solutions similar to solution (42) [8].

In the important case when the pumping wave $\vec{E}_4$ and the signal wave $\vec{E}_1$ are much stronger than the idler waves $\vec{E}_{2,3}$, the approximate solution can be obtained similarly to the solution (42) and (43) in the case of SLS [8–10]. It has been shown that this solution is stable in the case of FWM [8].

In the particular case when the fundamental waves (44) are counter-propagating, the phase conjugation is possible as a result of the nondegenerate FWM in SALC [8–10]. Optical phase conjugation (OPC) is the wavefront reversion property of a backward propagating optical wave with respect to a forward propagating wave [22]. The optical waves are phase conjugated to each other if their complex amplitudes are conjugated with respect to their phase factors [22]. Typically, OPC results from nonlinear optical processes such as FWM and SLS [20]. LC are commonly used for FWM and OPC [22].

Suppose that the waves $\vec{E}_{1,4}$ are phase-conjugate while the waves $\vec{E}_{2,3}$ are forward-going and backward-going pumping waves, which have the form [8]

$$\begin{aligned}
\vec{E}_1 &= \vec{e}_1\left\{A_1 \exp i\left[\left(\vec{k}_4\cdot\vec{r}\right) + \omega_1 t\right] + c.c.\right\} \\
\vec{E}_2 &= \vec{e}_2\left\{A_2 \exp i\left[\left(\vec{k}_2\cdot\vec{r}\right) - \omega_2 t\right] + c.c.\right\} \\
\vec{E}_3 &= \vec{e}_3\left\{A_3 \exp i\left[\left(\vec{k}_2\cdot\vec{r}\right) + \left(\Delta\vec{k}\cdot\vec{r}\right) + \omega_3 t\right] + c.c.\right\} \\
\vec{E}_4 &= \vec{e}_4\left\{A_4 \exp i\left[\left(\vec{k}_4\cdot\vec{r}\right) - \omega_4 t\right] + c.c.\right\}
\end{aligned} \tag{46}$$

Here, $\Delta\vec{k} = \Delta\vec{k}_{32}$ is the wave vector mismatch of the FWM process. In the case of OPC caused by SLS the frequency balance condition between the waves with the same vectors is necessary. We assume that $\omega_3 - \omega_1 = \omega_4 - \omega_2$. Suppose that the pumping waves $\vec{E}_{2,3}$ are much stronger than the probe wave $\vec{E}_4$ and the phase-conjugate wave $\vec{E}_1$ propagating in the negative direction as it is seen from Eq. (46). In such a case, using the constant pumping approximation (CPA) [14] where $A_{2,3} = const$, we obtain the following solution for the probe wave and the phase-conjugate wave SVA $A_{1,4}$ [8–10]

$$A_{1,4} = A_{01,4} \exp\left[gr \pm \frac{i}{2}\left(\Delta\vec{k}\cdot\vec{r}\right)\right] \tag{47}$$

Analysis of the truncated equations for $A_{1,4}$ shows that there exists the solution with the gain $\mathrm{Re}g < 0$ corresponding to the amplification of the phase-conjugate wave $\vec{E}_1$ [8–10]. Such a case can be characterized as a kind of the Brillouin-enhanced FWM (BEFWM) based on the optical nonlinearity related to the smectic layer normal displacement [8–10]. Numerical estimations show that the amplification of the phase-conjugate wave $\vec{E}_1$ is possible for the typical values of

SALC parameters and for the pumping wave intensity of about 100 $MWcm^{-2}$ [9], which is feasible [20, 21]. OPC in the homeotropically oriented SALC film with the thickness of 250 μm had been demonstrated experimentally [20].

The components of the nonlinear electric induction $\vec{D}^{NL}$, which are not phase matched to the fundamental waves (46) give rise to 12 doubly degenerate combination harmonics of the type.

$A_m A_p A_n^* \exp i\left[\left(\left(\vec{k}_m + \vec{k}_p - \vec{k}_n\right)\cdot \vec{r}\right) - (\omega_m + \omega_p - \omega_n)t\right]$ and 12 harmonics of the type $A_m^2 A_n^* \exp$ $i\left[2\left(\left(\vec{k}_m \cdot \vec{r}\right) - \omega_m t\right) - \left(\vec{k}_n \cdot \vec{r}\right) + \omega_n t\right]$ [8–10].

6. Nonlinear interaction of surface plasmon polaritons (SPP) in SALC

Integration of strongly nonlinear LC with plasmonic structures and metamaterials would enable active switching and tuning operations with low threshold [4]. LC may be also used in reconfigurable metamaterials for tuning the resonant frequency, the transmission/ reflection coefficient, and the refractive index [23]. Combination of metamaterials and active plasmonic structures with NLC has been investigated [4, 23]. In this section, we discuss the nonlinear optical effects caused by the SPP mixing in SALC, which is characterized by low losses and a strong nonlinearity related to the smectic layer normal displacement without the change of the mass density [11–13]. Consider the interface $z = 0$ between a homeotropically oriented SALC $(z > 0)$ and a metal $(z < 0)$ shown in **Figure 6** [11, 12]. The SALC optical Z axis and the X axis are chosen to be perpendicular and parallel to the interface $z = 0$, respectively.

SPP from the metal penetrate into SALC. The permittivity of the metal $\varepsilon_m(\omega)$ determined by the Drude model is given by $\varepsilon_m(\omega) = 1 - \omega_p^2 \omega^{-1}(\omega + (i/\tau))^{-1}$ where $\omega_p = \sqrt{n_0 e^2/(\varepsilon_0 m)}$ is the plasma frequency, n_0 is the free electron density in the metal, e, m are the electron charge and mass, respectively, ω, τ are the SPP angular frequency and lifetime, respectively [24, 25]. The efficient SLS in SALC takes place for the counter-propagating SPP with close frequencies

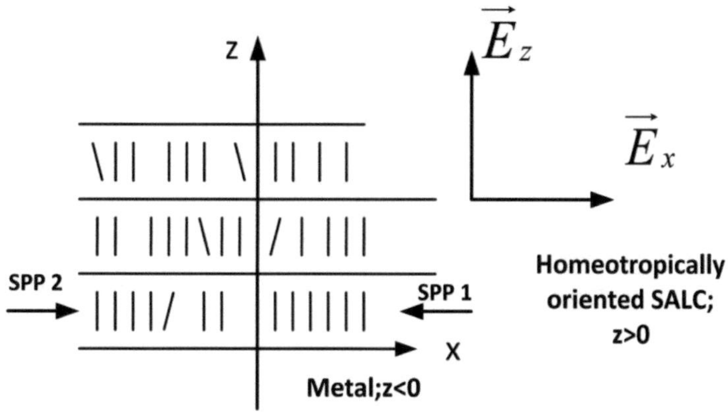

Figure 6. The counter-propagating SPP at the interface between a metal $(z < 0)$ and a homeotropically oriented SALC $(z > 0)$.

$\omega_1 > \omega_2$ such that $\Delta\omega = \omega_1 - \omega_2 \ll \omega_1$ [12]. The spatially localized electric fields of these SPP in SALC have the form [24, 25]

$$\vec{E}_{1,2} = \frac{1}{2}\left\{\vec{e}_{1,2}A_{1,2}(x,t)\exp\left[\pm ik_x(x \pm d) - k_z^S z - i\omega_{1,2}t\right] + c.c.\right\} \tag{48}$$

The SPP are polarized as transverse magnetic (TM) waves with the electric field components $E_{x,z}$ and the magnetic field component H_y [23, 24]. In an optically uniaxial SALC, SPP propagate as extraordinary waves [18]. The numerical estimations show that for the optical frequency range and the small frequency difference $\Delta\omega \approx (10^{-7} - 10^{-5})\omega_1$ the $SPP_{1,2}$ wave vectors are practically equal [11, 12]. They have the form [11, 12]

$$\begin{aligned} k_z^S &= \sqrt{k_x^2(\varepsilon_\perp/\varepsilon_\|) - \omega_1^2\varepsilon_\perp/c^2} \\ k_x &= (\omega_1/c)\sqrt{\varepsilon_m(\omega_1)[1 - (\varepsilon_m(\omega_1)/\varepsilon_\perp)]\left[1 - \left(\varepsilon_m^2(\omega_1)/(\varepsilon_\perp\varepsilon_\|)\right)\right]^{-1}} \end{aligned} \tag{49}$$

Numerical estimations show that for the typical values of $\omega_{1,2}, \omega_p, \tau$ the following relations are valid: $\mathrm{Re}k_z^S \gg \mathrm{Im}k_z^S$, $\mathrm{Re}k_x \gg \mathrm{Im}k_x$ [11, 12, 24, 25]. For the optical wavelength $\lambda_{opt} \approx (0.6 - 1.33)\mu m$, the SPP propagation length L_x and the wavelength λ_s are given by, respectively: $L_x = (\mathrm{Im}k_x)^{-1} \approx (84 - 550)\mu m$, $\lambda_s = 2\pi/(\mathrm{Re}k_x) \approx (0.33 - 0.77)\mu m \ll L_x$ [12]. The SPP localization length $L_z = \left(\mathrm{Re}k_z^S\right)^{-1} \sim 10^{-6}m$ belongs to the subwavelength scale: $\mathrm{Im}k_z^S \sim 10m^{-1} \ll \mathrm{Re}k_z^S$ and can be neglected [12].

Substituting the SPP fields (48) into the smectic layer equation of motion (9), we obtain the dynamic grating $u(x,z,t)$ given by [11, 12]

$$u(x,z,t) = 0.5\left[U\exp\left\{i(2\mathrm{Re}k_x)x - 2(\mathrm{Im}k_x)d - \left(2\mathrm{Re}k_z^S\right)z - i\Delta\omega t\right\} + c.c.\right] \tag{50}$$

Here,

$$U = -\frac{\varepsilon_0(2\mathrm{Re}k_x)^2 hA_1(x,t)A_2^*(x,t)}{\rho\left[-(2\mathrm{Re}k_x)^2 + \left(2\mathrm{Re}k_z^S\right)^2\right]G\left(k_x,k_z^S,\Delta\omega\right)} \tag{51}$$

$$\begin{aligned} h &= -\left(2\mathrm{Re}k_z^S\right)\left(-a_\perp|e_{1x}|^2 + a_\||e_{1z}|^2\right) - 4\varepsilon_a(2\mathrm{Re}k_x)\mathrm{Im}\left(e_{1z}e_{1x}^*\right); \\ G\left(k_x,k_z^S,\Delta\omega\right) &= (\Delta\omega)^2 - \frac{B(2\mathrm{Re}k_x)^2\left(2\mathrm{Re}k_z^S\right)^2}{\rho\left[-(2\mathrm{Re}k_x)^2 + \left(2\mathrm{Re}k_z^S\right)^2\right]} \\ &\quad -i\frac{\Delta\omega}{\rho}\left[-\alpha_1\frac{(2\mathrm{Re}k_x)^2\left(2\mathrm{Re}k_z^S\right)^2}{\left[-(2\mathrm{Re}k_x)^2 + \left(2\mathrm{Re}k_z^S\right)^2\right]} + \frac{(\alpha_4 + \alpha_{56})\left[-(2\mathrm{Re}k_x)^2 + \left(2\mathrm{Re}k_z^S\right)^2\right]}{2}\right] \end{aligned} \tag{52}$$

Unlike the dynamic grating (28) created by the interfering optical waves, the grating (50) caused by SPP is spatially localized both in the X and in the Z directions [11, 12]. The localized grating (50) can be characterized as an enhanced Rayleigh wave of SS [26]. Analysis of $G(k_x, k_z^S, \Delta\omega)$ (52) shows that the resonant case $\mathrm{Re}G(k_x, k_z^S, \Delta\omega) = 0$ cannot be achieved for the frequency difference $\Delta\omega \sim (10^7 - 10^8)s^{-1}$, and the spontaneous SS surface wave can be neglected [12]. The cubic susceptibility of the SALC-metal system $\chi_{ijkl}^{(3)}(\Delta\omega); i, j, k, l = x, z$ related to the localized grating (50) is essentially complex. For the typical values of the SALC parameters, SPP in silver, $\omega_1 = 1.4 \times 10^{15}s^{-1}$ and $\Delta\omega \sim (10^7 - 10^8)s^{-1}$ the numerical estimations yield $\left|\chi_{xxxx}^{(3)}\right| \approx \left|\chi_{zzzz}^{(3)}\right| \sim (10^{-20} - 10^{-19})m^2/V^2$ [11], which is larger by one-two orders of magnitude than the cubic susceptibilities of some organic liquids and solid materials [27].

We substitute the localized layer displacement $u(x, z, t)$ (50) into the expression of the SALC nonlinear permittivity (30), evaluate the nonlinear part of the electric induction $\vec{D}^{NL}$ for SPP (48), and by using the standard procedure, we obtain from Eq. (13) the truncated equations for the SPP SVA $A_{1,2}(t) = |A_{1,2}(t)|\exp i\gamma_{1,2}(t)$. The dependence of SVA $A_{1,2}$ on the x coordinate can be neglected in the central part of the dynamic grating (50) for the distances of several SPP wavelengths [12]. Integrating the SPP electric field and nonlinear electric induction over $z \in [0, \infty]$, we obtain the following truncated equations for the normalized SPP intensities $I_{1,2} = \left(|A_{1,2}|^2/\omega_{1,2}\right)I_0^{-1}$.

$$\frac{\partial I_{1,2}}{\partial t} = \mp g I_1 I_2 \tag{53}$$

Here, $\left(|A_1|^2/\omega_1\right) + \left(|A_2|^2/\omega_2\right) = I_0 = const$ is the integral of motion obtained from the Manley-Rowe relation [14], and the gain g has the form [12]

$$g = \frac{\varepsilon_0 (2\mathrm{Re}k_x)^2 hb \mathrm{Im}\{G(k_x, k_z^S, \Delta\omega)\} I_0 \omega_1 \omega_2 \exp(-2)}{6\rho\left(\varepsilon_\perp |e_x|^2 + \varepsilon_\parallel |e_z|^2\right)\left[(2\mathrm{Re}k_x)^2 - \left(2\mathrm{Re}k_z^S\right)^2\right]\left|G(k_x, k_z^S, \Delta\omega)\right|^2} > 0 \tag{54}$$

Here, $b = -2\mathrm{Re}k_z^S\left(-a_\perp |e_{1x}|^2 + a_\parallel |e_{1z}|^2\right) + 4\varepsilon_a(\mathrm{Re}k_x)\mathrm{Im}\left(e_{1z}e_{1x}^*\right)$. Solution of Eq. (53) has the form [12]

$$I_1(t) = \frac{I_1(0)}{I_1(0) + [1 - I_1(0)]\exp(-gt)}; I_2(t) = \frac{[1 - I_1(0)]\exp(-gt)}{I_1(0) + [1 - I_1(0)]\exp(-gt)} \tag{55}$$

It is easy to see from Eq. (55) that $I_1(t) + I_2(t) = 1$.

Expressions (55) show that the energy exchange between SPP takes place. In the limiting case $t \to \infty$, we obtain: $I_1(t) \to 1; I_2(t) \to 0$ [12]. The time dependence of the normalized SPP intensities $I_1(t), I_2(t)$ (55) is presented in **Figure 7**. The phases $\gamma_{1,2}(0)$ of the SPP SVA have the form

$$\gamma_1(t) - \gamma_1(0) = \frac{\mathrm{Re}\{G(k_x, k_z^S, \Delta\omega)\}}{2\mathrm{Im}\{G(k_x, k_z^S, \Delta\omega)\}} \ln[(1 - I_1(0))\exp(-gt) + I_1(0)] \tag{56}$$

$$\gamma_2(t) - \gamma_2(0) = -\frac{\mathrm{Re}\{G(k_x, k_z^S, \Delta\omega)\}}{2\mathrm{Im}\{G(k_x, k_z^S, \Delta\omega)\}} \ln[I_1(0)\exp(gt) + 1 - I_1(0)] \tag{57}$$

It is easy to see from Eqs. (56) and (57) that for $t \to \infty$ the phase $\gamma_1(t)$ of the amplified SPP I_1 tends to a constant value $\gamma_1(t) - \gamma_1(0) \to \frac{\mathrm{Re}\{G(k_x, k_z^S, \Delta\omega)\}}{2\mathrm{Im}\{G(k_x, k_z^S, \Delta\omega)\}} \ln[I_1(0)]$, while the phase of the depleted SPP I_2 $\gamma_2(t) - \gamma_2(0)$ for large time intervals such that $gt \gg 1$ takes the form $\gamma_2(t) - \gamma_2(0) \to -\frac{\mathrm{Re}\{G(k_x, k_z^S, \Delta\omega)\}}{2\mathrm{Im}\{G(k_x, k_z^S, \Delta\omega)\}} gt$ and $\gamma_2(t) \to -\infty$ for $t \to \infty$. The SVA of the depleted SPP I_2 undergoes strong XPM and rapidly oscillates in the time domain. The results (55)–(57) show that the Rayleigh stimulated scattering [27] of SPP on the smectic layer normal displacement localized grating is accompanied by XPM and the parametric energy exchange between SPP [12]. The rise time of the amplified SPP is about $1 - 2\mu s$ as it is seen from **Figure 7**. It is much faster than the thermal response time $\tau_R = 100\,\mu s$ and the purely orientational response time $\approx 25\,ms$ in NLC [4]. Numerical estimations show that for the SPP electric field of $10^7 V/m$ the rise time of about 10 ns can be achieved, which is much less than the Brillouin relaxation time $\tau_B \approx 200\,ns$ [4, 12].

Structures consisting of alternative conducting and dielectric thin films are capable of guiding SPP light waves [24, 25]. Each single interface can sustain bound SPP. When the distance between adjacent interfaces is comparable or smaller than the SPP localization length $L_z = \left(\mathrm{Re}k_z^S\right)^{-1}$, the coupled modes occur due to the interaction between SPP [24]. The following specific three-layer guiding systems can be considered: (i) an insulator/metal/insulator (IMI) heterostructure where a thin metallic layer is sandwiched between two infinitely thick dielectric claddings; (ii) a metal/insulator/metal (MIM) heterostructure where a thin dielectric core layer is sandwiched between two infinitely thick metallic claddings [24]. LC can be used as a tunable cladding material or as the guiding core material due to their excellent electro-optic properties

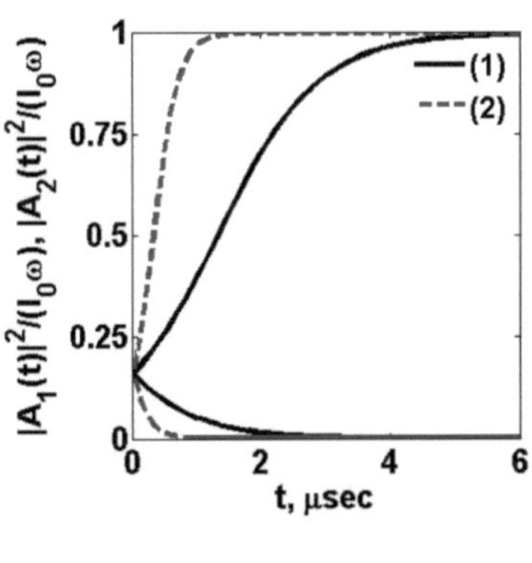

Figure 7. The temporal dependence of the SPP normalized intensities $I_{1,2}(t)$ for the input electric field of $10^6 V/m$ and optical wavelengths $\lambda_{opt1} = 0.6\mu m$ (curves 1) and $\lambda_{opt1} = 1.33\mu m$ (curve 2).

and large nonlinearity [28]. Photonic components based on plasmonic waveguides with NLC core have been theoretically investigated in a number of articles (see, for example, [28–31] and references therein).

We consider the nonlinear optical processes in an MIM waveguide with the SALC core [13]. The structure of such a waveguide is shown in **Figure 8**. SPP propagating in the metal claddings and in SALC core are TM waves [24, 25]. The SPP electric and magnetic fields in the metallic claddings $z > d; z < -d$ $\vec{H}_{1,2}(x,z,t), \vec{E}_{1,2}(x,z,t)$, and in the SALC core $|z| \le d$ $\vec{H}_{SA}(x,z,t), \vec{E}_{SA}(x,z,t)$ have the form, respectively [24]

$$\vec{H}_{1,2}(x,z,t) = \frac{1}{2}\vec{a}_y H_{1,20}\exp\left(\mp k_z^m z + ik_x x - i\omega t\right) + c.c., |z| > d \tag{58}$$

$$\vec{E}_{1,2}(x,z,t) = \frac{1}{2}\left[\vec{a}_x E_{1,2x0} + \vec{a}_z E_{1,2z0}\right]\exp\left(\mp k_z^m z + ik_x x - i\omega t\right) + c.c., |z| > d \tag{59}$$

$$\vec{H}_{SA}(x,z,t) = \frac{1}{2}\vec{a}_y\left[A\exp\left(k_z^S z\right) + B\exp\left(-k_z^S z\right)\right]\exp(ik_x x - i\omega t) + c.c., |z| \le d \tag{60}$$

$$\begin{aligned}&\vec{E}_{SA}(x,z,t)\\&= \frac{1}{2}\left\{\vec{a}_x\frac{k_z^S}{i\omega\varepsilon_0\varepsilon_\perp}\left[A\exp\left(k_z^S z\right) - B\exp\left(-k_z^S z\right)\right] - \vec{a}_z\frac{k_x}{\omega\varepsilon_0\varepsilon_\parallel}\left[A\exp\left(k_z^S z\right) + B\exp\left(-k_z^S z\right)\right]\right\}\\&\times\exp i(k_x x - \omega t) + c.c.; |z| \le d\end{aligned} \tag{61}$$

The complex wave number k_z^S of SPP in SALC in the linear approximation is determined by expression (49) and the SPP wave number in the metallic claddings is given by $k_z^m = \sqrt{k_x^2 - \varepsilon_m(\omega)\omega^2/c^2}$ [24]. Using the boundary conditions for the tangential components of the SPP fields (58)–(61) at the interface $z = d$ between the metallic cladding and the SALC core, we obtain the dispersion relation for the MIM modes [13, 24]

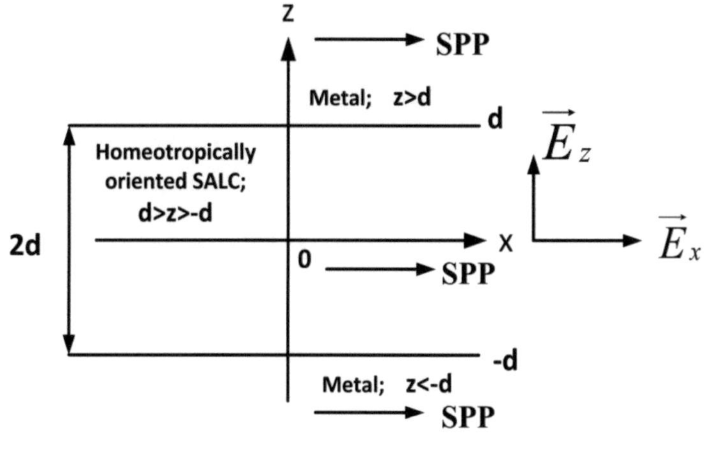

Figure 8. The MIM waveguide with the homeotropically oriented SALC as a core.

$$\exp\left(-4k_z^S d\right) = \left(\frac{k_z^m}{\varepsilon_r(\omega)} + \frac{k_z^S}{\varepsilon_\perp}\right)^2 \left(\frac{k_z^m}{\varepsilon_r(\omega)} - \frac{k_z^S}{\varepsilon_\perp}\right)^{-2} \tag{62}$$

Numerical estimations show that for the typical values of the SPP frequency ω, the plasma frequency ω_p, the SPP lifetime τ mentioned above, and the MIM thickness $2d = 1\mu m$ the following relationships take place:

$\mathrm{Re}k_z^S \sim 10^6 m^{-1} \gg \mathrm{Im}k_z^S \sim 10^4 m^{-1}$, $\mathrm{Re}k_x \sim 10^7 m^{-1} \gg \mathrm{Im}k_x \sim 10^3 m^{-1}$. The SPP wavelength in the Z direction is given by $2\pi\left(\mathrm{Im}k_z^S\right)^{-1} \sim 10^2 \mu m$ and can be neglected inside the MIM waveguide core with the thickness of $2d \sim 1\mu m$. Then, a single localized TM can exist in the MIM waveguide according to the dispersion relation (62). The even TM mode has the form [13]

$$\vec{E}_{SA} = E_0 \left[\vec{a}_x \cosh\left(k_z^S z\right) - \vec{a}_z i \frac{k_x \varepsilon_\perp}{k_z^S \varepsilon_\parallel} \sinh\left(k_z^S z\right)\right] \exp[i(k_x x - \omega t)] + c.c. \tag{63}$$

The distribution of the TM even mode normalized intensity $\left|\vec{E}_{SA}\right|^2 / |E_0|^2$ in the MIM waveguide core is presented in **Figure 9**. It is seen from **Figure 9** that the intensity is filling the MIM waveguide core due to the overlapping of SPP inserted from the metallic claddings $z < -d; z > d$. Substituting the SPP field (63) into equation of motion (9), we obtain the smectic layer normal strain [13].

$$\frac{\partial u}{\partial z} = \frac{\varepsilon_0 |E_0|^2}{B} \exp[-2(\mathrm{Im}k_x)x] \left\{ a_\perp \left|\cosh\left(k_z^S z\right)\right|^2 + a_\parallel \frac{|k_x|^2 \varepsilon_\perp^2}{\left|k_z^S\right|^2 \varepsilon_\parallel^2} \left|\sinh\left(k_z^S z\right)\right|^2 \right\} \tag{64}$$

The nonlinear polarization in the SALC core caused by the smectic layer strain (64) has the form [13]

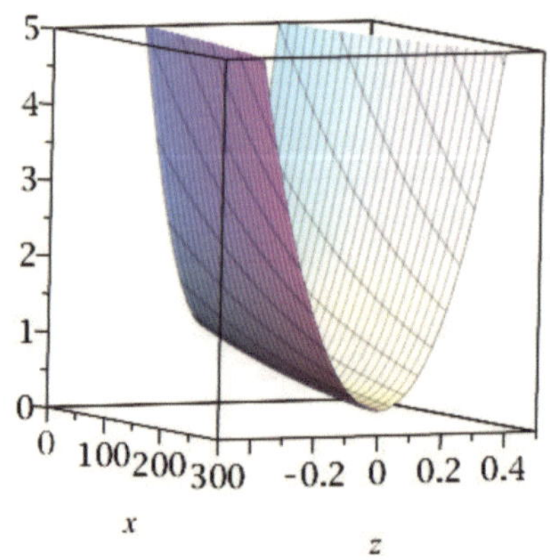

Figure 9. Distribution of the SPP normalized intensity $\left|\vec{E}_{SA}\right|^2 / |E_0|^2$ in the MIM waveguide core (arbitrary units).

$$\vec{D}_{SA}^{NL} = \varepsilon_0 \frac{\partial u}{\partial z} E_0 \left[\vec{a}_x a_\perp \cosh\left(k_z^S z\right) - \vec{a}_z i a_\parallel \frac{k_x \varepsilon_\perp}{k_z^S \varepsilon_\parallel} \sinh\left(k_z^S z\right) \right] \exp[i(k_x x - \omega t)] + c.c. \tag{65}$$

Substituting the SPP electric field (63) and nonlinear polarization (65) into Eq. (13) and separating linear and nonlinear parts, we obtain the truncated equation for SVA $E_0(t) = |E_0(t)|$ $\exp i\varphi(t)$.

$$\begin{aligned} &-2i \frac{\partial E_0}{\partial t} E_0^* \varepsilon_\perp \left[\left|\cosh\left(k_z^S z\right)\right|^2 + \frac{|k_x|^2 \varepsilon_\perp}{\left|k_z^S\right|^2 \varepsilon_\parallel} \left|\sinh\left(k_z^S z\right)\right|^2 \right] \\ &= \omega \frac{\varepsilon_0 |E_0|^4}{B} \exp[-2(\mathrm{Im} k_x) x] \left[a_\perp \left|\cosh\left(k_z^S z\right)\right|^2 + a_\parallel \frac{|k_x|^2 \varepsilon_\perp^2}{\left|k_z^S\right|^2 \varepsilon_\parallel^2} \left|\sinh\left(k_z^S z\right)\right|^2 \right]^2 \end{aligned} \tag{66}$$

At the distances $x \ll L_x = (\mathrm{Im} k_x)^{-1}$, the SVA dependence on x can be neglected since $\exp[-2(\mathrm{Im} k_x) x] \approx 1$ [13]. The dispersion effects can be neglected because the dispersion length $L_D \gg L_x$ [13]. Integrating both sides of Eq. (66) over $z \in [-d, d]$ and separating real and imaginary parts we obtain the following equations for the magnitude $|E(t)|$ and phase $\varphi(t)$ of the SPP SVA.

$$\frac{\partial |E_0|^2}{\partial t} = 0; \quad \frac{\partial \varphi}{\partial t} = \omega \frac{\varepsilon_0 |E_0|^2}{16B} \frac{F_2\left(k_z^S, k_x, d\right)}{F_1\left(k_z^S, k_x, d\right)} \tag{67}$$

Here,

$$\begin{aligned} F_1\left(k_z^S, k_x, d\right) &= \varepsilon_\perp \left[\left(1 + \frac{\varepsilon_\perp |k_x|^2}{\varepsilon_\parallel \left|k_z^S\right|^2} \right) \sinh\left[2\left(\mathrm{Re} k_z^S\right) d\right] + \left(1 - \frac{\varepsilon_\perp |k_x|^2}{\varepsilon_\parallel \left|k_z^S\right|^2} \right) 2\left(\mathrm{Re} k_z^S\right) d \right] \\ F_2\left(k_z^S, k_x, d\right) &= \left(a_\perp + a_\parallel \frac{\varepsilon_\perp^2 |k_x|^2}{\varepsilon_\parallel^2 \left|k_z^S\right|^2} \right)^2 \sinh\left[4\left(\mathrm{Re} k_z^S\right) d\right] \\ &+8 \left(a_\perp^2 - \left(a_\parallel \frac{\varepsilon_\perp^2 |k_x|^2}{\varepsilon_\parallel^2 \left|k_z^S\right|^2} \right)^2 \right) \sinh\left[2\left(\mathrm{Re} k_z^S\right) d\right] \\ &+8 \left[\frac{1}{2} \left(a_\perp + a_\parallel \frac{\varepsilon_\perp^2 |k_x|^2}{\varepsilon_\parallel^2 \left|k_z^S\right|^2} \right)^2 + \left(a_\perp - a_\parallel \frac{\varepsilon_\perp^2 |k_x|^2}{\varepsilon_\parallel^2 \left|k_z^S\right|^2} \right)^2 \right] \left(\mathrm{Re} k_z^S\right) d \end{aligned} \tag{68}$$

The solution of Eq. (67) has the form

$$|E_0|^2 = const; \; \varphi(t) = \left\{ \omega \frac{\varepsilon_0 |E_0|^2}{16B} \frac{F_2\left(k_z^S, k_x, d\right)}{F_1\left(k_z^S, k_x, d\right)} \right\} t \tag{69}$$

Eq. (69) shows that the strong SPM of the even SPP mode in the MIM wave guide occurs. It is enhanced by a large geometric factor $F_2\left(k_z^S, k_x, d\right) / F_1\left(k_z^S, k_x, d\right)$, which can achieve a value of $10^2 - 10^4$ for $2d = 1\mu m$ and $\mathrm{Re} k_z^S \sim \left(10^6 - 3 \times 10^6\right) m^{-1}$, $\mathrm{Re} k_x \sim \left(5 \times 10^6 - 10^7\right) m^{-1}$.

7. Conclusions

In SALC, there exists a specific mechanism of the optical nonlinearity related to the normal displacement $u(x, y, z, t)$ of smectic layers. This mechanism combines the properties of the orientational mechanism typical for LC and of the electrostrictive mechanism. In particular, the smectic layer oscillations occur without the mass density change. Under the resonant condition (11), the SS acoustic wave propagates in SALC in the direction oblique to the layer plane. The cubic nonlinearity related to this mechanism is characterized by a strong anisotropy, a short time response, a weak temperature dependence, a resonant frequency dependence, and a strong dependence on the optical wave polarization and propagation direction. The cubic susceptibility related to the smectic layer displacement is larger than the Kerr type susceptibility in ordinary organic liquids. It should be noted that the nonlinear optics of NLC has been mainly studied. However, SALC are promising candidates for nonlinear optical applications due to their low losses and higher degree of the long range order.

We derived the equation of motion (9) of the smectic layer displacement $u(x, y, z, t)$ in the electric field of optical waves. We investigated theoretically the nonlinear optical phenomena in SALC based on this specific mechanism. We solved simultaneously the equation of motion (9) and the wave Eq. (13) for the optical waves including the nonlinear polarization. The solution was based on the SVAA.

In an optically uniaxial SALC, an ordinary wave and an extraordinary one can propagate. Both the ordinary and extraordinary optical beams propagating in SALC undergo self-focusing and self-trapping and form spatial solitons. The optical wave self-trapping can occur at the interface between the linear medium and SALC. We obtained the analytical solutions for the SVA of the self-trapped beams.

SLS of two arbitrary polarized optical waves in SALC transforms into the partially frequency degenerate FWM because each optical wave splits into the ordinary and extraordinary waves. The coupled optical waves create a dynamic grating of the smectic layer normal displacement $u(x, y, z, t)$ and undergo the parametric energy exchange and XPM. The signal optical waves with the lower frequency are amplified up to a saturation level determined by the Manley-Rowe relation, while the pumping optical waves with higher frequency are depleted. It has been shown that the system of the coupled optical waves and the dynamic grating is stable. The analytical expressions for the magnitudes and phases of SVA have been obtained in the limiting case when the waves are mainly polarized either perpendicular to the propagation plane, or in it. The SLS gain coefficient is significantly larger than the one in the case of the Brillouin SLS in isotropic organic liquids. The SLS in SALC also results in the generation of the Stokes and anti-Stokes harmonics with the combination wave vectors.

The nondegenerate FWM in SALC results in the amplification of the signal optical wave with the lowest frequency and depletion of three other waves with higher frequencies. The polarization-decoupled FWM may take place when the polarizations of some optical waves are perpendicular to one another. If the coupled optical waves are counter propagating and their frequencies satisfy the balance conditions typical for OPC process then BEFWM takes place accompanied by the amplification of the phase-conjugate wave. The spectrum of the scattered harmonics consists of 24 Stokes and anti-Stokes terms with combination frequencies and wave vectors.

LC applications in nanophotonics, plasmonics, and metamaterials attracted a wide interest due to the combination of LC large nonlinearity and strong localized electric fields of SPP. Until now, NLC applications in nanophotonics and plasmonics have been investigated. We studied theoretically the nonlinear optical processes at the interface of a metal and a homeotropically oriented SALC. In such a case, SPP penetrating into SALC create the spatially localized surface dynamic grating of smectic layer normal displacement. We have shown that for optical frequencies of about 10^{15} s^{-1} and coupled SPP frequency difference of about 10^{8} s^{-1}, the SALC-metal system cubic susceptibility may be one to two orders of magnitude larger than the cubic susceptibility of isotropic organic liquids. We solved the wave Eq. (13) for the counter-propagating SPP in SALC with the spatially localized nonlinear polarization and obtained the explicit expressions (55)–(57) for the magnitudes and phases of the coupled SPP SVA. It has been shown that the Rayleigh stimulated scattering of SPP on the surface smectic layer oscillations occurs. The rise time of the amplified SPP of about 10 ns can be achieved, which is much faster than the Brillouin relaxation constant in NLC.

The plasmonic waveguides with NLC for nanophotonic and plasmonic have been theoretically investigated. We proposed an MIM waveguide with an SALC core. We evaluated the electric field of the strongly localized SPP even mode, the smectic layer normal strain and the nonlinear polarization in the MIM core. The evaluation of the SPP SVA shows that the strong SPM process takes place. The SPM is enhanced by the geometric factor caused by the strong SPP localization in the MIM core.

Author details

Boris I. Lembrikov*, David Ianetz and Yossef Ben Ezra

*Address all correspondence to: borisle@hit.ac.il

Department of Electrical Engineering and Electronics, Holon Institute of Technology (HIT), Holon, Israel

References

[1] De Gennes PG, Prost J. The Physics of Liquid Crystals. 2nd ed. New York, USA: Oxford Univeristy Press; 1993. 597 p. DOI: ISBN: 978-0198517856

[2] Khoo I-C. Liquid Crystals. 2nd ed. Hoboken, New Jersey, USA: Wiley; 2007. 368 p. DOI: ISBN: 978-0-471-75153-3

[3] Khoo IC. Nonlinear optics of liquid crystalline materials. Physics Reports. 2009;**471**:221-267. DOI: 10.1016/j.physrep.2009.01.001

[4] Khoo IC. Nonlinear optics, active plasmonics and metamaterials with liquid crystals. Progress in Quantum Electronics. 2014;**38**(2):77-117. DOI: 10.1016/j.pquantelec.2014.03.001

[5] Kventsel GF, Lembrikov BI. Two-wave mixing on the cubic non-linearity in the smectic A liquid crystals. Liquid Crystals. 1994;**16**(1):159-172. DOI: ISSN: 0267-8292

[6] Kventsel GF, Lembrikov BI. Stimulated light scattering in smectic A liquid crystals. Liquid Crystals. 1995;**19**(1):21-37. DOI: ISSN: 0267-8292

[7] Kventsel GF, Lembrikov BI. Self-focusing and self-trapping in smectic A liquid crystals. Molecular Crystals and Liquid Crystals. 1995;**262**(1):629-643. DOI: ISSN: 1542-1406

[8] Kventsel GF, Lembrikov BI. The four-wave mixing and the hydrodynamic excitations in smectic A liquid crystals. Molecular Crystals and Liquid Crystals. 1995;**262**(1):591-627. DOI: ISSN: 1542-1406

[9] Kventsel GF, Lembrikov BI. Second sound and nonlinear optical phenomena in smectic A liquid crystals. Molecular Crystals and Liquid Crystals. 1996;**282**(1):145-189. DOI: ISSN: 1542-1406

[10] Lembrikov BI. Light interaction with smectic A liquid crystals: nonlinear effects. HAIT Journal of Science and Engineering. 2004;**1**(2):306-347. DOI: ISSN: 1565-4990

[11] Lembrikov BI, Ben-Ezra Y. Surface plasmon polariton (SPP) interactions at the interface of a metal and smectic liquid crystal. In: 17th International Conference on Transparent Optical Networks (ICTON 2015); July 5–9, 2015; Budapest, Hungary. 2015. p. We.C4.4, 1-4. DOI: 978–1–4673-7879-6/15

[12] Lembrikov BI, Ben-Ezra Y, Ianetz D. Stimulated scattering of surface plasmon polaritons (SPPs) in smectic A liquid crystal. In: 18th International Conference on Transparent Optical Networks (ICTON-2016); July 10–14, 2016; Trento, Italy. 2016. p. We.B4.2, 1-4. DOI: 978–1–5090-1466-8/16

[13] Lembrikov BI, Ianetz D, Ben-Ezra Y. Metal/Insulator/Metal (MIM) plasmonic waveguide containing a smectic A liquid crystal (SALC) layer. In: 19th International Conference on Transparent Optical Networks (ICTON 2017); July 2–6, 2017; Girona, Catalonia, Spain. 2017. p. Tu.A4.3, 1-4. DOI: 978–1–5386-0858-6/17

[14] Shen YR. The Principles of Nonlinear Optics. Hoboken, New Jersey, USA: Wiley; 2003. 563 p. DOI: ISBN: 0-471-43080-3

[15] Liao Y, Clark NA, Pershan PS. Brillouin scattering from smectic liquid crystals. Physical Review Letters. 1973;**30**(14):639-641. DOI: 10.1103/PhysRevLett.30.639

[16] Ricard L, Prost J."Second sound" propagation and the smectic response function. Journal DE Physique Colloque C3. 1979;**40**(supplement au no. 4):C3–83-C3–86. DOI: 10.1051/jphyscol:1979318

[17] Ricard L, Prost J. Critical behaviour of second sound near the smectic a nematic phase transtion. Journal De Physique. 1981;**42**(6):861-873. DOI: 10.1051//jphys:01981004206086100

[18] Vagner ID, Lembrikov BI, Wyder P. Electrodynamics of Magnetoactive media. 1st ed. Heidelberg, Germany: Springer; 2004. 422 p. DOI: ISBN: 3-540-43694-4

[19] Newell AC. Solitons in Mathematics and Physics. 1st ed. Philadelphia, PA, USA: Society for Industrial and Applied Mathematics; 1985. 244 p. DOI: ISBN: 978-08-9871-1967

[20] Khoo I-C, Michael RR, Yan P-Y. Simultaneous occurence of phase conjugation and pulse shortening in stimulated scattering in liquid crystal mesophases. IEEE Journal of Quantum Electronics. 1987;**23**(8):1344-1347. DOI: 10.1109/JQE.1987.10734497

[21] Svetlana G. Lukishova. Liquid crystals under two extremes: (1) high-power irradiation, (2) single-photon level. Molecular Crystals and Liquid Crystals. 2012;**559**:127-157. DOI: 10.1080/154221406.2012.658703

[22] Guang S. He. Optical phase conjugation: principles, techniques, and applications. Progress in Quantum Electronics. 2002;**26**(3):131-191. DOI: ISSN: 0079-6727

[23] Oliveri G, Werner DH, Massa A. Reconfigurable electromagnetics through metamaterials -a review. Proceedings of the IEEE. 2015;**103**(7):1034-1056. DOI: 10.1109/JPROC.2015.2394292

[24] Maier SA. Plasmonics: Fundamentals and Applications. 1st ed. New York, USA: Springer Science; 2007. 223 p. DOI: ISBN: 978-0387-33150-8

[25] Sarid D, Challener WA. Modern Introduction to Surface Plasmons. 1st ed. New York, USA: Cambridge University Press; 2010. 371 p. DOI: ISBN 978-0-521-76717-0

[26] Fedorov DO, Romanov VP, Ul'yanov SV. Surface oscillations of smectic-A liquid crystals. Physical Review E. 2000;**62**(1):681-688. DOI: 1063-651X/2000/62(1)/681(8)

[27] Robert W. Boyd. Nonlinear Optics. 3rd ed. Amsterdam: Elsevier; 2008. 613 p. DOI: ISBN: 978-0-12-369470-6

[28] Beeckman J, James R, Fernandes FA, De Cort W, Vanbrabant PJM, Neyts K. Calculation of fully anisotropic liquid crystal waveguide modes. Journal of Lightwave Technology. 2009;**27**(17):3812-3819. DOI: 10.1109/JLT.2009.2016673

[29] Zografopoulos DC, Becceherelli R, Tasolamprou AC, Kriezis EE. Liquid-crystal tunable for integrated plasmonic components. Photonics and Nanostructures - Fundamentals and Applications. 2013;**11**:73-84. DOI: 10.1016/j.photonics.2012.08.004

[30] Zografopoulos DC, Beccherelli R. Liquid-crystal-tunable metal-insulator-plasmonic waveguides and Bragg resonators. Journal of Optics. 2013;**15**:1-5. DOI: 10.1088/2040-8978/15/5/055009

[31] Prokopidis KP, Zografopoulos DC, Kriezis EE. Rigorous broadband investigation of liquid-crystal plasmonic structures using finite-difference time-domain dispersive-anisotropic models. Journal of Optical Society of America B. 2013;**30**(10):2722-2730. DOI: 10.1364/JOSAB.30.002722

Chapter 8

Temperature Effects on Liquid Crystal Nonlinearity

Lamees Abdulkaeem Al-Quraainy and
Kais A.M. Al Naimee

Additional information is available at the end of the chapter

http://dx.doi.org/10.5772/intechopen.70414

Abstract

The effect of temperature variation on nonlinear refractive indices of several types of liquid crystal (LC) compounds has been reported. Five samples have been investigated: two pure components (E7, MLC 6241-000) and three mixtures are obtained by mixing the previous two in different proportions. Birefringence, the average refractive index and the temperature gradients of refractive indices of the LCs are determined. The variations in refractive indices and birefringence were fitted theoretically using the modified Vuks equation. Excellent agreement is obtained between the fitted values and experimental data. Finally, the bistability of nonlinear refractive indices with temperature of liquid crystal (LC) compounds has been studied. The bistability of liquid crystals based on temperature is clearly observed for all samples. Also, the extraordinary refractive index has larger bistability than the ordinary refractive index. The measurements are performed at 1550 nm wavelength using wedged cell refractometer method.

Keywords: extraordinary, refractive index, liquid crystal, birefringence, order parameter

1. Introduction

Liquid crystals exhibit optical anisotropy or birefringence (Δn). This is an essential physical property of liquid crystals and is a key element in how they are implemented in the display [1, 2], photonic devices [3], communications signal processing [4], and beam steering [5].

When light propagates through anisotropic media such as liquid crystals, it will be divided into two rays which travel through the material at different velocities, and therefore have different refractive indices, the ordinary index (n_o), and extraordinary index (n_e), and the difference is called as birefringence or double refraction ($\Delta n = n_e - n_o$). Depending on the values of n_e and n_o, birefringence can be positive or negative [6, 7].

Temperature plays a fundamental role in affecting the refractive indices of LCs. As the temperature increases, ordinary (n_o) and extraordinary (n_e) refractive indices of LCs behave differently from each other [8, 9].

Several techniques have been studied to describe the temperature effect on LC refractive indices. Horn measured the refractive indices as a function of temperature throughout the nematic phase of 4-n pentyl-4-cyanobiphenyl (5CB) and the smectic A and nematic phases of 4-n-octyl-4-cyanobiphenyl (8CB) using the method of Pellet and Chatelain at the wavelength (589 and 632.8 nm) [10]. Wu developed a single-band model and a three-band model for understanding the refractive index dispersions of liquid crystals. The three- and two-coefficient Cauchy equations based on the three-band model for the wavelength- and temperature-dependent refractive indices of anisotropic liquid crystals were derived by Jun Li and Wu. For low birefringence liquid crystal mixtures, the two-coefficient Cauchy model works equally well as the three-coefficient model in the off-resonance spectral region [11]. A four-parameter model for describing the temperature effect on the refractive indices of LCs based on the Vuks equation was derived by Jun Li et al. Four different LC materials with different birefringence were used to validate these parameters. An excellent agreement between theory and experiment was obtained [12]. Jun Li et al. measured the refractive indices of E7 LC mixture at six visible and two infrared ($\lambda = 1.55$ and $10.6\ \mu m$) wavelengths at different temperatures using Abbe and wedged cell refractometer methods [13]. In the present chapter, the temperature effect on nonlinear refractive indices of several types of liquid crystal (LC) compounds has been studied. In the beginning, the mathematical models are discussed. Second, using the refractive indices data, other parameters can be determined, such as birefringence (Δn), average refractive indices, and the temperature gradient of refractive indices (dn_e/dT, dn_o/dT). Finally, the bistability of nonlinear refractive indices with temperature of LC is also reported. The measurements performed at 1550 nm wavelength using wedged cell refractometer method. The variation in refractive indices was fitted theoretically using the modified four-parameter model, which is based on the Vuks equation.

2. Modeling

The classical Clausius-Mossotti equation correlates the dielectric constant (ε) of an isotropic media with molecular packing density (N), which means the number of molecules per unit volume and molecular polarizability (α) are as follows [6, 14, 15]

$$\frac{\varepsilon - 1}{\varepsilon + 2} = \frac{4\pi}{3} N\alpha \tag{1}$$

The Lorentz-Lorenz equation correlates the refraction index of anisotropic medium with molecular polarizability at optical frequencies and obtained the following equation [14, 15]

$$\frac{n^2 - 1}{n^2 + 2} = \frac{4\pi}{3} N\alpha \tag{2}$$

http://dx.doi.org/10.5772/intechopen.70414

For an isotropic LC, the ordinary and extraordinary refractive indices are determined by corresponding ordinary and extraordinary molecular polarizability α_o and α_e. Vuks modified the Lorentz-Lorenz equation by assuming that the internal field in a liquid crystal is equal in all directions, and therefore produce a semi-empirical equation correlating macroscopic refractive index with microscopic molecular polarizability [11, 15].

$$\frac{n_{e,o}^2 - 1}{\langle n^2 \rangle + 2} = \frac{4\pi N}{3}\alpha_{e,o} \tag{3}$$

n_e and n_o are the extraordinary and ordinary refractive indices, $<n^2>$ is defined as $<n^2> = (n_e^2 + 2n_o^2)/3$

The temperature-dependent LC refractive indices based on Vuks model can be expressed by the average refractive index <n> and birefringence Δn as [11, 16],

$$n_e = \langle n \rangle + \frac{2}{3}\Delta n \tag{4}$$

$$n_o = \langle n \rangle - \frac{1}{3}\Delta n \tag{5}$$

On the other hand, the LC birefringence is linearly proportional to the order parameter S. Through Haller's semi-empirical equation, the order parameter can be approximated as

$$S = \left(1 - \frac{T}{T_c}\right)^B \tag{6}$$

where T is the operating temperature, T_c is the clearing temperature of LC material, and *B* is the material constant, for many LC compounds and mixtures studied $\beta = (0.2–0.25)$. Thus, the temperature-dependent Δn can be written as

$$\Delta n(T) = (\Delta n)_o \left(1 - \frac{T}{T_c}\right)^\beta \tag{7}$$

where $(\Delta n)_o$ is the LC birefringence in the crystalline state at absolute zero (T = 0 K). From Eqs. (6) and (7), the order parameter can be written as:

$$S = \frac{\Delta \mathrm{n}}{(\Delta \mathrm{n})_\mathrm{o}} \tag{8}$$

The average refractive index decreases linearly with increasing temperature as [16]:

$$\langle n \rangle = A - BT \tag{9}$$

Substituting Eqs. (7) and (9) back to Eqs. (4) and (5), the modified four-parameter model for describing the temperature effect on the LC refractive indices is obtained [17–19]:

$$n_e(T) \approx A - BT + \frac{2(\Delta n)_o}{3}\left(1 - \frac{T}{T_c}\right)^{\beta} \tag{10}$$

$$n_o(T) \approx A - BT - \frac{(\Delta n)_o}{3}\left(1 - \frac{T}{T_c}\right)^{\beta} \tag{11}$$

Eqs. (10) and (11) contain four unknown parameters A, B, $(\Delta n)_o$, and β. The parameters A and B can be obtained by fitting the temperature-dependent average refractive index using Eq. (9), while $(\Delta n)_o$ and β can be obtained by fitting the birefringence data Δn as a function of temperature using Eq. (7). By taking temperature derivatives of Eqs. (10) and (11), the temperature gradient for n_e and n_o can be derived [9, 19–24].

$$\frac{dn_e}{dT} = -B - \frac{2\beta(\Delta n)_o}{3T_c\left(1 - \frac{T}{T_c}\right)^{1-\beta}} \tag{12}$$

$$\frac{dn_o}{dT} = -B + \frac{\beta(\Delta n)_o}{3T_c\left(1 - \frac{T}{T_c}\right)^{1-\beta}} \tag{13}$$

3. Experimental refractive index

Five samples have been investigated in this work; two of them are E7 and MLC 6241-000 samples, the clearing temperatures are 333 and 373.7 K, respectively. The other three have been obtained by mixing the previous two liquid crystals in different proportions (75% E7 with 25% MLC 6241-000, 50% E7 with 50% MLC 6241-000, and 25% E7 with 75% MLC 6241-000). The clearing temperature (T_c) of three mixture samples is measured using hot-stage optical microscope, and they are found to be 348, 353, and 368 K, respectively.

The effect of temperature variation on nonlinear refractive indices of these liquid crystal compounds has been reported using wedged cell refractometer method. The cell is made up of two glass substrates separated by two spacers that have different thickness as shown in **Figure 1**. The wedged angle was measured using an optical method [9], and the angle value of five samples is shown in **Table 1**.

The experimental setup used during this work is shown in **Figure 2**, and a more detailed description of this setup and the measurement method can be found in [9]. The laser beam is divided into two rays, when it passes through LC sample because of birefringence of LC, the values of two refractive indices (n_o and n_e) of the liquid crystal can be calculated by these equations

$$n_o = \frac{\sin(\theta + \delta_o)}{sin\theta} \tag{14}$$

$$n_e = \frac{\sin(\theta + \delta_e)}{sin\theta} \tag{15}$$

where θ is the angle of the wedged formed by two plates, δ_o and δ_e are the angles formed by two beams R_o and R_e.

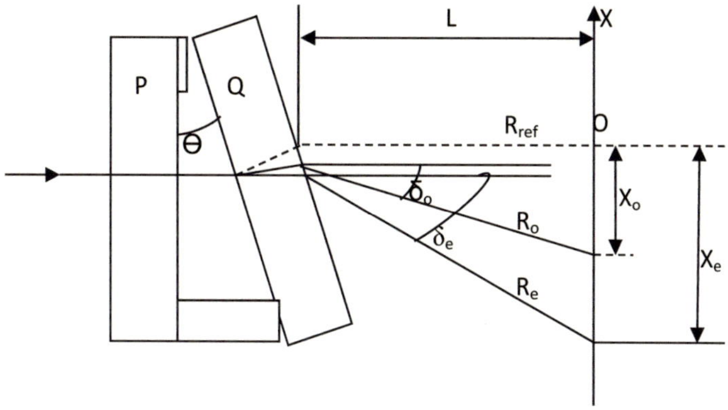

Figure 1. The used liquid crystal wedged cell.

Sample	Angle
E7	0.036215581
25% E7 with 75% MLC 6241-000	0.036635508
50% E7 with 50% MLC 6241-000	0.049305551
75% E7 with 25% MLC 6241-000	0.032724923
MLC 6241-000	0.052359877

Table 1. The angle's value formed by the two substrates of five samples.

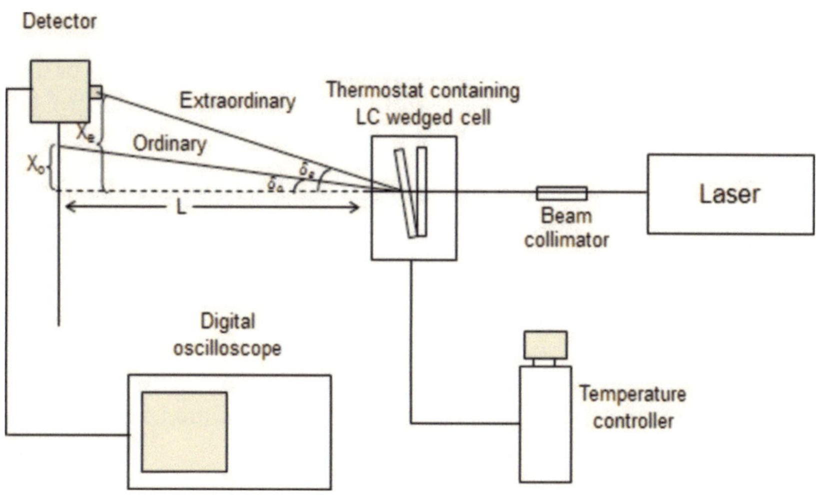

Figure 2. The experimental setup for the measurement of the refractive indices of LC.

$$\delta_o = \tan^{-1}\left(\frac{x_o}{D}\right) \tag{16}$$

$$\delta_e = \tan^{-1}\left(\frac{x_e}{D}\right) \tag{17}$$

So that

$$n_o = \frac{\sin\left(\theta + \tan^{-1}(x_o/D)\right)}{sin\theta} \tag{18}$$

$$n_e = \frac{\sin\left(\theta + \tan^{-1}(x_e/D)\right)}{sin\theta} \tag{19}$$

These measurements have been repeated many times by changing the temperature of the LC sample to measure the refractive indices at different temperatures. Also, this experiment has been repeated by decreasing the temperature of LC sample to study the bistability of liquid crystal due to temperature.

4. Results and discussions

4.1. Nonlinear refractive index

The refractive indices of the five liquid crystals were measured using wedged cell refractometer method at 1550 nm wavelength [9]. **Figure 3** shows the temperature dependence of the refractive indices n_o and n_e of E7 and MLC 6241-000. Red squares and blue circles are experimental data for refractive indices n_e and n_o of E7 and MLC 6241-000, respectively. The solid curves are fittings using the four-parameter model Eqs. (10) and (11). The fitting parameters are listed in **Table 2**.

The temperature dependence of refractive indices of three LC mixtures is shown in **Figure 4**. Blue squares, green triangles, and red circles represent the refractive indices of (50% E7 with 50% MLC 6241-000, 25% E7 with 75% MLC 6241-000, and 75% E7 with 25% MLC 6241-000), respectively. The solid curves are fittings using Eqs. (10) and (11).

Brugioni et al. also studied the temperature effect on nonlinear refractive indices of E7 liquid crystal at mid-infrared region (10.6 μm) using a wedge cell refractometer method as shown in **Figure 5** [13, 25].

Figure 6 shows the wavelength-dependent refractive indices of E7 at T = 25°C. Squares and circles represent the n_e and n_o of E7 in the visible region, while the downward and upward triangles stand for the measured data at λ = 1.55 and 10.6 μm, respectively [25]. **Figure 6** also indicates that the refractive indices will saturate in the far-infrared region, the agreement between experiment and theory is very good.

The temperature-dependent birefringence of E7 and MLC 6241-000 at 1550 nm is shown in **Figure 7**. Blue squares and red circles represent the birefringence of E7 and MLC 6241-000,

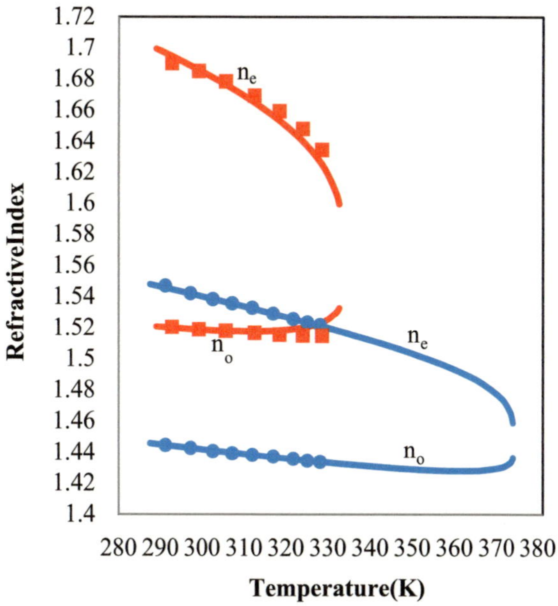

Figure 3. Temperature dependent refractive indices of E7 and MLC 6241-000 at λ = 1550 nm. Red squares and blue circles represent the refractive indices n_e and n_o of E7 and MLC 6241-000, respectively. The solid curves are the fitting using Eqs. (10) and (11).

LC materials	<n>		Δn	
	A	B (K^{-1})	$(\Delta n)_o$	β
E7	1.750172	5.88×10^{-4}	0.301812	0.259542
25% E7 with 75% MLC 6241-000	1.61647	4.17×10^{-4}	0.153497	0.246991
50% E7 with 50% MLC 6241-000	1.622715	4.18×10^{-4}	0.189193	0.22493
75% E7 with 25% MLC 6241-000	1.650715	4.49×10^{-4}	0.182306	0.204806
MLC 6241-000	1.6008	4.21×10^{-4}	0.147301	0.248339

Table 2. Fitting parameters for the average refractive index <n> and birefringence Δn of five LCs at 1550 nm.

respectively, while solid lines are fitting results using Eq. (7). The fitting parameters for these samples are also listed in **Table 2**. Through fittings, we obtain parameters $(\Delta n)_o$ and β. As shown in **Figures 3**, **4**, and **7**, the four-parameter model fits the experimental data of samples E7 and MLC 6241-000 very well.

Figure 8 shows the temperature-dependent birefringence of three mixing samples. Blue triangles, green circles, and red squares, represent the birefringence LCs of (50% E7 with 50% MLC 6241-000, 25% E7 with 75% MLC 6241-000, and 75% E7 with 25% MLC 6241-000), respectively. The solid lines are fitting curves using Eq. (7). The LC mixture (50% E7 with 50% MLC 6241-000) has higher birefringence.

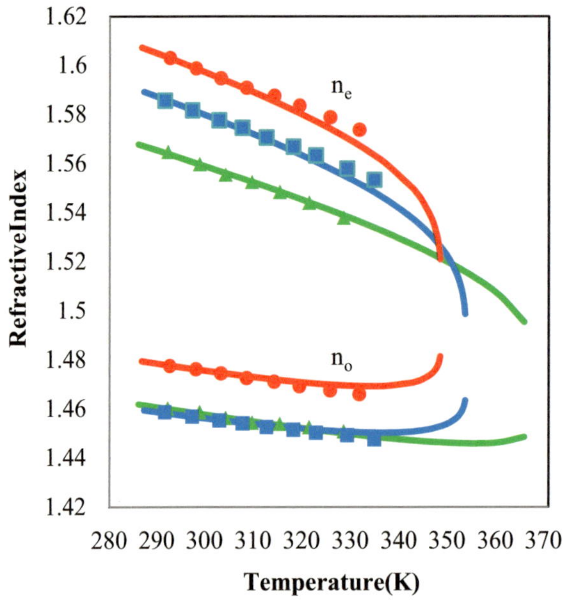

Figure 4. Temperature-dependent refractive indices of three mixture samples at λ = 1550 nm. Red circles, blue squares, and green triangles represent the refractive indices n_e and n_o of (75% E7 with 25% MLC 6241-000, 50% E7 with 50% MLC 6241-000, and 25% E7 with 75% MLC 6241-000), respectively. The solid curves are the fitting using Eqs. (10) and (11).

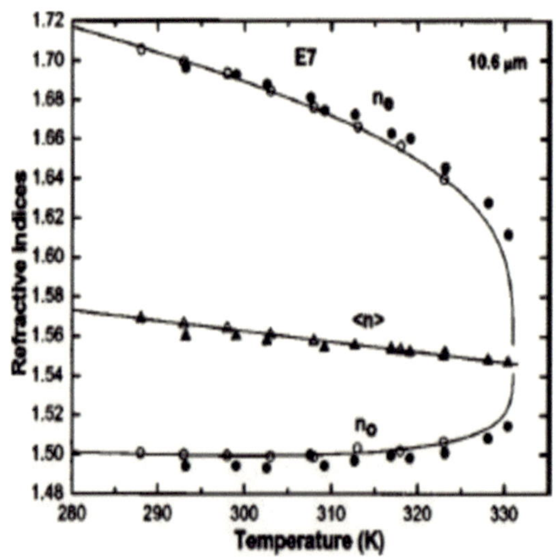

Figure 5. Temperature-dependent refractive indices of E7 at 10.6 μm. The filled circles are the refractive indices measured by the wedged LCcell refractometer method. The filled triangles are the average refractive index calculated by the experimental data. The open circles represent n_o and n_e extrapolated from the experimental data measured at visible spectrum using the extended Cauchy model. The open triangles are the average refractive index calculated by the extrapolated n_e and n_o, respectively. The solid curves are the fitting using Eqs. (10) and (11).

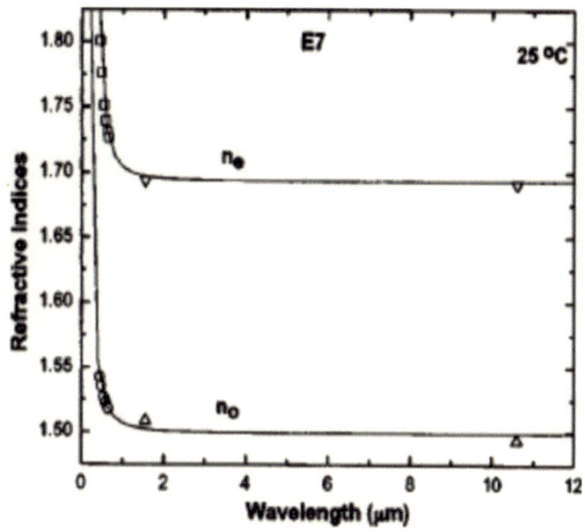

Figure 6. Wavelength-dependent refractive indices of E7 at T = 25°C. The open squares and circles are the n_e and n_o of E7 measured at the visible spectrum. The solid curves are the fittings to the experimental data measured at the visible spectrum using the extended Cauchy model. The downward and upward triangles are n_e and n_o of E7 measured at T = 25°C and λ = 1.55 and 10.6 μm, respectively.

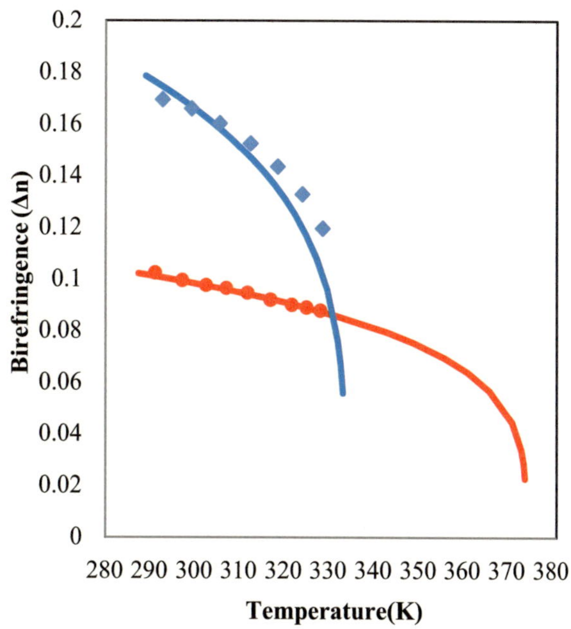

Figure 7. Temperature-dependent birefringence of E7 and MLC 6241-000 at λ = 1550 nm, blue squares and red circles represent the birefringence of E7 and MLC 6241-000, respectively. Solid lines are fitting results using Eq. (7).

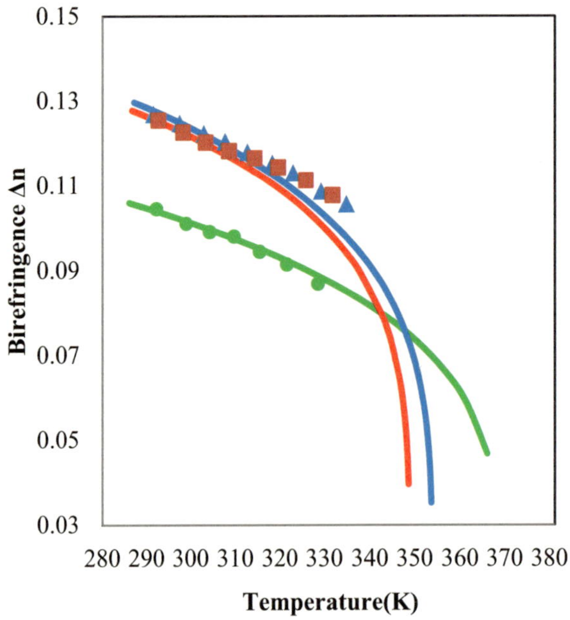

Figure 8. Temperature-dependent birefringence of three mixture samples at $\lambda = 1550$ nm. Red squares, blue triangles, and green circles represent the birefringence LCs (75% E7 with 25% MLC 6241-000, 50% E7 with 50% MLC 6241-000, and 25% E7 with 75% MLC 6241-000), respectively. Solid lines are fitting results using Eq. (7).

The temperature-dependent average refractive index of five liquid crystal samples is shown in **Figure 9**. Green triangles, red squares, black circles, brown stars, and blue monoclinic represent the average refractive index LCs of E7, 75% E7 with 25% MLC 6241-000, 50% E7 with 50% MLC 6241-000, and 25% E7 with 75% MLC 6241-000, and MLC 6241-000), respectively, while solid lines are fitting results using the four-parameter model (Eq. (9)). The fitting parameters A and B for these five samples are listed in **Table 2**. Average refractive index decreases linearly as the temperature increases.

4.2. Temperature gradient of refractive index

For practical applications, it is necessary to operate the LC device at room temperature. So, LC should be designed with crossover temperature (T_o) lower than 300 K to obtain a positive dn_o/dT at room temperature. The temperature dependence of $-dn_e/dt$ and dn_o/dt for liquid crystal can be found using Eqs. (12) and (13). These are shown in **Figure 10a–e**. The calculated values of crossover temperatures for LCs (E7, MLC 6241-000, 25% E7 with 75% MLC 6241-000, 50% E7 with 50% MLC 6241-000, and 75% E7 with 25% MLC 6241-000) are ~311.1, 360.54, 354.61, 335.8, and 333.56 K, respectively. Negative temperature gradient ($-dn_e/dt$) positive for all LC samples, that means the extraordinary refractive index decreases with increase in temperature, whereas the positive temperature gradient (dn_o/dt) changes its sign from negative to positive value. The dn_o/dt is negative when the temperature is below T_o, whereas it becomes positive when the temperature is above T_o. To achieve a high dn_o/dT, high birefringence and low clearing temperature are two important factors for this.

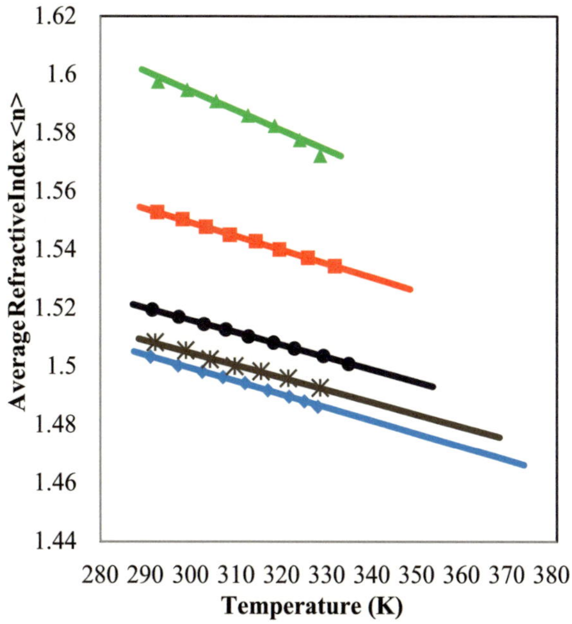

Figure 9. Temperature-dependent average refractive index > n < of five LCs at 1550 nm. Green triangles, red squares, black circles, brown stars, and blue monoclinic represent the average refractive index LCs of E7, 75% E7 with 25% MLC 6241-000, 50% E7 with 50% MLC 6241-000, and 25% E7 with 75% MLC 6241-000, and MLC 6241-000), respectively. Solid lines are fitting results using Eq. (9).

4.3. Studying bistability based on temperature

The refractive indices of the five liquid crystal samples were measured at λ = 1550 nm in the temperature range from 290 to 330 K. **Figure 11a–e** shows the temperature-dependent extraordinary refractive index of E7, MLC 6241-000, and three mixture samples (25% E7 with 75% MLC 6241-000, 75% E7 with 25% MLC 6241-000, and 50% E7 with 50% MLC 6241-000), respectively. From these figures, the bistability of LCs due to temperature for extraordinary refractive index is clearly observed. In the case of mixture (50% E7 with 50% MLC 6241-000), the two curves of increasing and decreasing liquid crystal's temperature intersect at 310 K to be equal until they reach to 291 K as shown in **Figure 11e**.

Figure 12a–e shows the temperature dependent ordinary refractive index of E7, MLC 6241-000, and three mixture samples (25% E7 with 75% MLC 6241-000, 75% E7 with 25% MLC 6241-000, and 50% E7 with 50% MLC 6241-000), respectively. Also, from these figures the bistability of LCs due to temperature is clearly observed for ordinary refractive index. As shown in **Figure 12e**, for mixture (50% E7 with 50% MLC 6241-000), the values of ordinary refractive index in the case of decreasing temperature become larger than its values in the case of increasing temperature case after the intersect point at 310 K.

Figure 13a–e shows the birefringence bistability of the five liquid crystal samples. The bistability is represented by the area between the temperature refractive indices and birefringence (increasing and decreasing). The measured areas are listed in **Table 3**.

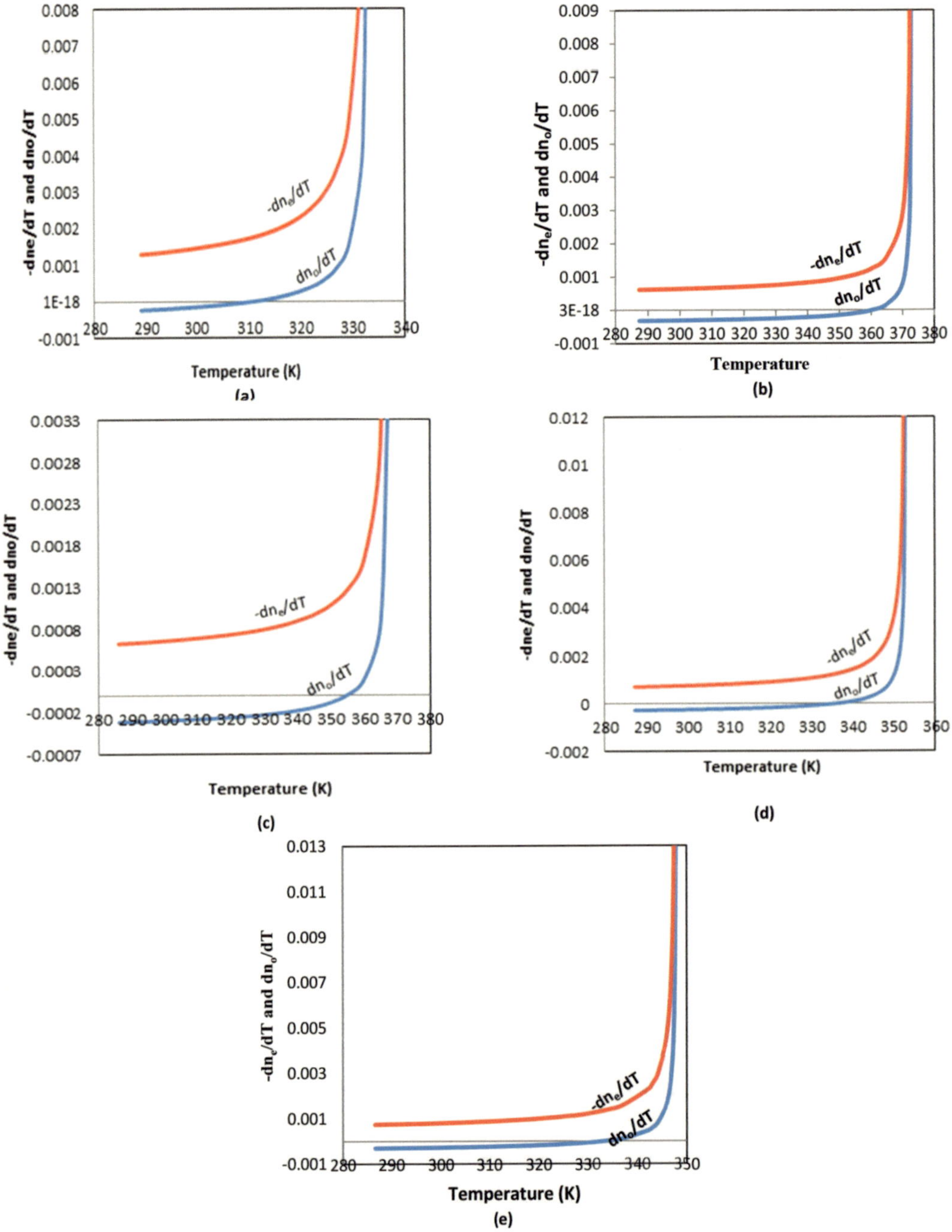

Figure 10. Temperature gradient for n_e and n_o of LC samples (a) E7, (b) MLC 6241-000, (c) 25% E7 with 75% MLC 6241-000, (d) 50% E7 with 50% MLC 6241-000, and (e) 75% E7 with 25% MLC 6241-000 at 1550 nm. Blue and red solid lines represent the calculated dn_o/dT and $-dn_e/dT$, respectively. The crossover temperature for these samples are around 311.2, 360.6, 354.7, 335.9, and 333.6 K for E7, MLC 6241-000, 25% E7 with 75% MLC 6241-000, 50% E7 with 50% MLC 6241-000, and 75% E7 with 25% MLC 6241-000, respectively.

http://dx.doi.org/10.5772/intechopen.70414

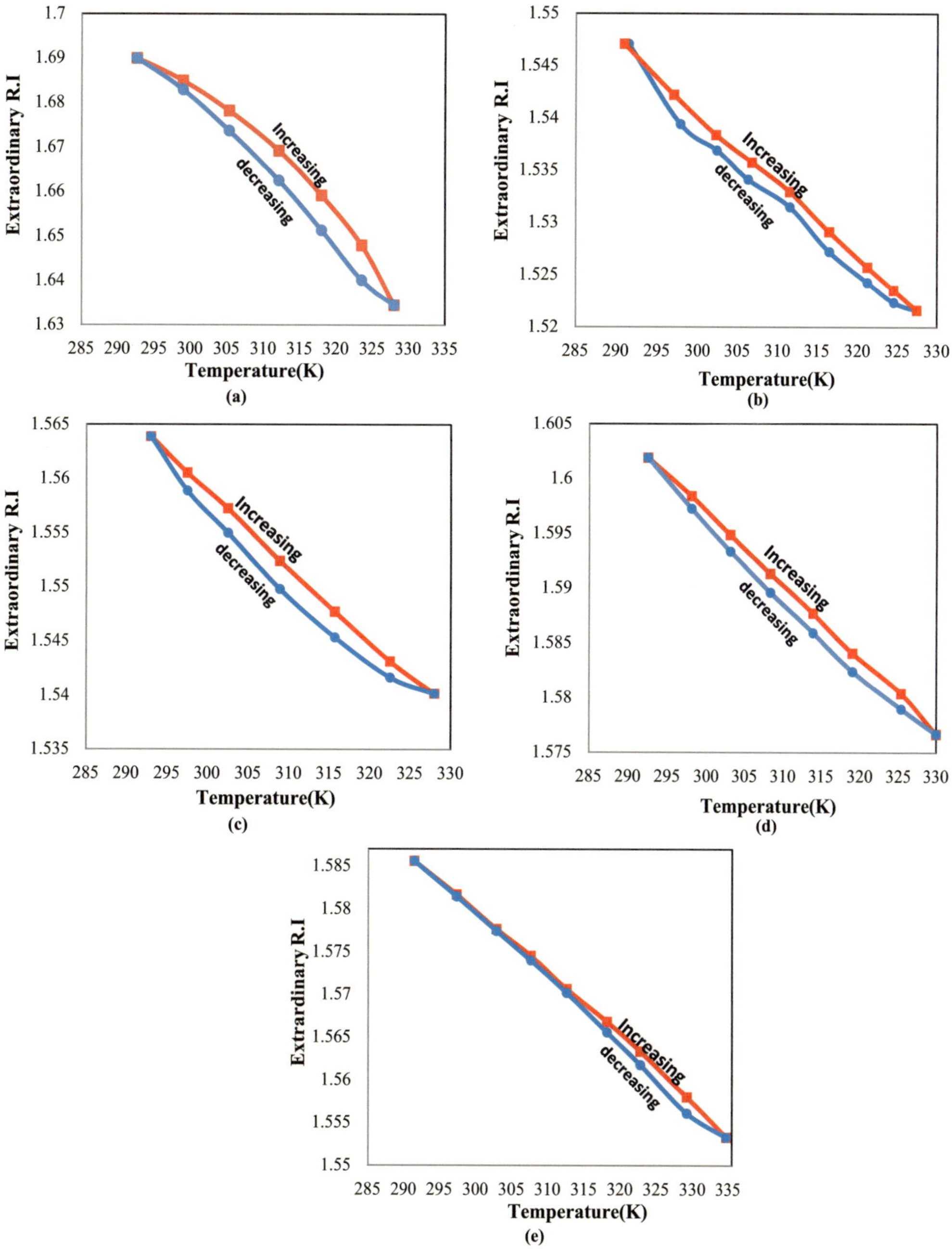

Figure 11. Temperature dependent extraordinary refractive index of (a) E7, (b) MLC 6241-000, (c) 25% E7 with 75% MLC 6241-000, (d) 75% E7 with 25% MLC 6241-000, and (e) 50% E7 with 50% MLC 6241-000, at 1550 nm. Squares and circles represent extraordinary refractive index for increasing (red line) and decreasing (blue line) temperature, respectively.

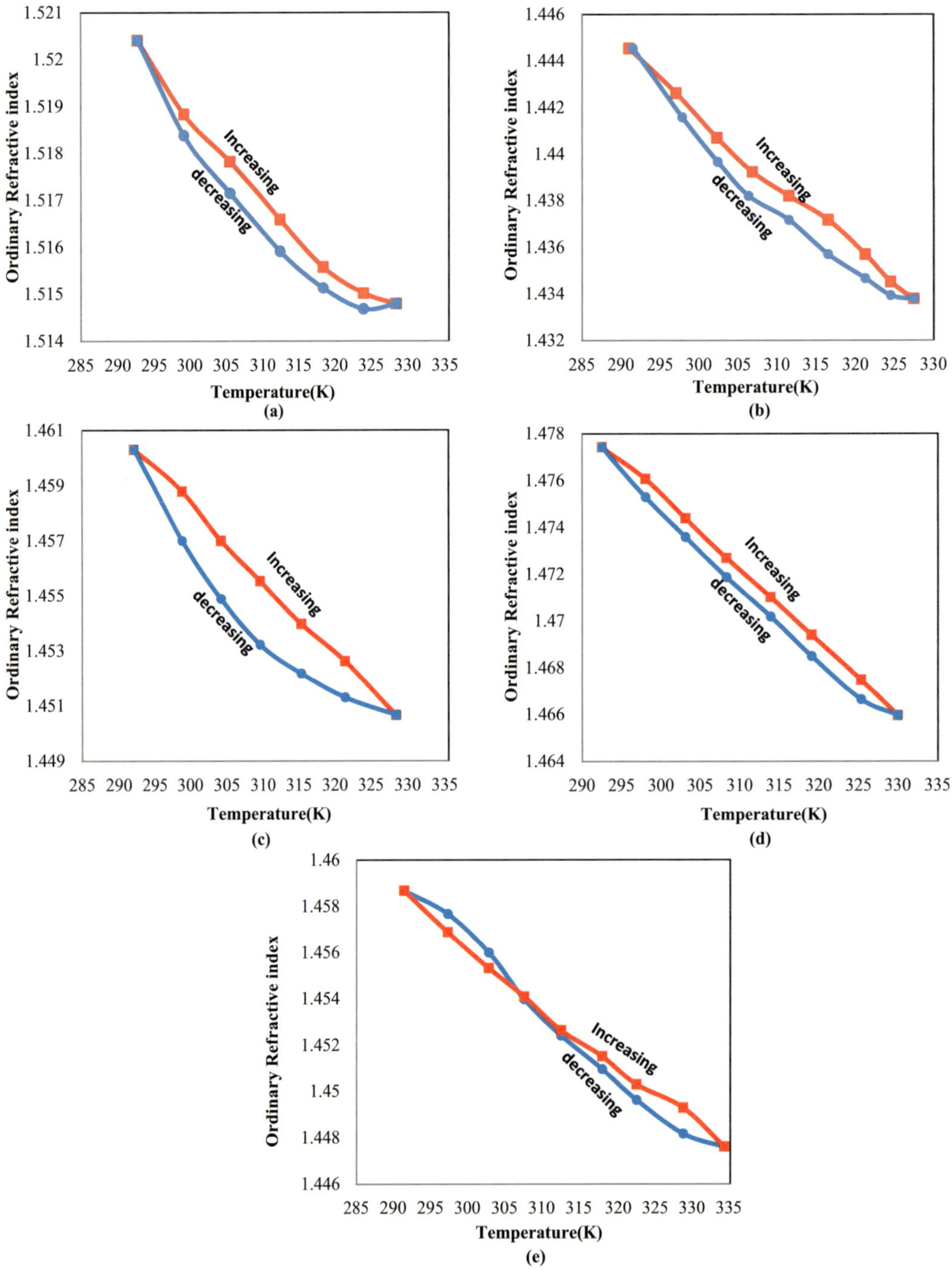

Figure 12. Temperature-dependent ordinary refractive index of (a) E7, (b) MLC 6241-000, (c) 25% E7 with 75% MLC 6241-000, (d) 75% E7 with 25% MLC 6241-000 and (e) 50% E7 with 50% MLC 6241-000 at 1550 nm. Squares and circles represent ordinary refractive index for increasing (red line) and decreasing (blue line) temperature respectively.

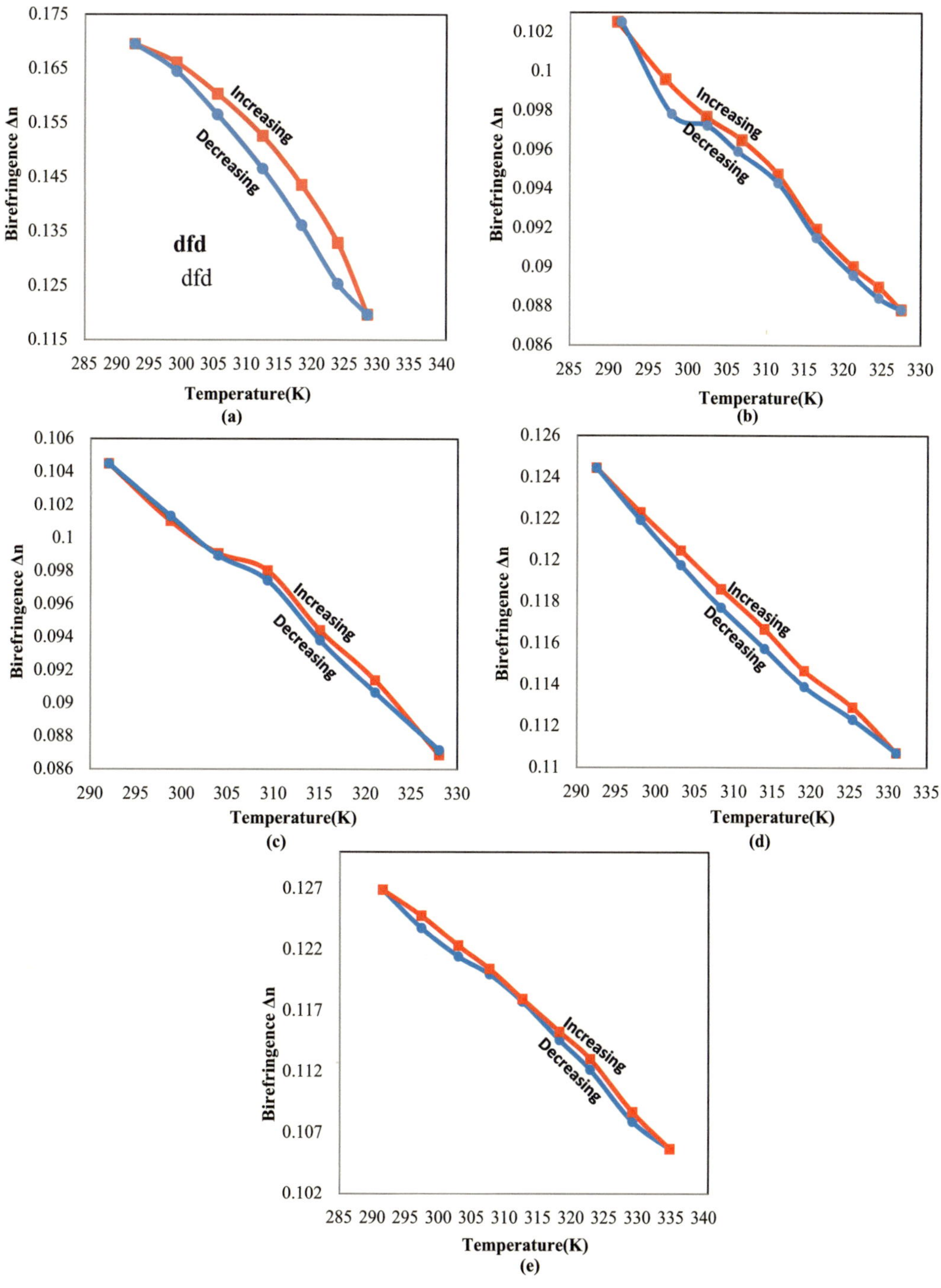

Figure 13. Temperature-dependent birefringence (Δn) of (a) E7, (b) MLC 6241-000, (c) 25% E7 with 75% MLC 6241-000, (d) 75% E7 with 25% MLC 6241-000, and (e) 50% E7 with 50% MLC 6241-000 at 1550 nm. Squares and circles represent birefringence refractive index for increasing and decreasing temperature.

Sample	Area		
	Extraordinary R.I	Ordinary R.I	Birefringence
E7	0.16882	0.01577	0.15304
25% E7 with 75% MLC 6241-000	0.06328	0.05379	0.00599
50% E7 with 50% MLC 6241-000	0.03333	0.00693	0.0264
75% E7 with 25% MLC 6241-000	0.05007	0.02704	0.02303
MLC 6241-000	0.05643	0.0337	0.02272

Table 3. The area between increasing and decreasing temperature curves of liquid crystal samples.

From these measurements, the extraordinary refractive index has larger bistability than the ordinary refractive index.

5. Conclusions

The ordinary and extraordinary refractive indices of five types of liquid crystals were measured at near-infrared region (1550 nm) and in temperature range from 290 to 330 K, using a wedged cell refractometer method. The variation in refractive indices, and average refractive index were fitted theoretically using a modified four-parameter model. Excellent agreement between the experimental data and fitted values by using four-parameter model is obtained. In addition, the birefringence of liquid crystal as a function of the temperature is calculated; high birefringence is obtained when the mixing is 50% E7 with 50% MLC 6241-000. The temperature gradients of liquid crystal refractive indices are presented; high birefringence and low clearing temperature are two important factors to achieve large dn_o/dT. The bistability of LCs due to temperature is also studied. The liquid crystal bistability based on the temperature is clearly observed for all samples. Also, the extraordinary refractive index has larger bistability than the ordinary refractive index.

Author details

Lamees Abdulkaeem Al-Qurainy[1*] and Kais A.M. Al Naimee[1,2]

*Address all correspondence to: lamees.alqurainy@ino.it

1 Physics Department, College of Science, University of Baghdad, Bagdad, Iraq

2 C.N.R. Istituto Nazionale di Ottica Applicata, Firenze, Italy

References

[1] Wu ST, Yang DK. Reflective Liquid Crystal Displays. New York: Wiley; 2001

[2] Abuleil MJ, Abdulhalim I. Birefringence measurement using rotating analyzer approach and quadrature cross points. Optical Society of America. 2014;**53**:2097-2104

[3] Wu ST, Hsu CS, Chuang YY. Room temperature Bis-tolane liquid crystals. Japanese Journal of Applied Physics. 1999;**38**

[4] Yang CSh, Lin ChJ, Pan RP, Que CT, Yamamoto K, Tani M, Pan CL. The complex refractive indices of the liquid crystal mixture E7 in the Terhertz frequency range. Optical Society of America. 2010;27

[5] McManamon PF, Dorschner TA, Corkum DL, Friedman LJ, Hobbs DS, Holz M, Liberman S, Nguyen HQ, Resler DP, Sharp RC, Watson EA. Optical phased Array technology. Proceedings of IEEE. 1996;84

[6] Scharf T. Polarized Light in Liquid Crystals and Polymers. John Wiley and Sons, Inc.; 2007

[7] Durbin SD, Arakelian SM, Cheung MM, Shen YR. Highly nonlinear optical effects in liquid crystals. Journal of Physique. 1983;**44**:161-169

[8] Thingujam KD, Sarkar SD, Choudhury B, Bhattacharjee A. Effect of temperature on the refractive indices of liquid crystals and validation of a modified four-parameter model. Acta Physica Polonica A. 2012;**122**

[9] Abdulkareem L, Abdalah SF, Al Naimee K, Meucci R. Temperature effect on nonlinear refractive indices of liquid crystals in visible and NIR. Optics Communications. 2016;**363**:188-194

[10] Horn RG. Refractive indices and order parameters of two liquid crystals. Journal of Physics. 1978;**39**:105-109

[11] Wu ST. A Semiempirical model for liquid crystal refractive index dispersions. Journal of Applied Physics. 1991;**69**:2080-2087

[12] Li J, Guaze S, Wu ST. Temperature effect on liquid crystal refractive indices. Journal of Applied Physics. 2004;**96**:19-24

[13] Li J, Wu ST, Brugioni S, Meucci R, Faetti S. Infrared refractive indices of liquid crystals. Journal of Applied Physics. 2005;**97**

[14] Blinov LM. Structure and Properties of Liquid Crystals. Springer; 2011

[15] Yang DK, Wu ST. Fundamentals of Liquid Crystal Devices. Wiley and Sons, Ltd.; 2006

[16] Li J, Wu ST. Extended Cauchy equations for the refractive indices of liquid crystals. Journal of Applied Physics. 2004;**95**

[17] Thingujam KD, Sarkar SD, Choudhury B, Bhattacharjee A. Effect of temperature on the refractive indices of liquid crystals and validation of a modified four- parameter model. Acta Physica Polonica A. 2012;**122**:754-757

[18] Li J, Guaza S, Wu ST. High dno/dT liquid crystals and their applications in a thermally tunable liquid crystal photonic crystal fiber. Molecular Crystals and Liquid Crystals. 2006;**453**:355-370

[19] Guaza S, Li J, Wu ST, Spadlo A, Dabrowski R, Tzeng Y, Cheng K. High birefringence and high resistivity Isothiocyanate-based Nematic liquid crystal mixtures. Liquid Crystals. 2005;**32**:1077-1085

[20] Li J, Gauzia S, Wu ST. High temperature-gradient refractive index liquid crystals. OSA. 2004;**12**

[21] Yang CSh, Lin ChJ, Pan RP, Que CT, Yamamoto K, Tani M, Pan CL. The complex refractive indices of the liquid crystal mixture E7 in the Terhertz frequency range. Optical Society of America. 2010;**27**

[22] Li J, Wen CH, Lu R, Wu ST. Refractive indices of liquid crystals for display applications. Journal of Display Technology. 2005;**1**

[23] Kumar A. Calculation of optical parameters of liquid crystals. Acta Physica Polonica A. 2007;**112**

[24] Ozbek H, Ustunel S, Kutlu E, Cetinkaya MC. A simple method to determine high-accuracy refractive indices of liquid crystals and the temperature behavior of the related optical parameters via high-resolution birefringence data. Journal of Molecular Liquids. 2014;**199**:275-286

[25] Brugioni S, Meucci R, Faetti S. Refractive indices of liquid crystals E7 and K15 in the mid- and near-IR regions. Opticheskii Zhurnal. 2006;**73**:15-17

Chapter 9

Micro/Nano Liquid Crystal Layer–Based Tunable Optical Fiber Interferometers

Haimei Luo, Changjing Wang, Yinghua Ji and Wen Yuan

Additional information is available at the end of the chapter

http://dx.doi.org/10.5772/intechopen.70413

Abstract

Miniaturization and integration are the main trends in modern photonic technology. In this chapter, two kinds of micro-/nano liquid crystal (LC) layer–based tunable optical fiber interferometers are proposed. One fiber interferometer is the optical fiber gratings (LPGs), and the other one is the locally bent microfiber taper (LBMT). The working principles of the devices are theoretically analyzed. The preparation process and the functional properties of the devices are experimentally investigated as well.

Keywords: liquid crystal device, fiber gratings, microfiber taper, interferometry, mode

1. Introduction

As optical materials that exhibit very large anisotropic properties, liquid crystal (LC) has been used in a variety of photonic applications with an eye toward enabling tunable optical responses, with stimuli including thermal [1], electrical [2], magnetic [3], and optical fields [4]. Miniaturization and integration are the main trends in modern photonic technology. With the help of micro-/nanofabrication technology, people can design and prepare various optical fiber micro-/nanostructures and devices, which exhibit significant difference in characteristics when the functional structure size decreases to micrometers and/or nanometers. These special physical properties have wide theoretical research prospects and practical applications. In this chapter, two kinds of novel micro-/nano LC layer–based tunable optical fiber interferometers are theoretically analyzed and experimentally studied, including the mode coupling properties and relevant micro-/nanofabrication technologies. The operation principle, preparation process, and functional properties of the proposed optical fiber devices are studied in detail. By comparison with that of the traditional sized optical fiber devices, the advantages in performance of the micro-/nano LC layer based ones are verified. The main contents of this chapter are as follows.

Two kinds of typical optical fiber–based modal interference devices and the relevant micro-/nanofabrication technology are proposed in Section 2.1. One typical fiber modal interference device is the optical fiber grating. The other typical device is the tapered optical fiber. Adopting optic fiber surface micro-/nanostructure technology, two kinds of micro-/nano liquid crystal layer–based wide range tunable optical fiber devices are achieved. The properties of the cladding modes in LPG coated with a high refractive index (HRI) micro-/nanometer overlay are theoretically studied in Section 2.2. The resonant wavelength and spectral characteristics of the four layer model long period grating are also analyzed based on the coupled-mode theory. Besides, the transmission spectra of LPG with different overlay thickness and refractive indices are numerically calculated. The tuning characteristics of locally bent microfiber taper (LBMT) covered with a nanosized HRI layer under different temperatures and electric field intensities have been theoretically analyzed in Section 2.3. The mathematical model for LBMT is established. The mode coupling and interference characteristics in a LBMT are described. In Section 2.4, a new structure LPG coated with nanosized HRI-LC layer is experimentally realized. The refractive indices of LC at different temperatures are measured. Using the sample brush coating technology, LC layers with different thicknesses are deposited on the surface of LPG. The sensitivity of the resonance wavelength to the change of the nanoscale overlay refractive index is experimentally observed. Experimental results show that the phenomenon of cladding mode reorganization in HRI-LC–coated LPG occurs when the high refractive index (HRI) of the nanosized LC overlay is changed from 1.477 to 1.515 resulting from temperature increasing from 20 to 65°C. The electro-optic tuning ability of LPG coated with the HRI-LC layer is also demonstrated. By choosing an appropriate operating point, the maximum tuning range can reach approximately 10 nm. The experimental results are in good agreement with the theoretical analysis in Section 2.2. The transmission characteristics of the nanosized HRI-LC layer–coated LBMT in response to the environmental temperature, and external electric fields have been experimentally investigated in Section 2.5. A microfiber taper with a diameter of ~3.72 μm is fabricated using the flame brushing technique. By bending the transition region of the taper and later by placing a ~200-nm LC layer over the uniform taper waist region, a high-efficiency thermal and electric tunable LC-coated LBMT interferometer is achieved. This suggests a potential application of this device as tunable all-fiber photonic devices, such as filters, modulators, and sensing elements. Finally, conclusions are drawn in section 3.

2. Micro-/nano LC layer–based tunable optical fiber interferometers

2.1. Two kinds of typical optical modal interferometers

2.1.1. Long period gratings (LPG)

In-fiber long period gratings (LPG), which have the ability to couple energy from the code mode to different cladding modes with the same propagation direction, have been widely investigated in the field of optical sensing and communication [5].

Up to date, most of the studies have concentrated on the analysis of the LPG response to the surrounding medium refractive index (SRI) smaller than that of silica [6]. The idea of coated LPG with a thin HRI layer was first put forward by Rees et al. [7]. Then, Wang et al. presented a detailed study to investigate the sensitivity of LPG to SRI when they are coated with a nano-sized HRI film [8]. In the same year, a comprehensive investigation of mode transition in HRI layer–coated LPG was reported by Cusano et al. [9]. Their analysis indicated that the cladding modes reorganization occurred for a fixed overlay thickness and refractive index by increasing the SRI. In fact, changing the HRI could also result in cladding modes reorganization. The theoretical and experimental study proposed by Del Villar et al. showed that by selecting an appropriate overlay thickness, the highest sensitivity of the resonance wavelengths to HRI change of the overlay can be obtained [10, 11].

2.1.2. Tapered optical fiber interferometer

Fiber tapers, which have a variety of functions including filtering, light coupling, and sensing, consist of a waist between two tapered sections fabricated by heating and pulling technology [12, 13]. Different length and shape of the first tapered section have different effects on the input mode [13].

Recently, we reported a compact fiber interferometer named LBMT interferometer [14–16]. This kind of interferometer has unique features of low insertion loss and ultrathin taper waist, so it is promising for high-sensitivity sensing. It is interesting to integrating a LBMT interferometer and functional material due to physical effect of functional material and the high sensitivity of the microfiber to the surroundings.

2.2. Mode transition in the nanosized HRI overlay–coated LPG

2.2.1. Theory background

Based on the phase-matching condition between the core and the cladding modes, the center wavelength λ_i of the *j*th attenuation band can be expressed as [17]

$$\lambda_i = \left(n_{co} - n_{cl}^i\right)\Lambda \tag{1}$$

where n_{co} is the effective refractive index of the core mode and n_{cl}^i is the effective refractive index of the *i*th cladding mode. Λ is the grating period [5].

Figure 1 shows the structure of HRI-coated LPG with four layers. Using the transfer matrix method proposed by Anemogiannis et al., the cladding mode effective index can be calculated [18].

2.2.2. Numerical analysis of HRI overlay–coated LPG

The analysis used the standard Corning SMF-28 optical fiber parameters: numerical aperture 0.14, refractive index difference 0.36% [9], cladding, and core diameter 125 and 8.2 μm, respectively. The HRI-LC overlay is with refractive index of 1.47–1.55 close to the experimental testing [19] and with different thickness changing from 600 to 900 nm.

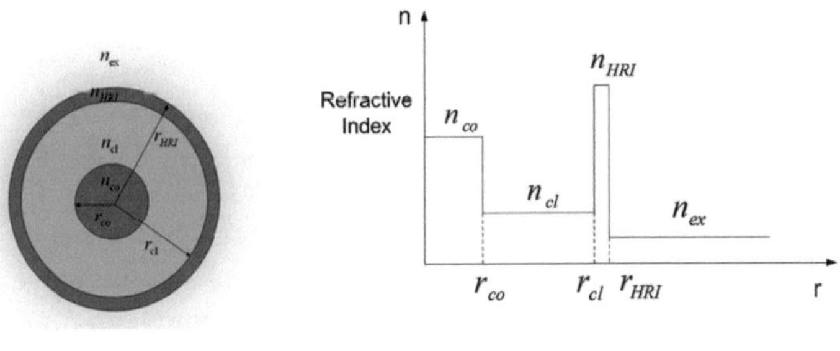

(a) (b)

Figure 1. (a) Illustrative schematic of LPG with an nm-thick thin-film coating. (b) Index profile of the thin-film coated LPG.

Each effective refractive index of the first six cladding modes as a function of the HRI is represented in **Figure 2(a)** for a 600-nm HRI overlay. From the figure, we know that each effective refractive index of the first six cladding modes goes up as the HRI increases, until a critical point is reached, when a significant shift in the effective refractive index occurs. There is a specific value of HRI that makes the lowest order cladding mode to be guided within the overlay for a fixed overlay thickness.

Figures 2(b)–(d) show the effective refractive index for a HRI overlay of 700, 800, and 900 nm, respectively. It can be seen that the transition point moves to a lower HRI as the overlay thickness increases.

2.3. Tuning effect of nanosized HRI-LC overlay–coated LBMTs

2.3.1. Mode coupling and interference in LBMTs

The fiber taper can be divided into two zones: (1) the taper waist with a constant diameter d_0 and (2) the transition region with a diameter continuously varying from d_0 to 125 μm. A LBMT is fabricated by bending the transition regions of the taper to form a modal interferometer.

To study the modal characteristics in the bent transition region, we assume the bent fiber taper as a sequence of straight segments of the same length l with an angle of θ [20] (see **Figure 3(a)**) [15]. The complex amplitude, $a_{pq}^{(i+1)}$, of the modes in the (i+1)th region is given by [20]

$$a_{pq}^{(i+1)} = \sum_{n=0}\sum_{m=0} \int_0^{\infty} \int_0^{2\pi} \Psi_{nm}^{i} \exp\left(-j\beta_{nm}^{i} l^{i}\right) \times \exp\left(j\beta_{nm}^{i} \theta r \cos \varnothing\right) \Psi_{nm}^{(i+1)*}\, r dr d\varnothing, \mathrm{i} = 1, 2, \ldots \quad (2)$$

where l^i is the length of the ith region, β_{nm}^{i} and ψ_{nm}^{i} are the propagation constant and the normal field of the LP_{mn} mode in the ith region, respectively [15]. $\psi_{pq}^{(i+1)*}$ is the complex conjugate of the mode field of the LP_{pq} mode in the (i+1)th region [15].

The local-mode power evolution along the LBMT under various bending curvatures was theoretically examined. The LBMT parameters are as follows: d_0 = 3.7 μm, L_0 = 6 mm, and

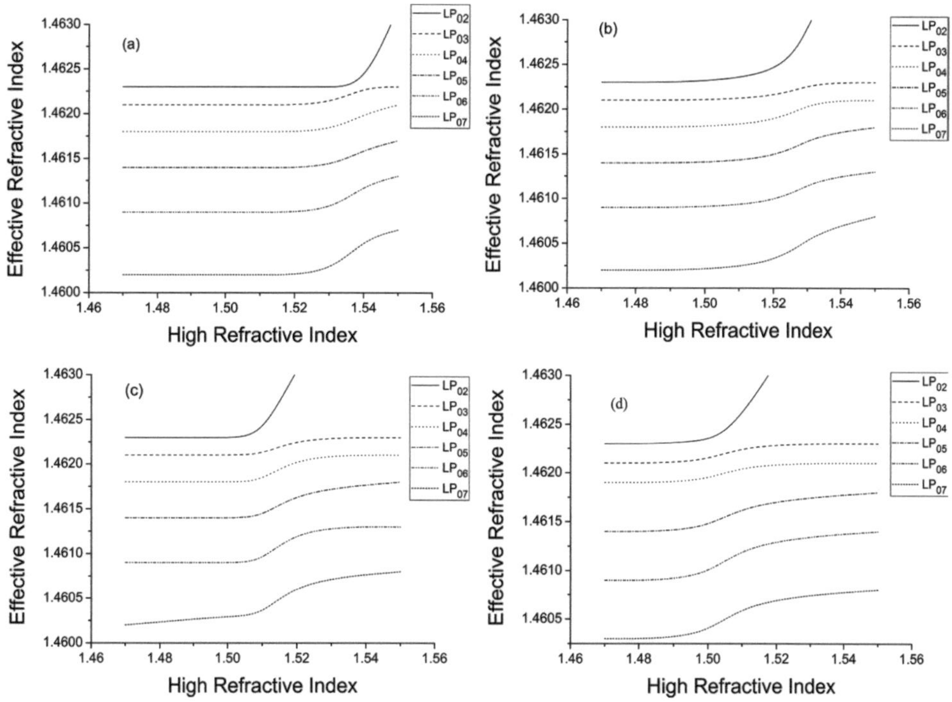

Figure 2. Effective refractive index of the $LP_{02} - LP_{07}$ cladding modes versus HRI-coated fiber with (a) 600, (b) 700, (c) 800, and (d) 900 nm film.

L_t = 3 mm at λ = 1.550 μm. Each bent transition region was divided into 100 steps with the same length (L_t/M). The 6-mm long uniform waist region has been considered as a step. From the modal shown in **Figure 3(b)** [15], we can calculate the appropriate values of the angle θ in the bent transition region. **Figure 4** shows the evolution of the first four modes (LP_{01}, LP_{11}, LP_{21}, and LP_{02}). As we can see that there is no power transfer from the fundamental mode to other high-order modes when the bending curvature $1/R = 0$. As the bending curvature goes up, the LP_{11}, LP_{21}, and LP_{02} modes are successively excited with their energy originated from the LP_{01} mode. The power of each mode remains almost constant in the central uniform taper waist. When $1/R$ increases, the coupling between the fundamental mode and the first higher order mode, the decisive factor in the interference extinction ratio is strengthened. The optimized status can be obtained at a certain bending curvature (e.g. 0.455 mm^{-1}) with the maximum extinction ratio and the relatively low loss.

2.3.2. Spectral tuning characteristics of a LBMT with a nanosized HRI-LC overlay

We calculated Poynting vectors of 200-nm LC layer–coated silica microfiber with different diameters ranging from 1 to 5 μm operating at 1.55 μm wavelength by solving Maxwell's equations of a three-layer structured cylindrical wavelength numerically (see **Figure 5(a)–(c)**)

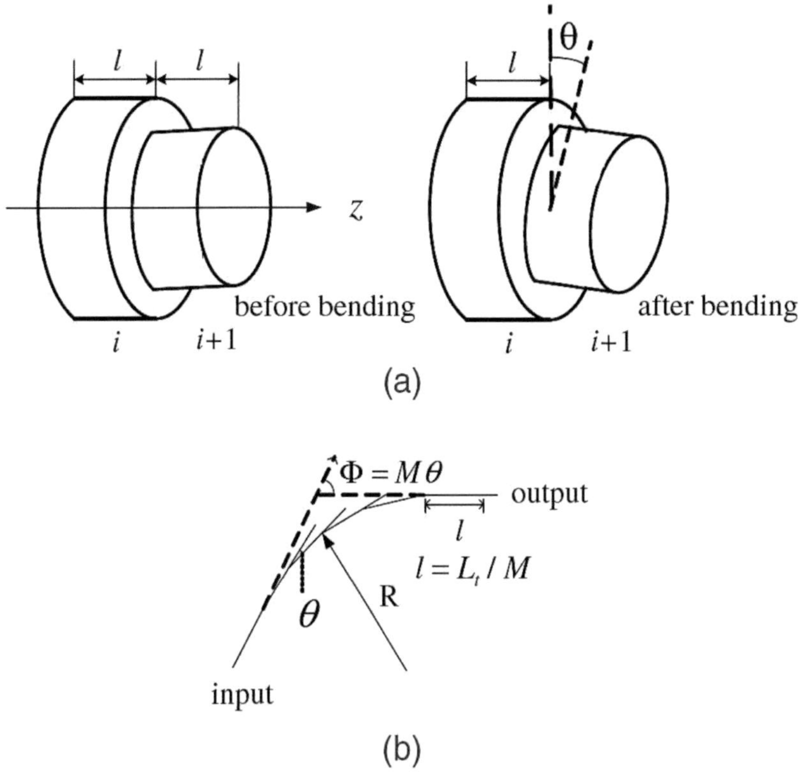

Figure 3. (a) Two adjacent sections of a fiber before and after bending. (b) Geometry used for the theoretical analysis of the bending effect on the fiber taper [15].

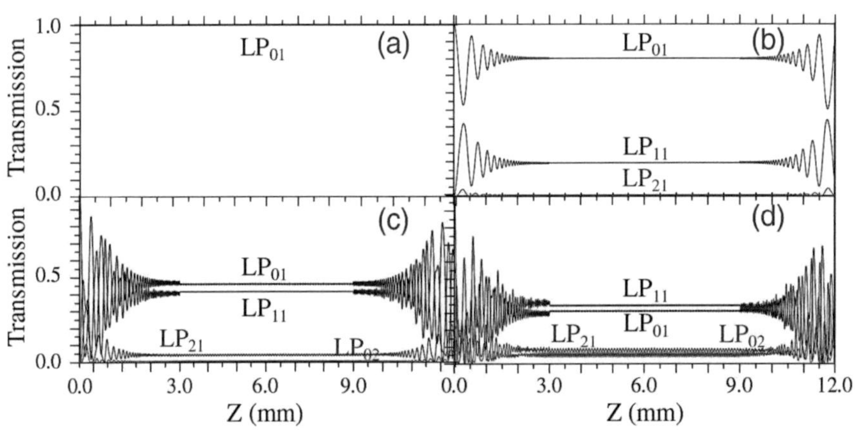

Figure 4. LP_{nm} local-mode power longitudinal evolution of the LBMT with d_0 = 3.7 μm and L_0 = 6 mm under bending curvature $1/R$ being (a) 0, (b) 0.2, (c) 0.445, and (d) 0.526 mm^{-1} at λ = 1.555 μm.

[16, 21]. In our calculation, the refractive indices of LC and silica microfiber are 1.50 and 1.44, respectively. We also calculated the amount of optical power guided in the LC layer and the air as a function of microfiber diameter in **Figure 5(d)** and **(e)**, respectively [16]. When the diameter of microfiber goes up from 1 to 5 μm, the amount of optical power is changed from 18.4 to 0.98% in the LC layer and from 10.5 to 0.31% in the air, respectively, which results in smaller tuning efficiency and lower transmission loss.

High-order modes are excited successively from the fundamental mode after the light injects into the uniform taper waist from a locally bent transmission region. In the output bent transition region, these modes will couple back into the fundamental mode and the wavelength-dependent transmission spectrum could be expressed as [16]

$$I = \sum_m I_m + 2\sum_{m>n} \sum \sqrt{I_m I_n} \cos(2\pi \Delta n_{eff} L_0/\lambda) \quad (3)$$

where I_m is the amplitude of the propagation mode, Δn_{eff} is the effective index difference, and L_0 is the length of the taper waist. The attenuation peak wavelength λ_N of the interferometer can be expressed as [16]

$$\lambda_N = 2\Delta n_{eff} L_0/(2N+1) \quad (4)$$

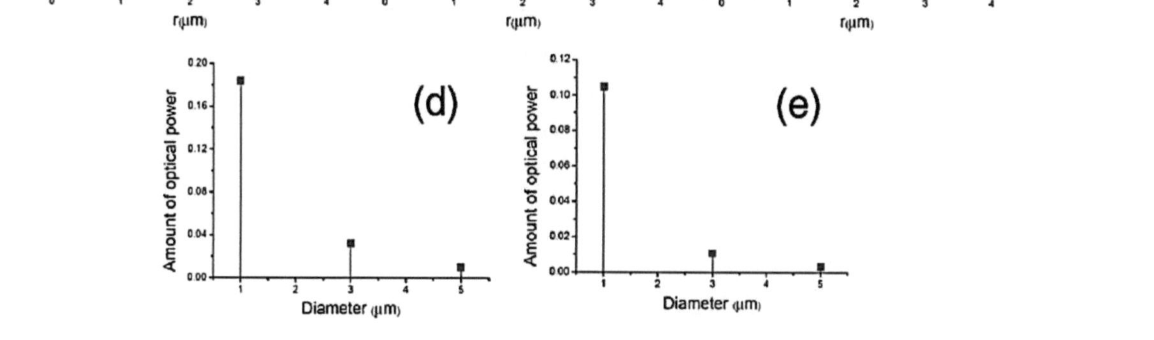

Figure 5. Calculated Poynting vector of a 200-nm thick LC-coated microfiber with diameter being (a) 1, (b) 3, and (c) 5 μm. The amount of optical power guided in (d) LC overlay, and (e) air as a function of microfiber diameter [16].

where N is the interference order. According to Eq. (4), Δn_{eff} changes when the effective index of the propagation mode of the microfiber taper waist is altered by the surrounding refractive index, and as a result, a wavelength shift is obtained. **Figure 6** shows the calculated Poynting vectors of a 3.72-μm diameter silica microfiber coated with different thickness HRI-LC overlay when wavelength is at 1.55 μm. As we can see, thicker LC films assure more optical power guided in the overlay for the fundamental mode and the first high-order mode. More optical power is converted to the radiation mode when the overlay thickness increases, which will result in a higher transmission loss.

The calculated transversal mode distributions for the first two modes with different 200-nm LC overlay HRIs are plotted in **Figure 7**. **Figure 8** depicts the effective index difference Δn_{eff} of the fundamental mode and the first high-order mode as a function of the refractive index of LC, when the overlay thickness being 200, 400, and 600 nm. We can see that more light energy concentrates in the outside LC layer with higher refractive index of the LC overlay, leading to an increased tuning range. The microfiber taper interferometer with higher overlay refractive index or thicker overlay thickness exhibits higher sensitivity. The higher the refractive index of the overlay, the smaller the effective index difference Δn_{eff}.

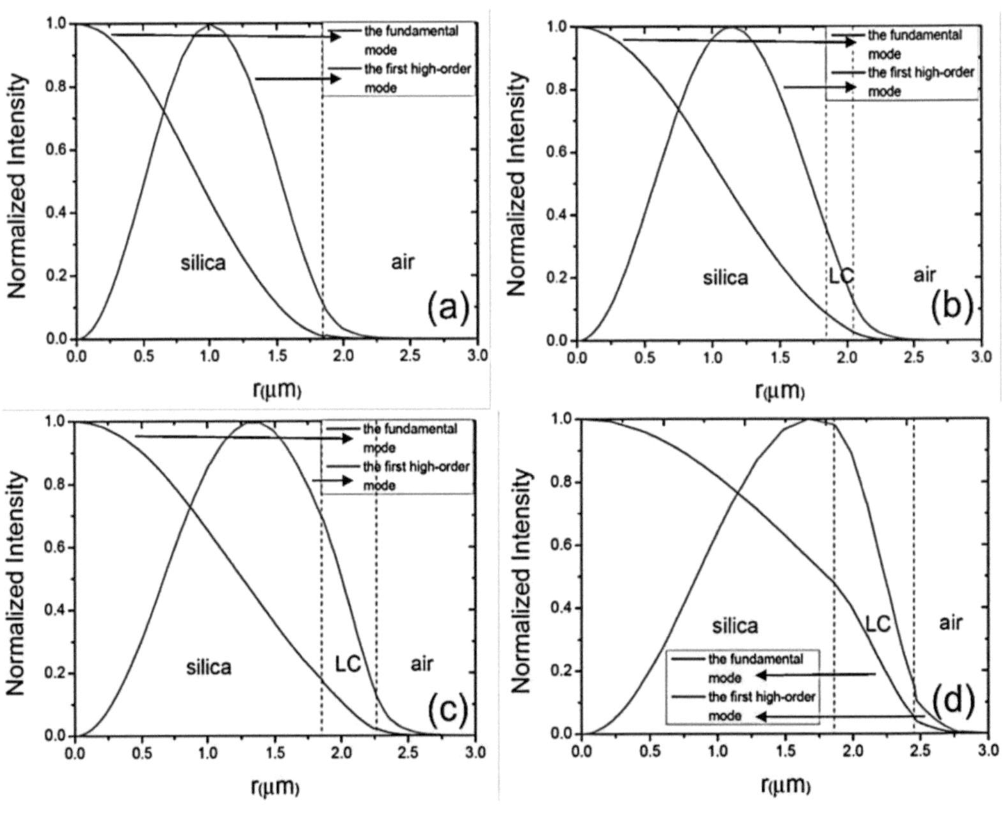

Figure 6. Distribution of the fundamental mode and the first high-order mode for three LC overlay thickness values: (a) 0, (b) 200, (c) 400, and (d) 600 nm.

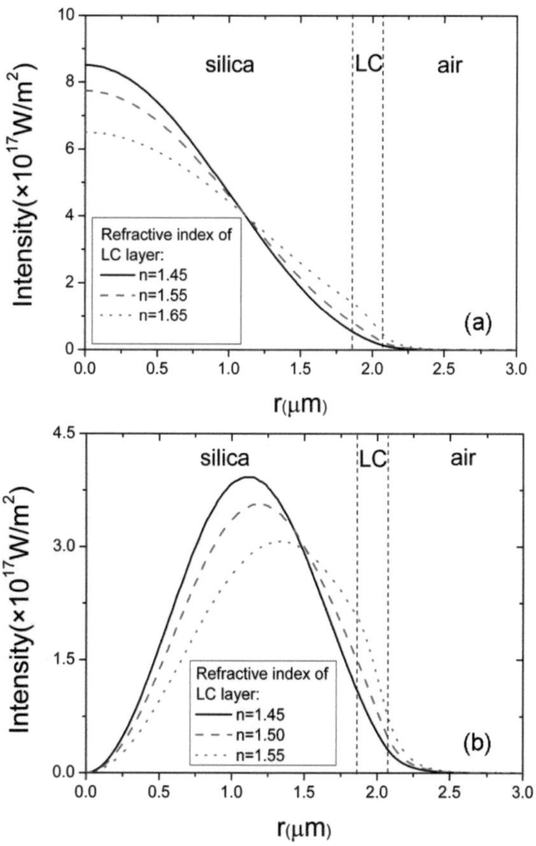

Figure 7. Transversal electric field distributions of the fundamental mode and the first high-order mode guided in the microfiber taper waist coated with a 200 nm LC overlay. (a) The fundamental mode and (b) the first high-order mode.

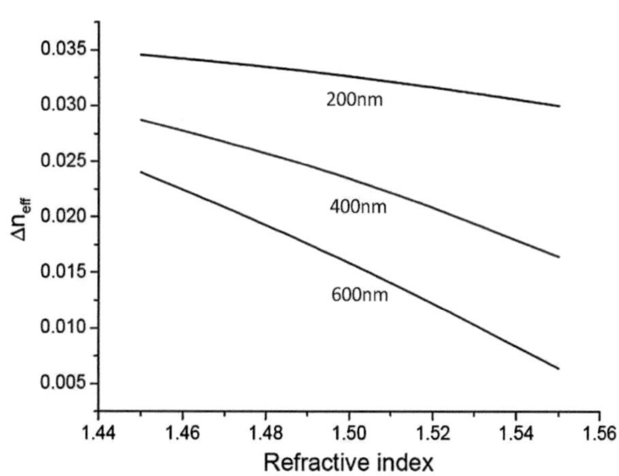

Figure 8. Calculated dependence of Δn_{eff} on the refractive index and the thickness of the LC overlay at the wavelength of 1559 nm.

2.4. Experimental study of nanosized HRI-LC layer–coated LPG

2.4.1. Temperature-dependent mode transition in HRI-LC–coated LPG

The LPG was fabricated on Corning SMF-28 fibers through CO_2 irradiation. The grating period is 620 μm, and the grating region is 50 mm. A white light source and an optical spectrum analyzer are used to record the spectral response. Attention bands were focus on the LP_{02} and LP_{03} modes in the range of 1400–1700 nm.

LC material MDA-98-3699 is from Merck. The refractive index of LC is higher than that of silica. The coating process is simple. We dipped a tampon with the liquid crystal, and daubed it on the fiber grating evenly. The surface tension of the slimy liquid crystal makes itself uniformly coated. **Figure 9** shows the CCD photographs of the bare and LC-coated LPG taken by

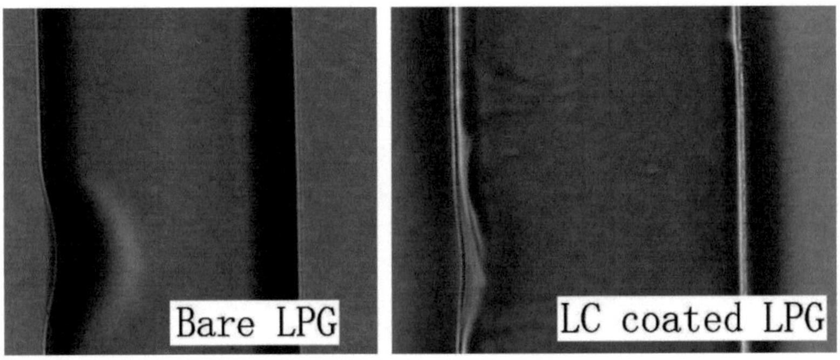

Figure 9. CCD photographs of bare and LC-coated LPG.

OLYMPUS STM6 measuring microscope. The thickness of the LC layer is controlled by the times of daubing. **Figure 10** shows the CCD photographs of overlay thickness of about (a) 400 and (b) 800 nm. A heater box was used to change the temperature of the liquid crystal from 20 to 65°C.

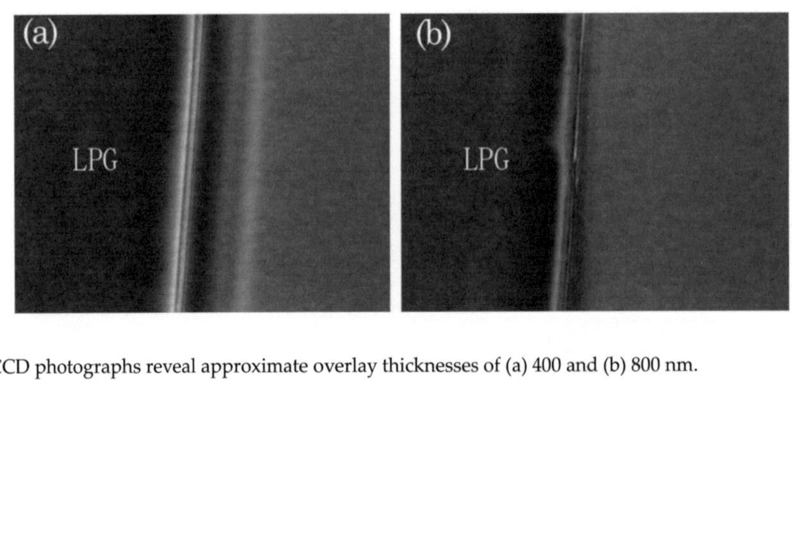

Figure 10. CCD photographs reveal approximate overlay thicknesses of (a) 400 and (b) 800 nm.

Figure 11 shows the transmission spectrum of the ultrathin LC-coated LPG for different temperatures between 20 and 65°C. From the figure, we know that when temperature is increased from 20 to 58°C, wavelength blueshifts slightly (the LP_{0i} mode has been marked with *i*). A large wavelength shift is obtained in the attenuation band of LP_{03} mode when the temperature goes up to about 59°C. The phenomenon of cladding modes reconfiguration appears when temperature in changed from 58 to 60°C.

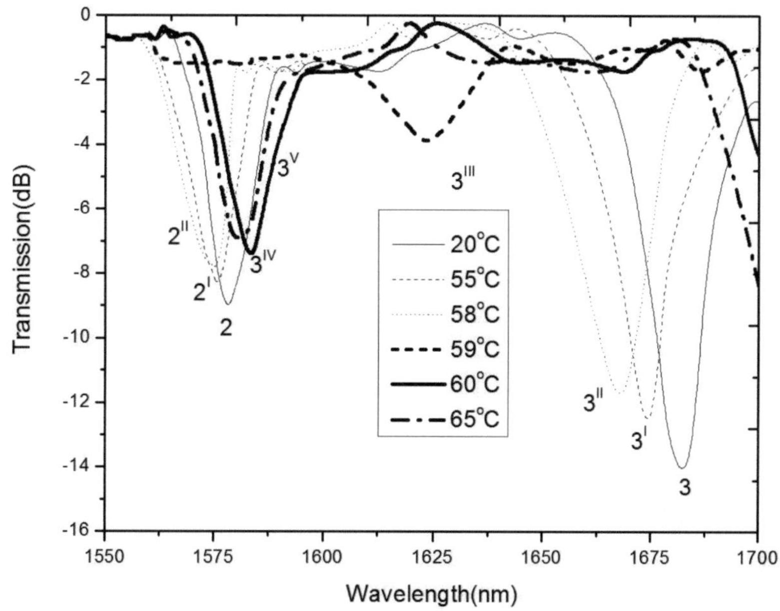

Figure 11. Transmission spectra of an ultrathin HRI-LC–coated LPG for different temperatures in the 20–65°C range.

Figure 12 shows a clearer wavelength shift. As we can see, outside the transition region, the center wavelength of the LP_{02} mode shifts from 1579.7 to 1579.2 nm, and the LP_{03} mode shifts from 1687.5 to 1665.7 nm. However, only LP_{03} mode can be observed within the transition region with its center wavelength changing from 1665.7 to 1583.7 nm, which shifts ~82 nm.

2.4.2. Measurement of the refractive index of LC at different temperatures

To clarify the experimental observations, the temperature dependent refractive index of the LC overlay was also measured. The methods for measuring the refractive index of liquids can be classified into refraction technique and reflection techniques including total reflection [22]. **Figure 13** shows the schematic diagram of the experiment setup with an efficient method developed by Pu et al. [23].

We immerse a detecting tip into air, water, and LC under different magnetic fields. The reflected light power under every condition is measured. **Figure 14** shows the thermo-dependent HRI of LC in the infrared light region based on the experimental data.

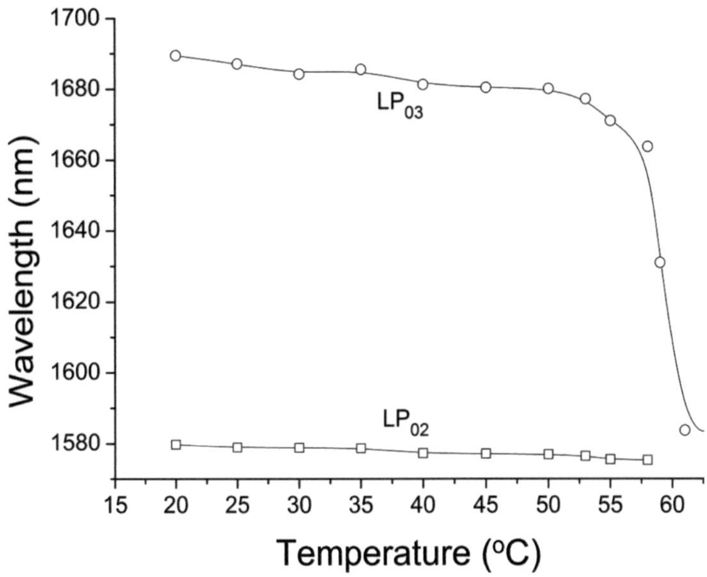

Figure 12. Wavelength shift of LP_{02} and LP_{03} cladding modes for the LPG coated with a HRI layer versus temperature.

Figure 15 shows the theoretically calculated center wavelength shifts with the measured HRI data. The experimental results are in good agreement with the theoretical analysis. For LP_{03} mode, the sensitivity to HRI in the transition region was approximately six times the sensitivity outside the transition region, which means that we can choose this region as high-sensitivity operating region for sensing application.

2.4.3. Electrical spectral tuning of LC-coated LPG at the highly sensitive operating point

A simple scheme for active electrically controlled tunable fiber gratings is illustrated schematically in **Figure 16**. Two parallel substrates are used to fix the LC-coated LPG separated by two 125-μm spacers. The external electric fields are applied through a pair of electrodes under stabilized temperature at 20, 55, 58, 59, 60, and 65°C. The electrical spectra tunability of the LC-coated LPG under different temperatures is shown in **Figure 17**. As we can see that the most sensitive operating point for the device is at 60°C, the maximum tuning range is about 10 nm.

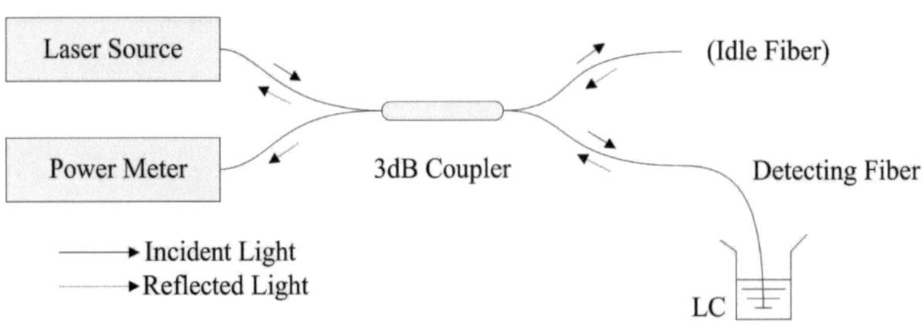

Figure 13. Schematic diagram of experimental setup for measuring the refractive index of the LC.

http://dx.doi.org/10.5772/intechopen.70413

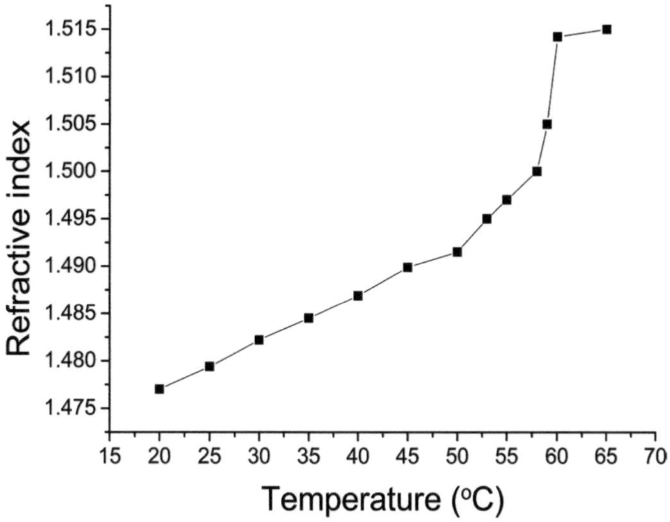

Figure 14. Refractive index versus temperature (°C) of LC MDA-98-3699.

2.5. Experimental investigation of nanosized LC-coated LBMT interferometer

Figure 18 shows the experimental setup. The microfiber taper integrated with the nanosized LC layer was pulled straightly first, and then, the transition regions of the microfiber taper were locally bent to form a fiber interferometer. Two pieces of strip electrodes parallel to each other were pressed against the microfiber with a 4-μm gap maintained by spacers. The profile parameters of the microfiber taper are as follows: the waist diameter $d_0 \approx 3.72$ μm, waist length $L_0 \approx 6.3$ mm, and bent transition region length $Lt \approx 3$ mm. These parameters are given in **Figure 18(b)** together with the critical diameter corresponding to the bent transition region. **Figure 18(c)** shows the SEM image of the taper waist.

The LC material MDA-98-3699 is from Merck [16, 24]. CCD photographs of the bare and LC-coated microfiber taper waist are shown in **Figure 19**. By rubbing the strip electrodes directly on both sides of the microfiber, the planar anchoring of the LC can be obtained.

A comparison of transmitted spectra of a bare and LC-coated LBMT interferometer at 25°C is given in **Figure 20**. Compared with the bare one, a redshift of ~4.9 nm and a decrease of the transmission loss peak of ~2 dB can be observed.

We first measured the transmission spectra of the bare microfiber interferometer at temperatures ranging from 25 to 50°C, and the temperature responses for the attenuation peak at the wavelength of 1529.3 nm are shown in **Figure 21**. Here, we can see from **Figure 21** that the temperature response is only 0.026 nm/°C.

As to thermal tuning, we use the temperature dependence of the ordinary effective index of LC. We measured the ordinary refractive indices of the LC under different temperatures at a wavelength of 1.55 μm [25]. The ordinary refractive index of the LC increases linearly from 1.4794 to 1.4845 when the temperature goes up from 25 to 35°C.

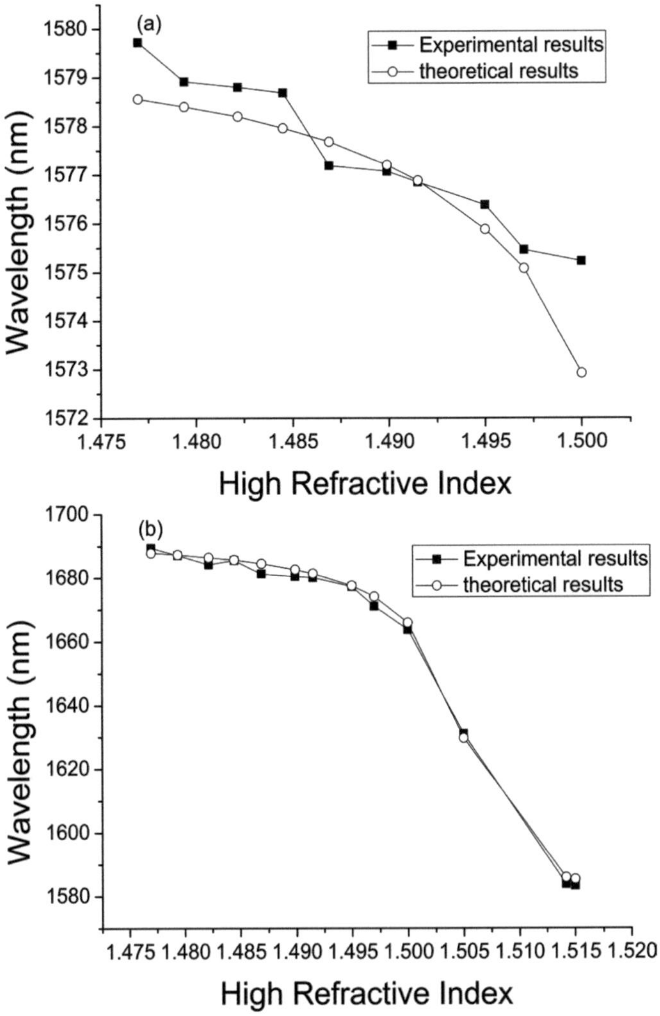

Figure 15. Theoretical and experimental center wavelength of (a) LP_{02} and (b) LP_{03} shift dependence on HRI.

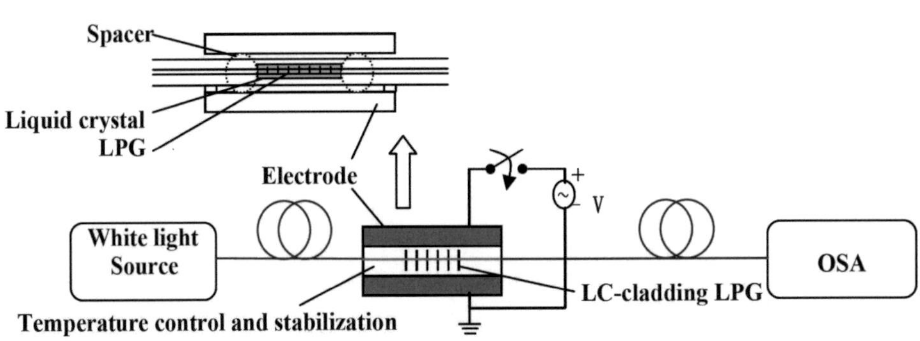

Figure 16. Experimental setup and schematic representation of the LC cladding LPG.

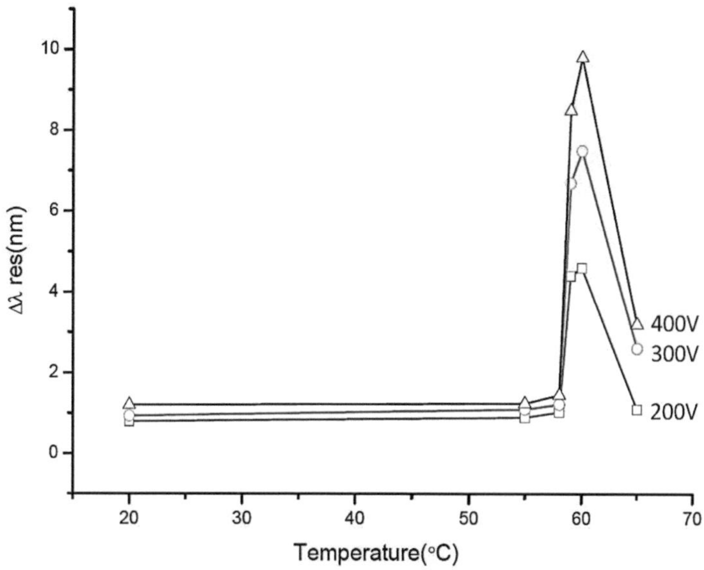

Figure 17. Electrical spectra tenability of the LC cladding LPG at different temperatures.

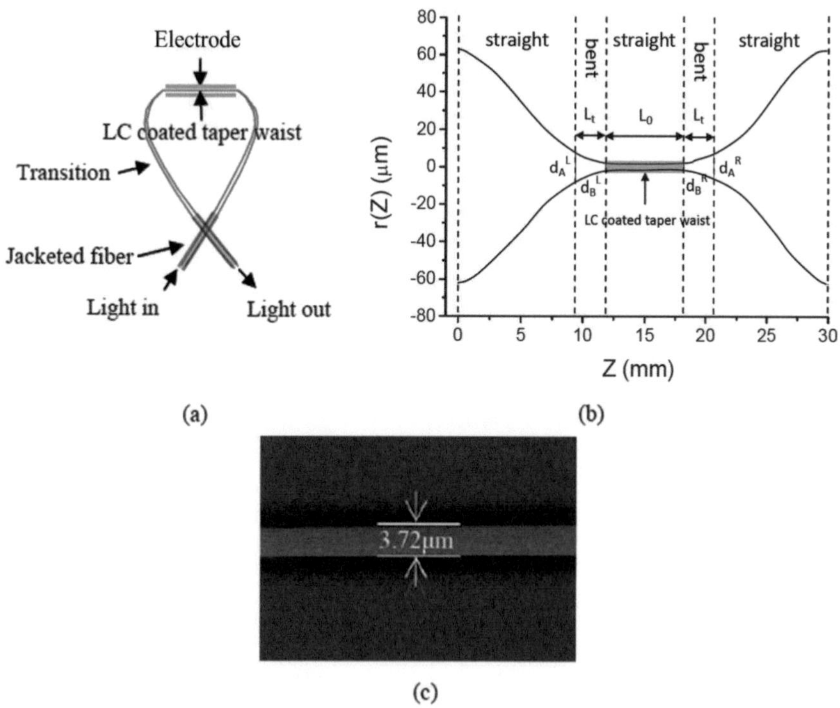

Figure 18. (a) Schematic diagram of the experimental setup for magnetic field tenability test; (b) measured profile of the LBMT having $d_0 \approx 3.72$ μm, $L_0 \approx 6.3$ mm with d_A and d_B corresponding to the bent transition region; (c) SEM image of the microfiber taper waist with a diameter of ~3.72 μm.

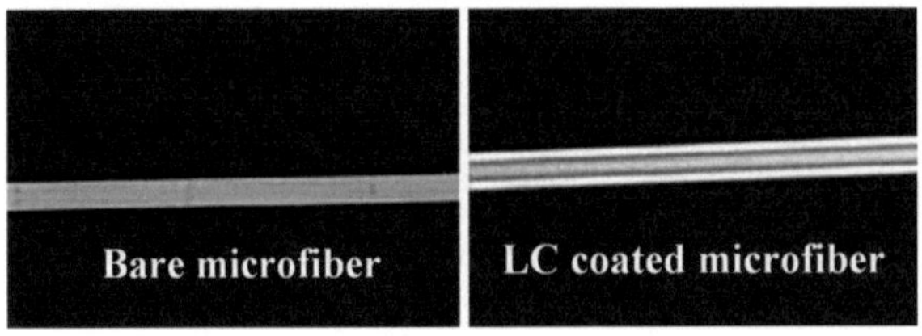

Figure 19. CCD photographs of bare and LC-coated microfiber taper waist.

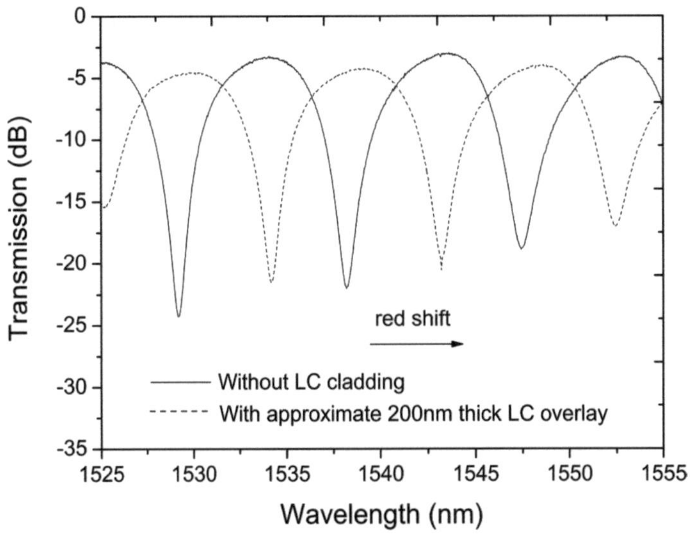

Figure 20. Comparison of the transmission spectra of a bare and an approximate 200 nm LC-coated LBMT interferometer.

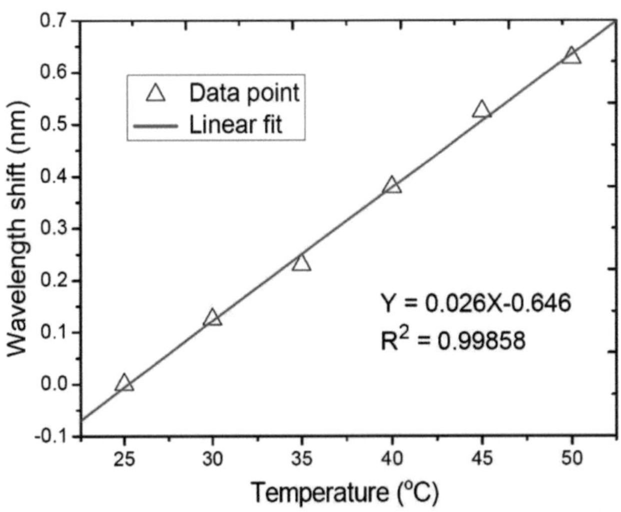

Figure 21. Temperature responses of bare microfiber interferometer for the attenuation peak at the wavelength of 1529.3 nm.

http://dx.doi.org/10.5772/intechopen.70413

A redshift of 10.5 nm for the attenuation peak wavelength of 1534.2 nm was achieved as temperature increased from 25 to 35°C, as shown in **Figure 22**. The tuning efficiency is ~1.05 nm/°C, which is approximately 40 times higher than that of the bare locally bent microfiber interferometer.

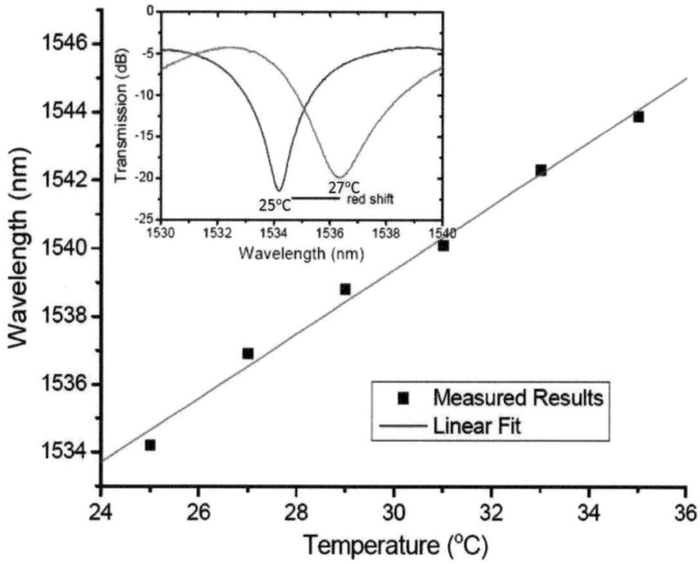

Figure 22. Measured wavelength shifts of a certain attenuation peak of the interferometer at different temperatures.

As to electric tuning, electric fields of 1.25, 2.50, and 3.75 V/μm were applied to the electrodes in our experiment by adjusting the tunable voltage source with the temperature stabilized at 25°C [16]. **Figure 23(a)** is the measured transmission spectra of the same attenuation peak under different electric fields. **Figure 23(b)** is the wavelength of attenuation peak as a function

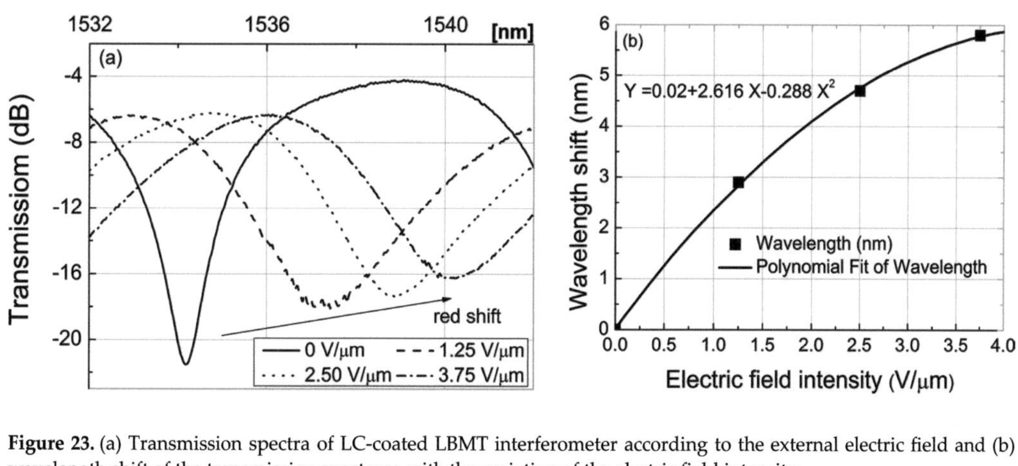

Figure 23. (a) Transmission spectra of LC-coated LBMT interferometer according to the external electric field and (b) wavelength shift of the transmission spectrum with the variation of the electric field intensity.

of the electric field. From the figures, we can see that the wavelength dip experiences some redshift by ~5.8 nm, when the electric field changes from 1.25 to 3.75 V/μm. Due to the uniaxial birefringent properties, the optic axis of nematic LC can be realigned by electric field. The effective refractive index is given by $n_{\text{eff}} = 1/\sqrt{(\cos^2\theta/n_0^2)^2 + (\sin^2\theta/n_e^2)^2}$, when the light polarization direction has an angle θ with respect to the average alignment of the molecules [19]. These experimental results are in good accordance with the above theoretical analysis.

3. Conclusions

In this chapter, two kinds of micro/nano LC layer based tunable optical fiber interferometers are reported. For interferometer 1, experimental results show that shifts of greater than 80 nm in the attenuation bands of the transmission spectrum of LPG can be obtained by changing the refractive index of the LC overlay through the thermo-optic effect. For interferometer 2, due to the high sensitivity of the microfiber taper waist to the LC overlay refractive index, the thermal sensitivity as high as 1.05 nm/°C has been achieved, and the electrical tuning range is ~ 5.8 nm when the external electric field is applied up to 3.75 V/μm. The results suggest that the potential application of the micro/nano LC layer based fiber interferometers as tunable all-fiber photonic devices, such as filter and all-optical switch.

Acknowledgements

This work was supported in part by National Science Foundation of China (NSFC) (51567011, 11264016, 61505073, 61363012); China Scholarship Council (CSC) (201608360081); Science and Technology Project of Jiangxi Province (20151BDH80060, 20151BBG70062, 20161ACB21011, 20161BAB212045); Project of Jiangxi Province Department (160273, 150313).

Author details

Haimei Luo*, Changjing Wang, Yinghua Ji and Wen Yuan

*Address all correspondence to: nclhm2002@hotmail.com

Jiangxi Normal University, Nanchang, Jiangxi, China

References

[1] Hu DJJ, Shum P, Lu C, Sun X, Ren GB, Yu X, Wang GH. Design and analysis of thermally tunable liquid crystal filled hybrid photonic crystal fiber coupler. Optics Communications. 2009;**282**(12):2343-2347. DOI: 10.1016/j.optcom.2009.03.023

[2] Song L, Lee WK, Wang XS. AC electric field assisted photo-induced high efficiency orientational diffractive grating in nematic liquid crystals. Optics Express. 2006;**14**(6):2197-2202. DOI: 10.1364/OE.14.002197

[3] Zyryanov VY, Myslivets SA, Gunyakov VA, Parshin AM, Arkhipkin VG, Shabanov VF, Lee W. Magnetic-field tunable defect modes in a photonic-crystal cell. Optics Express. 2010;**18**(2):1283-1288. DOI: 10.1364/OE.18.001283

[4] Chen Z, Hsiao VKS, Li XQ, Li Z, Yu JH, Zhang J. Optically tunable microfiber-knot resonator. Optics Express. 2011;**19**(15):14217-14222. DOI: 10.1364/OE.19.014217

[5] Luo HM, Wang CJ, Wang XP, Wang YF, Yuan W. The effects of the nanosized high refractive index overlay on tunable long period gratings with normal and reduced cladding diameters. Optik. 2016;**127**:5230-5234. DOI: 10.1016/j.ijleo.2016.02.083

[6] Patrick HJ, Kersey AD, Bucholtz F. Analysis of the response of long period fiber gratings to external index of refraction fraction. Journal of Lightwave Technology. 1998;**16**(9): 1606-1612.

[7] Rees ND, James SW, Tatam RP, Ashwell GJ. Optical fiber long-period gratings with Langmuir–Blodgett thin-film overlays. Optics Letters. 2002;**27**(9):686-688. DOI: 10.1364/OL.27.000686

[8] Wang Z, Heflin JR, Stolen RH, Ramachandran S. Analysis of optical response of long period fiber gratings to nm-thick thin-film coatings. Optics Express. 2005;**13**(8):2808-2813. DOI: 10.1364/OPEX.13.002808

[9] Cusano A, Iadicicco A, Pilla P, Contessa L, Campopiano S, Cutolo A. Mode transition in high refractive index coated long period gratings. Optics Express. 2006;**14**(1):19-34. DOI: 10.1364/OPEX.14.000019

[10] Del Villar I, Matias IR, Arregui FJ. Long-period fiber gratings with overlay of variable refractive index. IEEE Photonics Technology Letters. 2005;**17**(9):1893-1895. DOI: 10.1109/LPT.2005.853283

[11] Del Villar I, Corres JM, Achaerandio M, Arregui FJ, Matias IR. Spectral evolution with incremental nanocoating of long period fiber gratings. Optics Express. 2006;**14**(25):11972-11981. DOI: 10.1364/OE.14.011972

[12] Goure JP, Verrier I. Optical Fibre Devices. 1st ed. London: CRC Press; 2010. 269 p.

[13] Mas S, Marti J, Palaci J. Curvature investigation in tapered fibers and its application to sensing and mode conversion. Optics and Lasers in Engineering. 2015;**74**:109-113. DOI: 10.1016/j.optlaseng.2015.05.011

[14] Luo HM, Li XW, Zou WW, Li X, Hong ZH, Chen JP. Temperature-insensitive micro-displacement sensor based on locally bent microfiber taper modal interferometer. IEEE Photonics Journal. 2012;**4**(2):772-778. DOI: 10.1109/JPHOT.2012.2197606

[15] Luo HM, Li XW, Zou WW, Jiang WN, Chen JP. Modal interferometer based on a C-shaped ultrathin fiber taper for high-sensitivity refractive index measurement. Applied Physics Express. 2012;**5**:012502. DOI: 10.1143/APEX.5.012502

[16] Luo HM, Wang CJ, Ji YH, Yuan W, Zhang GP, Wang YF, Hong ZH. Spectral tuning of a locally bent microfiber taper interferometer with a nanosized liquid crystal overlay. Applied Optics. 2016;**55**(26):7393-7397. DOI: 10.1364/AO.55.007393

[17] Vengsarkar AM, Lemaire PJ, Judkins JB, Bhatia V, Erdogan T, Sipe JE. Long-period fiber gratings as band-rejection filters. Journal of Lightwave Technology. 1996;**14**(1):58-65. DOI: 10.1109/50.476137

[18] Anemogiannis E, Glysisv EN, Gaylord TK. Transmission characteristics of long-period fiber gratings having arbitrary azimuthal/radial refractive index variation. Journal of Lightwave Technology. 2003;**21**(1):218-227. DOI: 10.1109/JLT.2003.808637

[19] Luo HM, Li XW, Wang SS, Chen JP. Temperature stabilized electrically tunable long period gratings coated with nanosized liquid crystal layer. In: Optical Society of America, editor. OFC/NFOEC; 22-26 March 2009; San Diego, CA, USA. IEEE. 2009. p. JThA22

[20] Shankar PM, Bobb LC, Krumboltz HD. Coupling of modes in bent biconically tapered single-mode fibers. Journal of Lightwave Technology. 1991;**9**(7):832-837. DOI: 10.1109/50.85782

[21] Lou JY, Tong LM, Ye ZZ. Dispersion shifts in optical nanowires with thin dielectric coatings. Optics Express. 2006;**14**(16):6993-6998. DOI: 10.1364/OE.14.006993

[22] Schmid K and Penzkofer A. Refractive-index measurements with a Pellin-Broca prism apparatus. Applied Optics. 1983;**22**(12):1824-1827. DOI: 10.1364/AO.22.001824

[23] Pu S, Chen X, Chen Y, Liao W, Chen L, Xia Y. Measurement of the refractive index of a magnetic fluid by the retroreflection on the fiber-optic end face. Applied Physics Letters. 2005;**86**:171904. DOI: 10.1063/1.1905808

[24] Luo HM, Li XW, Li SG, Chen JP. Analysis of temperature-dependent mode transition in nanosized liquid crystal layer-coated long period gratings. Applied Optics. 2009; **48**(25):F95-F100. DOI: 10.1364/AO.48.000F95

[25] Yin SZ, Chung KW, Zhu X. A highly sensitive long period grating based tunable filter using a unique double-cladding layer structure. Optics Communications. 2001;**188** (5-6):301-305. DOI: 10.1016/S0030-4018(00)01172-X

Chapter 10

Liquid-Crystal-Based Phase Gratings and Beam Steerers for Terahertz Waves

Ci-Ling Pan, Chia-Jen Lin, Chan-Shan Yang,
Wei-Ta Wu and Ru-Pin Pan

Additional information is available at the end of the chapter

http://dx.doi.org/10.5772/intechopen.70449

Abstract

We review our theoretical and experimental studies on a class of liquid crystal (LC) photonic devices, i.e., terahertz (THz) phase gratings and beam steerers by using LCs. Such gratings can function as a THz polarizer and tunable THz beam splitters. The beam splitting ratio of the zeroth-order diffraction to the first-order diffraction by the grating can be tuned from 10:1 to 3:5. Gratings with two different base dimensions were prepared. The insertion loss is lower by approximately 2.5 dB for the one with the smaller base. The response times of the gratings were also studied and were long (tens of seconds) as expected because of the thick LC layer used. Accordingly, the devices are not suitable for applications that require fast modulation. However, they are suitable for instrumentation or apparatuses that require precise control, e.g., an apparatus requiring a fixed beam splitting ratio with occasional fine tuning. Schemes for speeding up the device responses were proposed. Based on the grating structure, we also achieved an electrically tunable THz beam steerer. Broadband THz radiation can be steered by 8.5° with respect to the incident beam by varying the driving voltages to yield the designed phase gradient.

Keywords: liquid crystals, liquid crystal devices, diffraction, phase grating, grating arrays, polarizer, beam splitter, submillimeter wave, THz radiation, tunable circuits and devices, ultrafast optics, beam steering

1. Introduction

Terahertz (THz) science and technology have advanced significantly over the last 3 decades. Applications are abundant in topics such as material characterization, data communication, biomedicine, 3D imaging, and environmental surveillance [1–5]. These developments were hampered as crucial quasi-optic components such as phase shifters [6–9], phase gratings [10–12], modulators [13, 14], attenuators [15], polarizers [16, 17], and beam splitters [18–21] in the THz range are still relatively underdeveloped.

To control the properties of electromagnetic waves at all wavelengths, periodic structures such as gratings are frequently employed. In the THz frequency range, gratings with various periods have been used for tailoring few-cycle pulses [22]. Gratings have also been used as couplers and filters [23]. Tunable THz devices based on an optically and electrically controlled carrier concentration in quantum-well structures have been demonstrated. However, these devices have a limited range of tunability and must be operated at cryogenic temperatures far below room temperature [24–26]. The potential of gratings with liquid-crystal-enabled functionalities was recognized 2 decades ago [27]. Recently, the focus has been on various tunable THz devices, such as phase shifters, filters, and switches that are controlled electrically or magnetically, employing liquid crystals, primarily nematic liquid crystals (NLCs) [6–10, 15, 16, 18, 27–33]. Previously, we demonstrated a magnetically controlled phase grating for manipulating THz waves [10]. This is based on magnetic-field-induced birefringence of the NLCs employed [34]. Nonetheless, electrically controlled phase gratings are generally regarded as desirable for many applications. Therefore, we also proposed and demonstrated an electrically controlled phase grating involving NLCs for THz waves [11]. However, the theoretical analysis was not described in detail and the issue of insertion loss was not touched upon in the previous communication.

Besides, there is an urgent need for THz beam steering devices for scanning the THz beam over the surface of targets to get full topological and spectral information, a metamaterial-based beam steerer has been demonstrated and achieved a maximal deflection angle of 6° [35]. Other groups employed highly-doped semiconductors, such as Indium antimonide (InSb) [36] and GaAs [37], so that the propagation properties of surface plasmons mode in highly-doped semiconductor slits can be tailored by changing the carrier density there [36]. On the other hand, the development of a reconfigurable THz antenna [38], which can electrically steer the THz beam or vary the beam shapes, are useful for applications, such as adaptive wireless, satellite communication networks, and automobile radar systems. The use of LC to construct a phase array for beam steering in millimeter wave range has also been reported recently [39].

In this chapter, we report our comprehensive experimental studies on a phase grating for THz waves. In particular, we analyzed the insertion loss in such gratings and devised an approach for improving the loss by 2.5 dB over existing designs. Further, we demonstrated an electrically tunable phase shifter array to modulate the phase of THz beam. By applying different voltages on each part of the phase array, we can achieve a gradient in phase shift. Finally, it is shown that the incident THz wave can be steered toward a selected direction.

2. Theoretical and experimental methods

2.1. Operation principles of the phase grating

We designed a binary phase grating consisting of alternating sections of two materials (fused silica and LCs) with different refractive indices. **Figure 1** shows the schematic of a generic

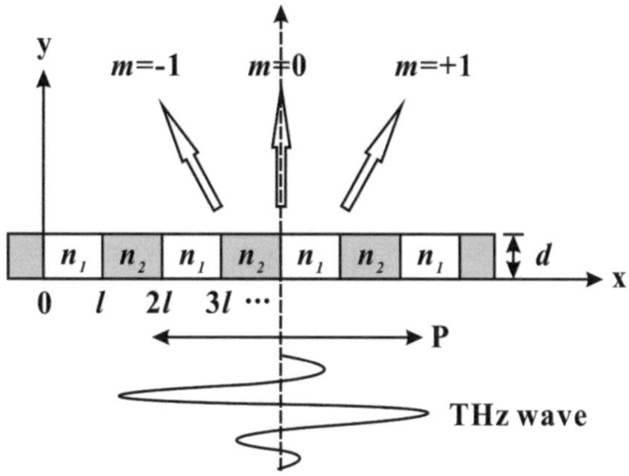

Figure 1. Schematic of a generic binary phase grating consisting alternating sections with refractive indices of n_1 and n_2. The width and height of each section of the phase element is respectively, l and d. P is the polarization direction of the THz wave.

binary phase grating. The grating is periodic along the x-direction. The THz wave is assumed to be polarized along the x-axis and propagates along the y-direction. Each section of the grating can be considered a retarder that introduces a phase shift. The Jones matrix [40] associated with a particular retarder can be written as

$$\widehat{R} = \begin{bmatrix} e^{i\delta_x} & 0 \\ 0 & e^{i\delta_y} \end{bmatrix}, \tag{1}$$

where δ is the phase retardation and a function of x. The Jones vector associated with the incident THz field is given by

$$E_i = E\begin{pmatrix} 1 \\ 0 \end{pmatrix}, \tag{2}$$

where E is a constant amplitude factor. The transmitted field E_t emerging from the retarder is then

$$E_t = \widehat{W} \cdot E_i = E\begin{pmatrix} e^{i\delta_x} \\ 0 \end{pmatrix}. \tag{3}$$

For our design, we set $\delta_x = n_N kd$, where $n_N = n_1 + \kappa_1$ or $n_2 + \kappa_2$ is the complex refractive index of the corresponding section, k is the wave number, and d is the thickness of the binary phase grating. The total transmitted field E_T is then the superposition of the field transmitted through all the alternating phase elements. We further write $E = E_0 e^{iky\sin\phi}$, where ϕ is the diffraction angle. Therefore, the total transmitted field can be written as

$$\begin{aligned} E_T(\varphi) = & \sum_{m=0}^{even} \int_{ml}^{(m+1)l} E_0 e^{iky\sin\varphi} e^{i(n_1+\kappa_1)kd} dy \\ & + \sum_{m=1}^{odd} \int_{ml}^{(m+1)l} E_0 e^{iky\sin\varphi} e^{i(n_2+\kappa_2)kd} dy. \end{aligned} \tag{4}$$

In Eq. (4), l is the width of each section (all sections are assumed to have identical widths) of the grating. The diffraction intensity $I(\phi)$ can then be expressed as

$$I(\varphi) = E_T(\varphi) \cdot E_T^{*}(\varphi). \tag{5}$$

For an ideal binary phase grating, the diffraction efficiency η_m of the mth-order diffracted wave, defined as the intensity ratio of the diffracted beam to that of the incident beam, is given by

$$\eta_m = \frac{1}{\Lambda^2}\left|\int_{-\Lambda/2}^{\Lambda/2} e^{i\delta} e^{-i(2\pi m y/\Lambda)} dy\right|^2, \tag{6}$$

where δ is the x-dependent phase shift of the grating with grating period Λ [18, 41]. Following [41], we can write

$$\eta_m = \begin{cases} \cos^2(\Delta\Gamma/2) \ \text{if } m = 0 \\ [(2/m\pi)\sin(m\pi/2)]^2 \sin^2(\Delta\Gamma/2) \ \text{if } m \neq 0 \end{cases}, \tag{7}$$

where $\Delta\Gamma$ is the relative phase difference between two adjacent sections in the phase grating. For $\Delta\Gamma = (2\,N + 1)\pi$ (where N is an integer), the diffraction efficiencies of the odd orders ($m = \pm1, \pm3, \pm5, \ldots$) are maximal. Eq. (7) reveals that the diffraction efficiency of the third order $\eta_{\pm3}$ (4.5%) is nine times smaller than that of the first-order $\eta_{\pm1}$ (40.5%). Therefore, in this study, we considered only the zeroth and the first orders of the diffracted beam. Eqs. (6) and (7) were used as guides for designing the parameters of the grating. In practice, insertion loss causes the experimentally observed efficiencies to be lower than expected.

Because of the THz wavelength and THz beam size, the grating can have only a finite number of grooves. Further, the number of grooves N affects the angular resolution of the diffracted beam [42]. The angular width $\Delta\phi$ is given by

$$\Delta\varphi = \frac{\lambda}{N\Lambda\cos\varphi}. \tag{8}$$

Let the frequency of the THz be centered at 0.3 THz or a wavelength of 1 mm, we designed Λ to be 2 mm in our devices. The relative phase difference between adjacent sections is designed to be π. Therefore, the diffraction angle for the first order is 30°. **Figure 2** shows a plot of the diffraction efficiency as a function of the diffraction angle (for the first order). It illustrates clearly the angular resolution achievable for a grating with 10 periods and a grating with 40 periods. According to Eq. (8), the angular widths for the first-order diffracted wave for 10 and 40 periods are 3.3 and 0.8°, respectively.

When the relative phase difference between adjacent groves is tuned between π and 2π, the diffracted signals between the zeroth order and the first order have the maximal tunable range. This is illustrated in **Figure 3** for a grating with 10 periods. For $\Delta\Gamma = \pi$, the diffracted signal lies mainly in the first-order, and $\eta_{\pm1} \approx 0.4$. By contrast, for $\Delta\Gamma = 2\pi$, the diffracted signal mostly

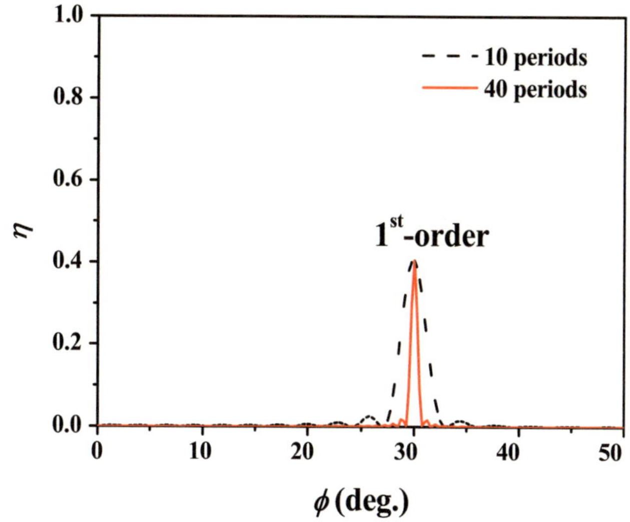

Figure 2. The diffraction efficiency of a phase grating with 10 (black dashed curve) and 40 periods (red solid line) is plotted as a function of the diffraction angle (for the first order).

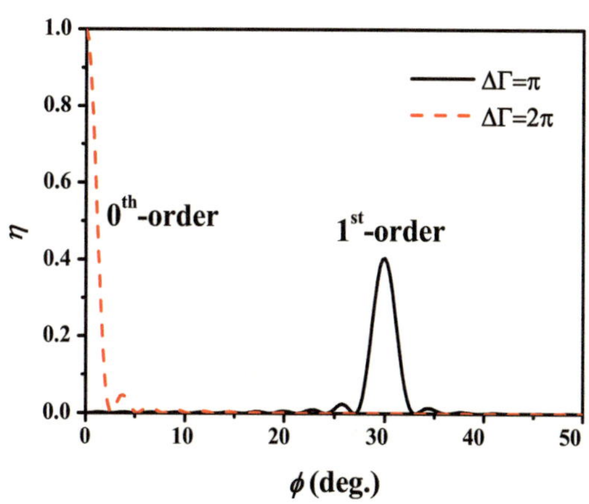

Figure 3. Diffraction efficiencies of a 10-period phase grating are plotted as a function of the diffraction angle for relative phase differences of π and 2π.

concentrated in the zeroth order, and $\eta_0 \approx 1$. These predictions are valid provided the insertion losses can be ignored.

2.2. Construction of the grating

The design of the grating was based on the structure of the electrically controlled THz phase grating reported in our previous study [11]. This is shown schematically in **Figure 4**. The incident THz wave was assumed to be polarized in the y-direction. Orientations of the LC molecules for two possible configurations are shown (See **Figure 4**). The device was designed

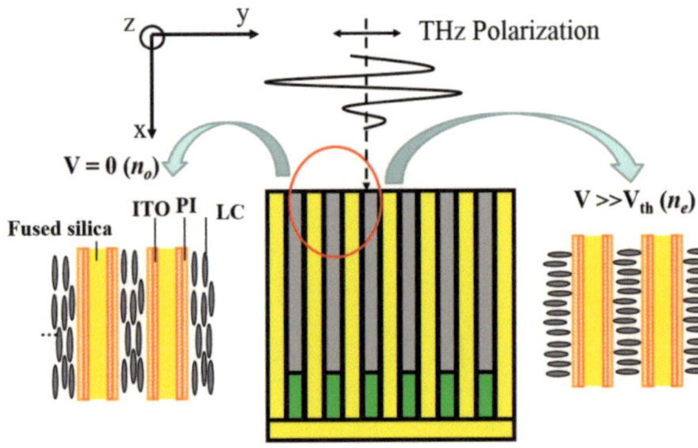

Figure 4. Structure of the electrically controlled THz phase grating using nematic liquid crystals. ITO: Indium tin oxide; PI: Polyimide; LC: Liquid crystal molecules; n_o: Ordinary index of refraction; n_e: Extraordinary index of refraction; V_{th}: Threshold voltage.

such that the frequency band of 0.3–0.5 THz would exhibit the highest zeroth-order diffraction efficiency.

Parallel grooves with a period of 2.0 mm, width of 1.0 mm, and groove depth of 2.5 mm were formed by stacking indium tin oxide (ITO)-coated fused silica substrates; the refractive index of these substrates is 1.95 in the sub-THz frequency region (0.2–0.8 THz). The surfaces of the fused silica substrates were coated with polyimide (SE-130B, Nissan) and then rubbed for homogenous alignment. The grooves were filled with NLCs (E7, Merck) and sealed with a sheet of fused silica coated with N,N-dimethyl-N-octadecyl-3-aminopropyltrimethoxysilyl chloride. At room temperature, E7 is a birefringent material with positive dielectric anisotropy. The LC molecules tend to be aligned parallel to the direction of the applied electric field when the applied voltage is greater than a threshold voltage. The effective refractive index of E7 [43], n_{eff} can be tuned from the refractive index for ordinary waves ($n_o = 1.58$) to that for extraordinary waves ($n_e = 1.71$) by varying the applied voltage. A stack of ITO-coated fused silica plates with dimensions identical to those of the grating was prepared as a reference. Bases of the phase grating with two different dimensions ($h_1 = 17.5$ mm and $h_2 = 7.5$ mm) were fabricated for analyzing the effect of dimension of the base on insertion losses of the phase grating.

2.3. Transmission measurements

A photoconductive (PC) antenna-based THz time-domain spectrometer (THz-TDS) [32, 44], was used for measuring the zeroth-order diffraction spectra of the device. Briefly, the pump beam from a femtosecond mode-locked Ti:sapphire laser was focused on a dipole antenna fabricated on LT-GaAs for generating a broadband THz signal, which was collimated and collected through the THz phase grating by using off-axis parabolic gold mirrors. A pair of parallel wire-grid polarizers (GS57204, Specac) was placed before and after the device under test. The zeroth-order diffraction of THz radiation was coherently detected by another PC antenna of the same type as that of the THz-TDS and grated by ultrafast pulses from the same laser.

In the second set of experiments, the broadband THz signal was filtered by using a metallic hole array to obtain a quasi-monochromatic wave centered at 0.3 THz and with a line width of 0.03 THz [45]. The diffraction pattern of this beam produced by a grating with various nematic LC orientations was detected and mapped by a liquid-helium-cooled Si bolometer, which was at a distance of 20 cm from the device and located on a rotating arm that could be swung with respect to the fixed grating. The bolometer had an aperture with a diameter of approximately 2.5 cm.

2.4. Insertion loss

To estimate the insertion loss of the THz grating, we regarded the device as a stack of parallel-plate waveguides. The ITO conductive film was not an ideal conductor. We recently showed that for a conductivity of 1500–2200 $\Omega^{-1}\cdot\text{cm}^{-1}$, the complex refractive indices of ITO are 20−70 for n and 20−70 for κ in the THz range [46]. On the basis of the manner of waveguide excitation, we can assume the mode of the propagating THz wave is transverse. The cutoff frequency of the parallel-plate waveguides can be written as $f_c = c/2\,nl$, where c is the velocity of light, l is the distance between two conductive layers, and n is the refractive index of the dielectric material within a waveguide. The attenuation constants α_c and α_d corresponding to conductor loss and dielectric loss, respectively, are given by [47, 48]

$$\alpha_c = \sqrt{\frac{\pi f \varepsilon}{\sigma_c}} \frac{2}{l\sqrt{(f_c/f)\left[1-(f_c/f)^2\right]}}, \tag{9}$$

and

$$\alpha_d = \frac{\pi f \tan\delta\sqrt{\varepsilon_r}}{c\sqrt{1-(f_c/f)^2}}, \tag{10}$$

where f is the frequency, ε is the permittivity of the dielectric material, σ_c is the conductivity of the electrode, tanδ is the loss tangent, and ε_r is the relative permittivity of the dielectric material. In practical applications, the insertion loss of the THz grating should be minimal.

2.5. Electrically controlled steering of the THz beam

We have also designed an electrically tunable phase shifter array which can function as the THz beam steerer. **Figure 5** shows the structure of the phase shifter array, which is constructed by alternately stacking a number of NLC layers and electrodes. Voltage sources are connected to the electrodes to apply control voltages to each NLC layer. The effective refractive index, n_{eff} (V), of each NLC layer can be electrically tuned by applying appropriate voltages. The polarization of the THz wave was assumed to be along the z-direction while the wave was normally incident to the device.

The device was designed such that, when no voltage was applied, the NLC molecules are aligned along the y-direction. In this case, the effective refractive index equals to that of the

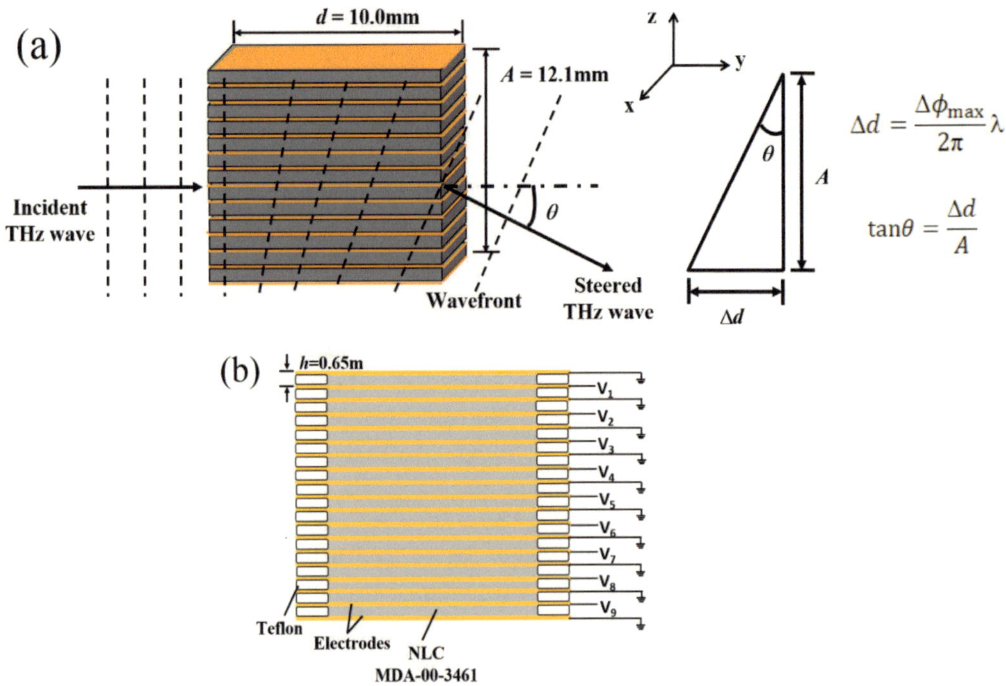

Figure 5. Schematic structure of the electrically controlled THz phase shifter array for beam steering. (a) Set-up of the beam steering experiment. The relationship between the steering angle, θ, optical delay length, Δd, and aperture, A, are shown. (b) Structure of the device with arrangement for voltage applied to each layer. Dimensions of the structure are also shown.

ordinary component of light in the LC, n_o. If sufficient control voltage is applied, the NLC molecules will orientate toward the direction parallel to the polarization direction of the THz wave (z-direction). The effective refractive index then equals to that of the extraordinary component of light in the LC, n_e. The traversing time, which the THz wave takes to pass through the NLC layers, can be changed by applying voltages. The corresponding phase shift $\Delta\phi(V)$ in the applying voltage V is given by

$$\Delta\phi(V) = kd\left(n_o - n_{eff}(V)\right), \tag{11}$$

where k is the wave number in free space and d is the propagation length of the NLC layer. When a certain phase gradient was created across the aperture of the device by adjusting the phase shift in each NLC layer, the wavefront of the transmitted wave was inclined against the aperture. According to the limited voltage source channels, we divided two NLC layers as a block. The steering angle θ can be determined by the aperture size A and the optical length delay Δd between the top NLC block and the bottom NLC block. The optical length delay Δd was according to the phase shift between the top and bottom NLC blocks $\Delta\phi_{max}$. Therefore, the relationship between steering angle θ, optical length delay Δd, and the phase shift $\Delta\phi_{max}$, was shown in **Figure 5(a)**, and can be written as,

$$\Delta d = \frac{\Delta\phi_{\max}}{2\pi}\lambda, \text{ and } \tan\theta = \frac{\Delta d}{A}, \quad (12)$$

where λ is the corresponding wavelength of THz wave. Accordingly, the THz wave can be steered by the control voltage. The phase shift $\Delta\phi_i$ in the certain ith NLC block, in our case can be written as,

$$\Delta\phi_i = (i-1)k(2h)\tan\theta, \quad (13)$$

where $2h$ was the thickness of the ith NLC block and θ was the steering angle against the normal of the aperture. We can see the construction of the device in **Figure 5(b)**, in which each NLC block includes two NLC layers and two electrodes.

In this work, we used the 550-μm-thick Teflon sheet as the spacer and the 100-μm-thick copper foil as the electrode. The copper foil was coated with PI Nissan SE-130B on both sides and rubbed for homogeneous alignment along y-direction before applying the voltage. The 18 NLC layers and 19 electrodes were stacked up alternately. The total thickness of the device was 12.1 mm, which corresponded to the size of the aperture, A, along z-direction. The size of the aperture along x-direction was designed to be 20.0 mm, and the propagation length, d, of the THz wave was designed to be 10.0 mm. Control voltage sources connected to the electrodes provided 1 kHz-sinusoidal waves to the NLC layers.

The threshold voltage V_c can be estimated by $V_c = \pi\ (k/\varepsilon_0\Delta\varepsilon)^{1/2} = 1.20\ V_{rms}$. The complex refractive indices of NLC MDA-00-3461 for ordinary and extraordinary in THz range are $n_o = 1.54$, $n_e = 1.72$, $\kappa_o = 0.03$, and $\kappa_e = 0.01$, respectively [49]. At a frequency of 0.3 THz, the

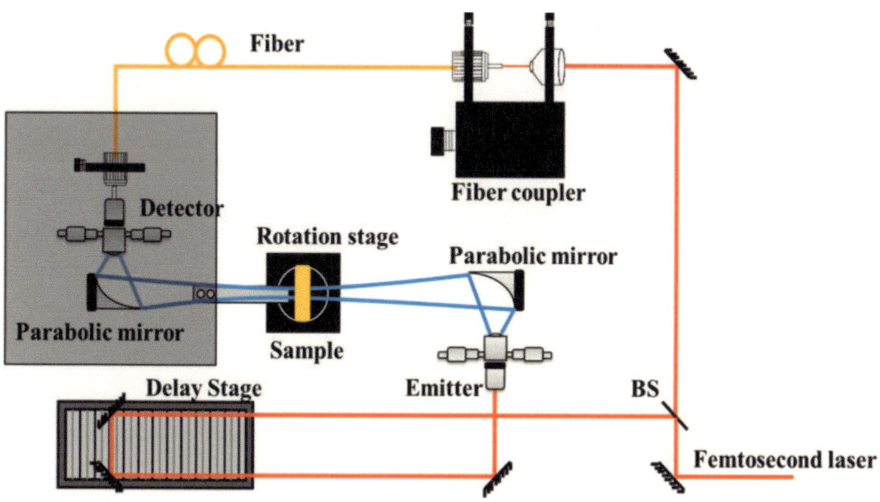

Figure 6. Improved THz-TDS. Probe beam is guided with a 1 m long optical fiber directly to the antenna. The detection assembly is located on a rotatable arm and can be moved without changing the optical path.

estimated maximum phase shift applied to a propagation wave passing through NLC layer would be 11.31 rad, which is calculated by Eq. (11) and the refractive indices of NLC MDA-00-3461. The maximum steering angle was estimated to be approximately 9° from Eq. (13) taking the maximum phase shift and aperture size into account.

For studying THz beam steering, we modified the THz-TDS by employing a 1 m-long single mode fiber (F-SF-C-1FC, from Newport Corp.) to guide the femtosecond laser directly to the detecting antenna. This way, the optical path remained fixed when the detecting arm was moved as the THz beam was steered. The schematic diagram of the setup is shown in **Figure 6**. The detection assembly was 20 cm away from the device and located on a rotation arm that can be swung with respect to the fixed device. This system was much more stable and convenient to use than the one employing the bolometer.

3. Results and discussions

3.1. Phase grating

We studied zeroth-order diffracted THz pulses by the phase grating for both ordinary and extraordinary waves were described in Ref. [11].

Experimentally, the diffraction efficiency of diffracted signals, η, in the frequency domain was determined by normalizing the diffracted signals in the frequency domain to the diffracted signals of the reference phase grating. To compare, a finite-difference time-domain (FDTD) algorithm (RSoft Design Group, Inc.) was used for simulating the diffraction of THz waves by a phase grating.

In the FDTD simulation, we analyzed the grating structure as a stack of rectangular-shaped waveguides. Neglecting conductive and magnetic loss of the materials involved, the Maxwell-Faraday and Maxwell-Ampere equations can be expanded in the Cartesian coordinates as

$$\begin{aligned} \frac{\partial H_x}{\partial t} &= \frac{1}{\mu_0}\left(\frac{\partial E_y}{\partial z} - \frac{\partial E_z}{\partial y}\right), \\ \frac{\partial H_y}{\partial t} &= \frac{1}{\mu_0}\left(\frac{\partial E_z}{\partial x} - \frac{\partial E_x}{\partial z}\right), \\ \frac{\partial H_z}{\partial t} &= \frac{1}{\mu_0}\left(\frac{\partial E_x}{\partial y} - \frac{\partial E_y}{\partial x}\right), \end{aligned} \tag{14}$$

$$\begin{aligned} \frac{\partial E_x}{\partial t} &= \frac{1}{\varepsilon_0}\left(\frac{\partial H_z}{\partial y} - \frac{\partial H_y}{\partial z}\right), \\ \frac{\partial E_y}{\partial t} &= \frac{1}{\varepsilon_0}\left(\frac{\partial H_x}{\partial z} - \frac{\partial H_z}{\partial x}\right), \\ \frac{\partial E_z}{\partial t} &= \frac{1}{\varepsilon_0}\left(\frac{\partial H_y}{\partial x} - \frac{\partial H_x}{\partial y}\right), \end{aligned} \tag{15}$$

where H_x, H_y, H_z, E_x, E_y, and E_z are components of the magnetic field and electric field, respectively. We set the refractive indices of the waveguides as those of fused silica and LCs.

http://dx.doi.org/10.5772/intechopen.70449

The two types of waveguides were made to alternate (black and yellow sections in **Figure 4**) in the structure. The grating device extended from $z = 0$ to $z = 2500$ μm. The incident wave was set to have a Gaussian shape, and the beam size was 19.0 mm, as large as the device aperture. The wave was normally incident in the z-direction from a source at $z < 0$ and was polarized in the x-direction. The dimensions of the grid in the FTD analysis were 10 μm × 10 μm in the xy-plane, while the time step was 1.67×10^{-14} s. **Figure 7** shows the simulation results for an incident THz wave that is yet to enter the device (a), THz waves at two positions in the grating (b and c), and a THz wave that has emerged from the grating. The false color (in 256 levels) indicates the strength of the electric field at a given spatial point. In **Figure 7(b, c)**, we show THz waves transmitted through the grating with different velocities at different x-positions because of the difference in the refractive indices. After the THz signal emerged from the device (**Figure 7(d)**), we set the time monitor to obtain the superposition signal.

To illustrate performance of the grating, experimental and FDTD simulation results of the zeroth-order diffraction efficiencies of the phase grating operated at four values of applied voltage are plotted as a function of frequency in **Figure 8** (reproduced from [11] with permission). Note that the experimentally measured diffraction efficiency was the highest near 0.3 THz, in agreement with the designed frequency. For an ordinary wave at 0.3 THz, the phase difference between fused silica and E7 was close to 2π. Therefore, the transmission of the

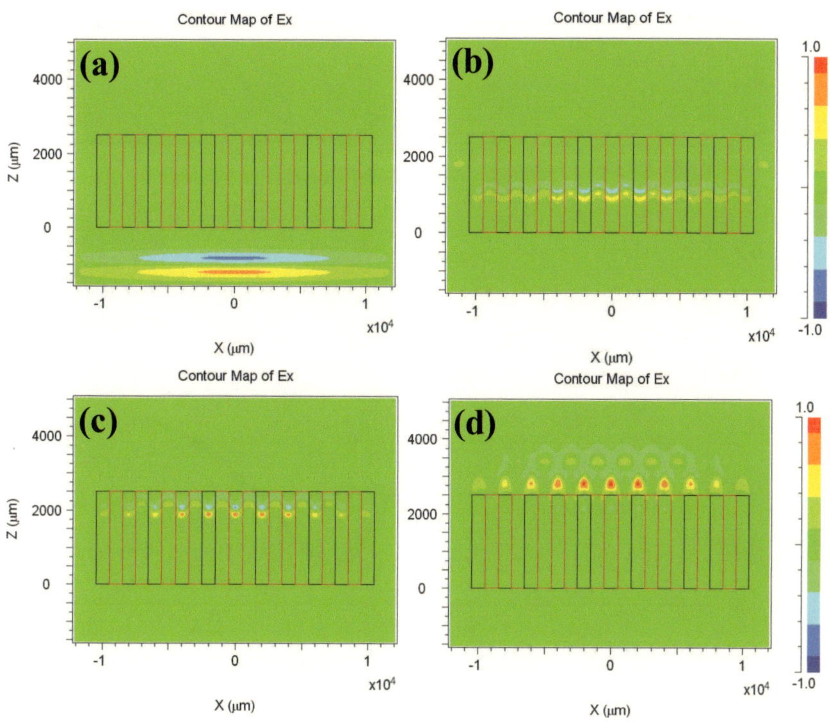

Figure 7. Simulation results for (a) an incident THz wave that is yet to enter the device, (b) and (c) THz waves at two positions in the grating, and (d) a THz wave that has emerged from the grating.

grating was higher. The THz wave was mainly concentrated in the zeroth order. By contrast, for extraordinary waves, the phase difference was close to π. Furthermore, the diffraction efficiency was lower for the zeroth order because the THz wave was mostly diffracted into the first order.

The experimental and FDTD simulation results are in general agreement. In **Figure 8(a, b)**, there are, however, some discrepancies in efficiencies and peak positions. This is expected as the thickness of the fused silica plates in the grating assembly varies by ± 0.1 mm. To check, we calculated diffraction efficiencies of gratings with dimensions of 2.4, 2.5, and 2.6 mm using the

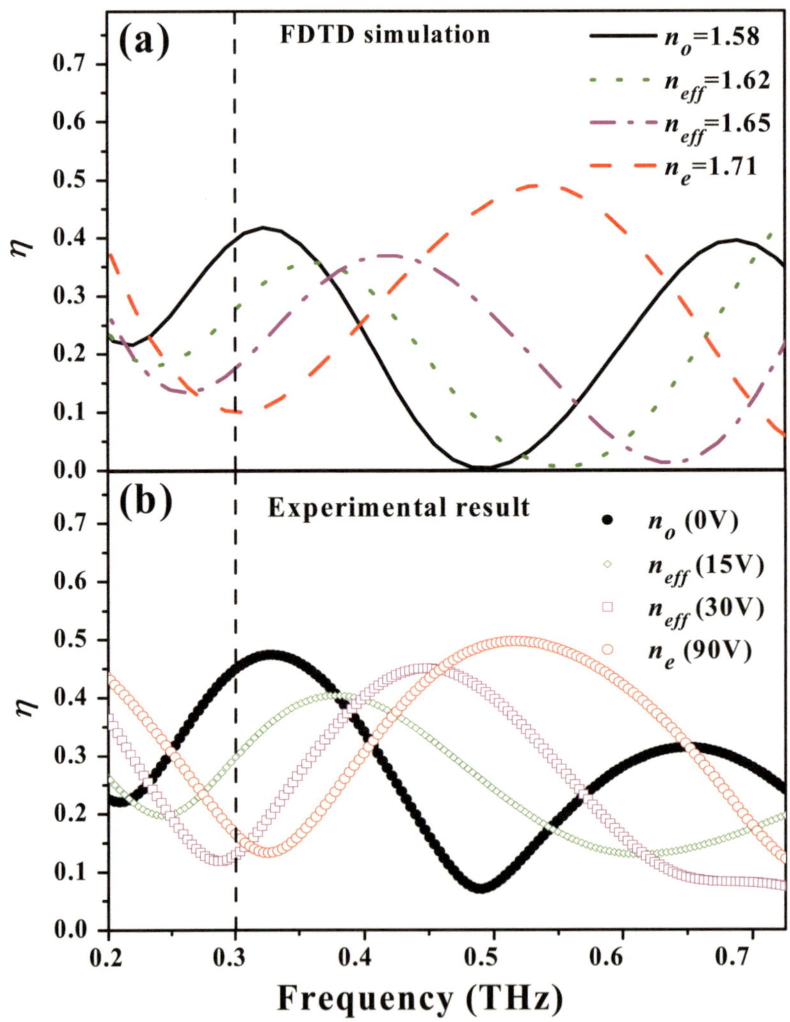

Figure 8. (a) FDTD simulation and (b) experimental results of the frequency dependence of the zeroth-order diffraction efficiencies of the phase grating operated at four values of applied voltages. (**Figure 3** of Ref. [11], reproduced by permission of the authors and IEEE).

FDTD software for the o-ray and e-ray, respectively. Further, a structure with random arrangement of sections with deviations of 0.1 mm centered around 2.5 mm was also studied. The results are shown in **Figure 9**. Clear shifts are observed in the curves. Therefore, we inferred that the experimental results are reliable.

Because of the periodically arranged ITO films in the grating, our device could be considered a wire-grid polarizer for the THz wave. Only a THz wave polarized perpendicular to the grooves could pass through the electrically tuned phase grating. The measurement result is shown in **Figure 10**. The extinction ratio of the device shows the ratio of the transmitted THz signals polarized parallel and perpendicular to the grooves is better than 1:100 at ~ 0.3 THz.

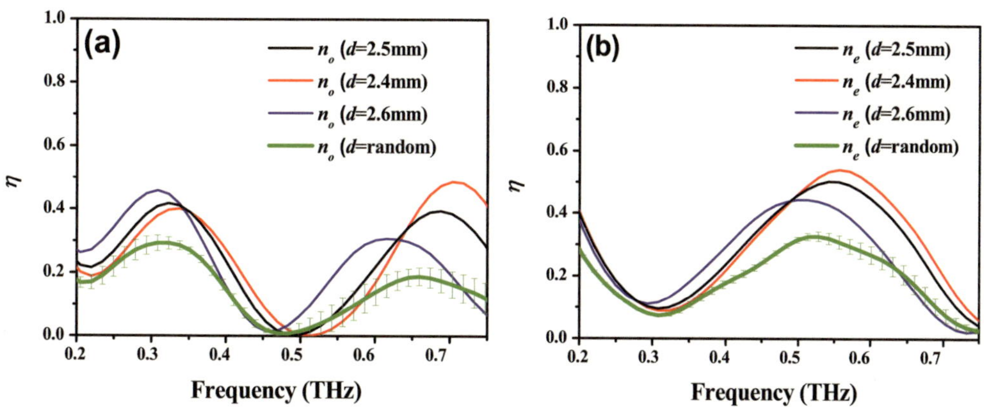

Figure 9. FDTD simulation result showing the diffraction efficiency as a function of the frequency for the phase grating of different thicknesses (see text): (a) o-wave and (b) e-wave.

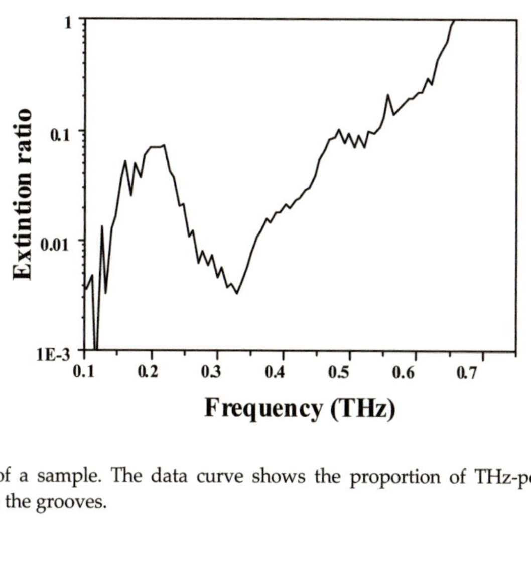

Figure 10. Extinction ratio of a sample. The data curve shows the proportion of THz-polarized transmitted signals parallel and perpendicular to the grooves.

3.2. Bolometer measurement results

In **Figure 11**, we present the intensity profiles of the diffracted 0.3 THz beam polarized in the y-direction. Data are shown for the grating biased from 0 to 90 V. The corresponding effective indices of refraction vary from 1.58 to 1.71. A diffraction maximum was detected at ϕ = 30°, which corresponds to the first-order diffracted beam predicted by the grating equation $\Lambda \sin\phi = m\lambda$, where Λ is 2.0 mm and the wavelength λ is 1.0 mm for the 0.3 THz wave. The measured diffraction efficiencies for the zeroth and first orders are in accord with the predictions of Eq. (5), considering the finite dimensions of the grating and the acceptance angle of the bolometer (±3°).

When the E7 molecules were aligned such that the refractive index was n_o, the phase difference was close to 2π. Most of the THz signal propagated in the direction of the zeroth-order diffraction. Experimentally, the diffraction efficiencies were determined to be 0.62 and 0.06 for the zeroth and first orders, respectively. The diffraction efficiencies were tuned by increasing the applied voltage (V_{appl}) gradually. When the refractive index of E7 was varied from 1.58 to 1.71, the diffraction efficiency of the zeroth order decreased; by contrast, the diffraction efficiency of the first order increased. When the E7 molecules were aligned such that the refractive index was n_e, the phase difference was close to π. The THz wave propagated mostly as a first-order diffracted beam. The diffraction efficiencies were 0.16 and 0.26 for the zeroth and first orders, respectively.

Figure 12(a, b) show the diffraction efficiencies of the zeroth and first orders as a function of the refractive index of E7 and V_{appl}. The experimental results, shown by dot symbols, are in good agreement with the theoretical predictions. Such results indicate that the grating functions as a variable beam splitter. The beam splitting ratio of the zeroth order to the first order can be tuned by varying the applied voltage.

Alternatively, these results indicate that the beam splitting ratio can be tuned and varied as a function of the refractive index of the nematic liquid crystal, E7. This is illustrated in **Figure 13**.

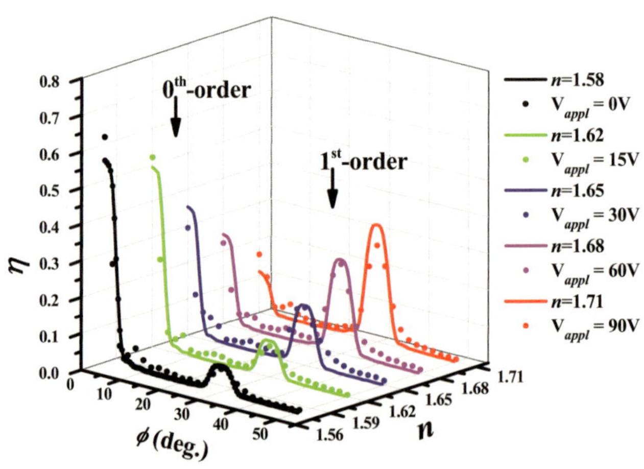

Figure 11. Diffraction efficiencies of the grating biased at several values of applied voltages are plotted as a function of the diffraction angle for the 0.3 THz beam. Solid lines are theoretical curves.

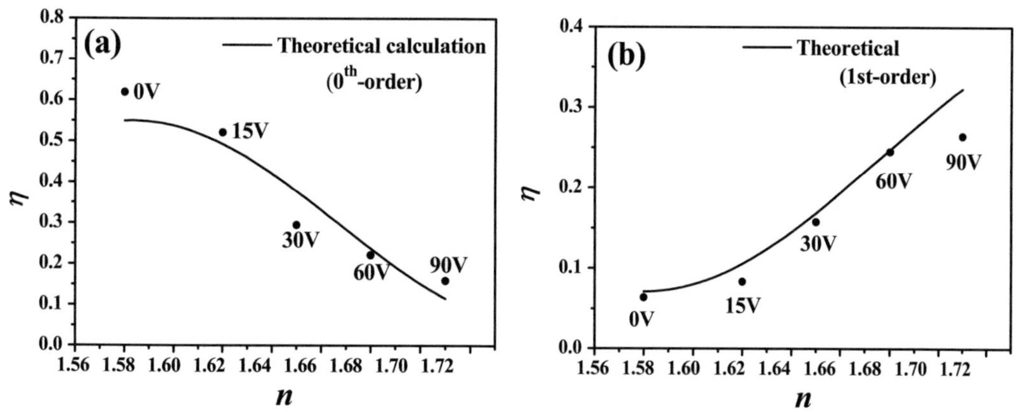

Figure 12. Diffraction efficiency as a function of the refractive index of E7 and the applied voltage for the (a) zeroth order and (b) first order. The theoretical calculation results and experimental results are shown by the curves and dots, respectively.

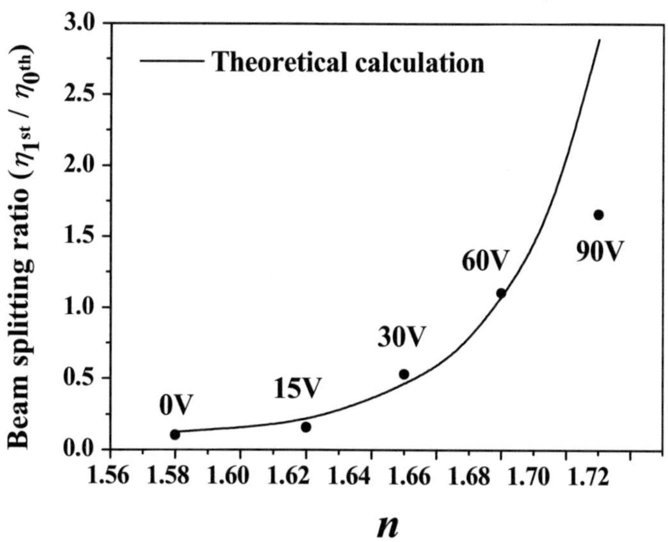

Figure 13. Beam splitting ratio as a function of the refractive index of E7. The theoretical calculation results and experimental results are shown by the curve and dots, respectively.

The theoretical calculation results and the experimental results are shown by the curve and dot symbols, respectively in **Figure 13**. The results indicate that the beam splitting ratio of the zeroth order to the first order can be tuned from 10:1 to 3:5.

3.3. Insertion loss

Insertion loss is a critical parameter for THz devices. We have experimentally and theoretically studied the insertion loss of two classes of devices. **Figure 14(a)** shows the diffraction efficiency

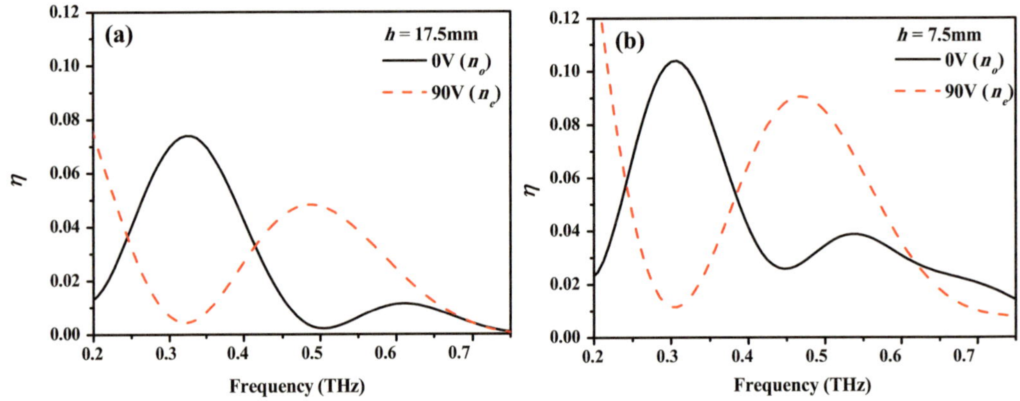

Figure 14. Diffraction efficiency of the devices with bases of (a) h = 17.5 mm and (b) h = 7.5 mm.

of a grating with a thicker base (h = 17.5 mm), obtained by normalizing the diffracted signals for the o-ray and e-ray to the reference THz signal for which the grating was removed. The experimentally measured diffraction efficiency for the o-ray at 0.3 THz is approximately 0.07. The diffraction efficiency for the o-ray at 0.3 THz predicted by the classic diffraction theory or evaluated by performing an FDTD simulation was approximately 0.45. The loss of the device was thus −8.0 dB for the o-ray at 0.3 THz. Similarly, the diffraction efficiency for the e-ray at 0.5 THz was approximately 0.046, whereas the theoretical prediction was approximately 0.45. The loss value of the grating for the e-ray at 0.5 THz was therefore −10 dB. A grating device with a smaller base component (h = 7.5 mm) was prepared to compare the insertion loss (**Figure 14(b)**). The experimentally measured diffraction efficiency for the o-ray at 0.3 THz was approximately 0.11, and the loss was −6.1 dB. The diffraction efficiency for the e−ray at 0.5 THz was approximately 0.083 and the loss value was −7.4 dB. The diffraction efficiency of the device with a smaller base is obviously higher than that of the device with a larger base.

The thickness of the ITO film we used was approximately 200 nm. According to [46], the conductivity σ of the film was $1.5 \times 10^3\ \Omega^{-1}\cdot\text{cm}^{-1}$. The parameters of fused silica and LC with different refractive indices in the frequency range of 0.2−0.8 THz are used for the insertion loss calculations shown in **Table 1**.

The estimated loss value was obtained from Eqs. (9) and (10). For the grating with a larger base component (h = 17.5 mm), the total loss for the o-wave at 0.3 THz and for the e-wave at 0.5 THz were estimated to be −9.0 and −13 dB, respectively. Similarly, for the grating with a smaller base component (h = 7.5 mm), the total loss for the o-wave at 0.3 THz and for the e-wave at

Material	Fused silica	E7 (n_o)	E7 (n_e)
ε_r	3.80	2.50	2.92
ε_i	0.008	0.095	0.041
tanδ	0.0021	0.038	0.014

Table 1. Parameters of fused silica and the NLC E7.

http://dx.doi.org/10.5772/intechopen.70449

Phase grating	h = 17.5 mm		h = 7.5 mm	
Driving voltage	0 V (n_o) at 0.3 THz	90 V (n_e) at 0.5 THz	0 V (n_o) at 0.3 THz	90 V (n_e) at 0.5 THz
(Estimated insertion loss)				
Conductor loss	4.7 dB	7.8 dB	2.3 dB	3.8 dB
Dielectric loss	4.3 dB	4.9 dB	3.2 dB	3.1 dB
Total	9.0 dB	13 dB	5.5 dB	6.9 dB
Measured value	8.0 dB	10 dB	6.1 dB	7.4 dB

Table 2. Insertion loss of phase gratings.

0.5 THz were estimated to be −5.5 and −6.9 dB, respectively. **Table 2** shows the estimated and measured values of the two devices. The discrepancy between the estimated insertion loss and measured data could be due to the finite collection efficiency of the detection system and nonideal assembly of the grating. The experimental results indicate that by reducing the thickness of the thick fused silica plates in the base of the device by10 mm, the insertion loss can be reduced by approximately 2.5 dB.

Therefore, if a grating device without a base (h = 2.5 mm) can be fabricated, the insertion loss can be as low as −2.5 dB. Such a device would be more attractive for practical use. Furthermore, the performance of the grating can be improved by using electrodes with higher conductivity. The thickness of the electrodes affects the conductor loss, as detailed in Ref. [50].

3.4. Device response times

For gratings using LCs, the response time of the device is a concern. The voltage-on and voltage-off times were measured by subjecting the device to a pulse signal. **Figure 15(a, b)** shows the normalized power as a function of the driving voltage in the voltage-on and voltage-off states, respectively. We defined the rise time as the duration for which the driving voltage was turned on for reducing the power to 37% of the maximum. The fall time was defined as the duration for which the driving voltage was turned off for increasing the power

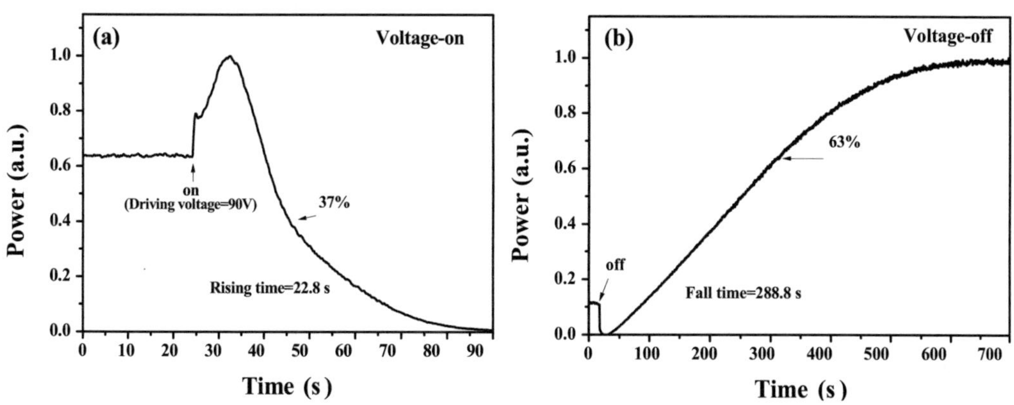

Figure 15. Response times of a phase grating: (a) voltage-on state and (b) voltage-off state.

to 63% of the maximum. The rise and fall times of the grating were found to be approximately 23 and 290 s, respectively. The phase grating responded slowly because of the thick LC layer used. Consequently, the present device is not suitable for applications that require fast modulation. However, the device is appropriate for instrumentation or apparatuses that require, for example, a fixed beam splitting ratio with occasional fine tuning.

The response time of the voltage-off state depended only on the material properties and cell thickness. Therefore, it cannot be shortened by applying a higher electric field. To shorten the response time, LCs with birefringence than E7 can be used. Alternatively, dual-frequency LCs can be employed; the use of dual-frequency LCs has been discussed in previous papers [51–54]. Dual-frequency LCs show high dielectric dispersion, and their dielectric anisotropy is frequency dependent, resulting in a change in sign at the crossover frequency. Dual-frequency materials in which the crossover frequency is a few kilohertz and changes markedly over the range are commercially available. Dual-frequency LCs would enable the operation of phase gratings in a nonzero applied voltage state.

3.5. Phase shifting and beam steering

We have studied the phase shift experienced by the THz wave propagating through the grating in which the control voltages were applied equally to all NLC layers. **Figure 16** shows the measured THz waveforms for biasing voltages varied from 0 to 28.8 V_{rms}. It is obviously that the pulses delay increase as applying voltages increased, as the NLC molecules re-orientate gradually from ordinary to extraordinary refractive index.

By applying Fourier transform on the waveforms in **Figure 16**, we obtained the phase shift as a function of frequency. This is shown in **Figure 17**. The phase shift increased with increasing

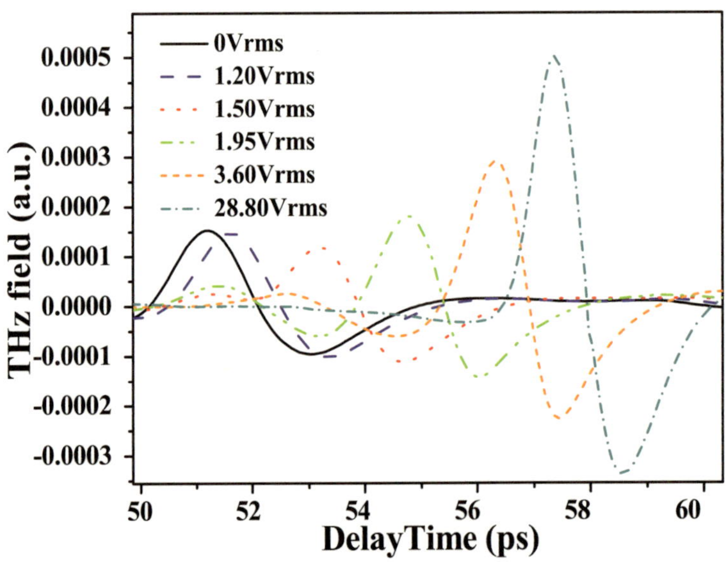

Figure 16. THz signal delay in time domain. Delay time increases as applying voltage increases.

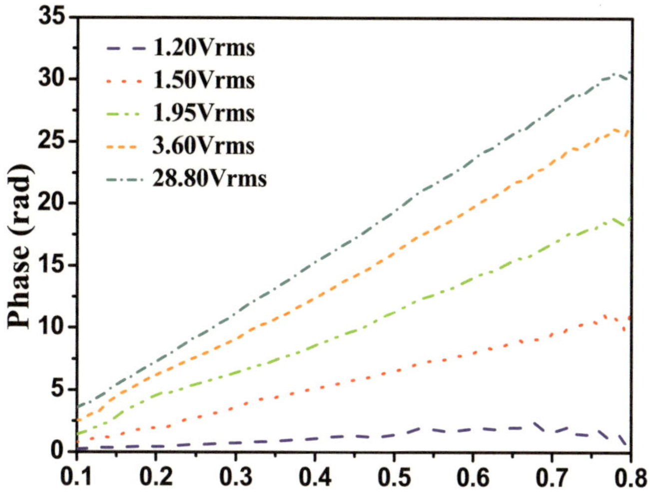

Figure 17. Spectra of phase shift of THz signal. Phase increases as applying voltage increases.

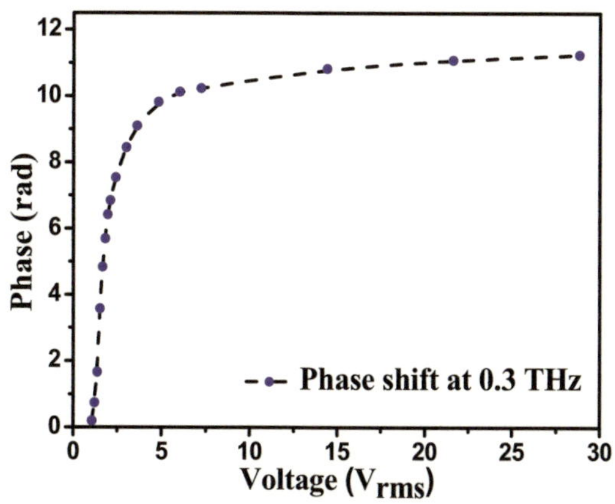

Figure 18. Phase shift at 0.3 THz as a function of the control voltage.

applying voltages as expected. **Figure 18** is a plot the phase shift at 0.3 THz as a function of the control voltage. Above the threshold voltage, 1.20 V_{rms}, the phase shift rapidly increases with the applying voltage. The maximum phase shift reached approximately 11.24 rad. This value is in close agreement with the calculated value.

We measured the beam steering characteristics of the phase-shifting array with the modified THz-TDS shown in **Figure 6**. Although the applying voltage should be adjusted layer-by layer for beam steering, only nine values of control voltages were available to be applied to each NLC block consisting two NLC layers. As the phase shift $\Delta\phi_i$ in each NLC block needed for beam steering is given by Eq. (12), the control voltage corresponding to phase shift can be determined from the experimental results in **Figure 18**, and are tabulated in **Table 3**.

	Applied voltage (V_{rms})	Phase shift (rad) at 0.3 THz
V_1	0	0
V_2	1.32	1.41
V_3	1.44	2.81
V_4	1.57	4.22
V_5	1.77	5.62
V_6	2.18	7.03
V_7	2.96	8.43
V_8	4.88	9.84
V_9	28.80	11.24

Table 3. Control voltage and corresponding phase shift at 0.3 THz.

The experimental results demonstrating beam steering are shown in **Figure 19**.

In the above figure, (a) shows the THz signal before transmitted to the device, and (b) and (c) show the THz signal transmitted through the device with ordinary and extraordinary refractive indices at $\theta = 0°$, respectively. The main beam was steered in the direction of $\theta = 8.5°$ as the control voltages were varied to yield the phase gradient as shown in (d). The signal vanishes as we removed the device as shown in (e). Applying FFT analysis, the corresponding THz spectra

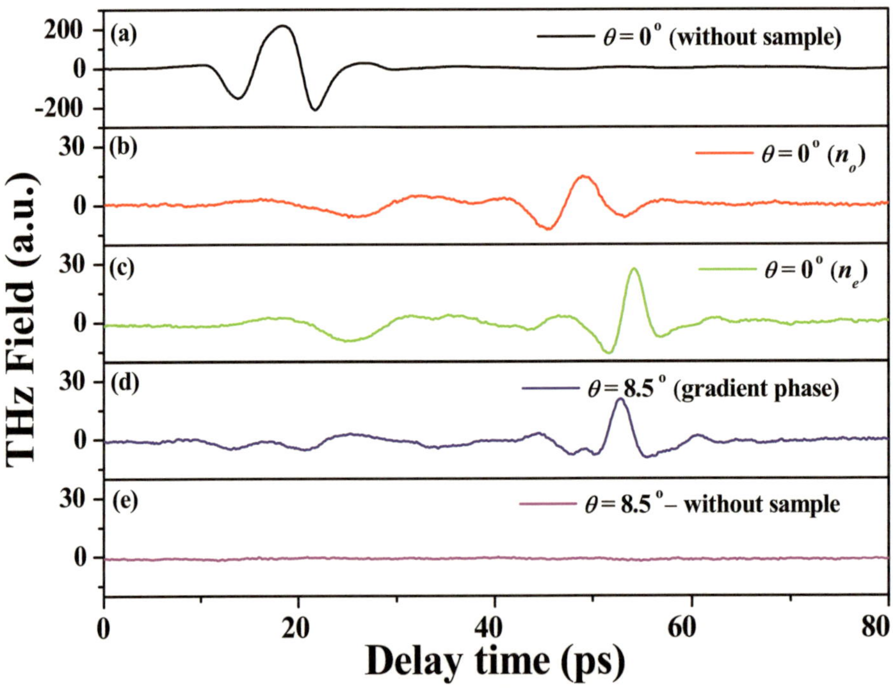

Figure 19. THz signals before steered at $\theta = 0°$: (a) without sample; with sample (b) at no-state, (c) at ne-state; steered signals at $\theta = 8.5°$: (d) sample applied voltage to yield gradient phase, (e) without sample.

http://dx.doi.org/10.5772/intechopen.70449

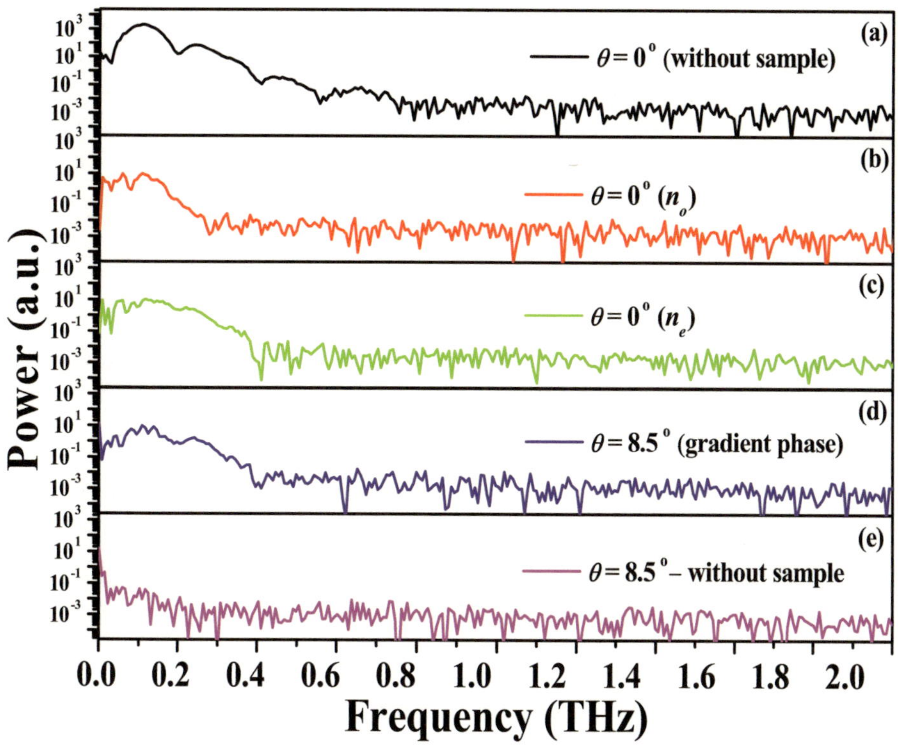

Figure 20. Spectra of THz signals before steered at $\theta = 0°$: (a) without sample; with sample (b) at no-state, (c) at ne-state; steered signals at $\theta = 8.5°$: (d) sample applied voltage to yield gradient phase, (e) without sample.

in frequency domain are shown in **Figure 20**. According to the results, the device can steer the broadband THz signal up to about 0.5 THz.

4. Summary

In this work, we review our theoretical and experimental studies on electrically controlled LC-based phase gratings for manipulating THz waves. This device can be used as a tunable THz beam splitter, and the beam splitting ratio of the zeroth-order diffraction to the first-order diffraction can be tuned from 10:1 to 3:5. An FDTD simulation was performed to investigate the diffraction effect of the phase grating. The experimental and simulation results were in general agreement. The signal losses of the device were discussed. It was observed that the insertion loss could be reduced by reducing the thickness of the fused silica plates in the base component of the device. The rise and fall times of the grating are approximately 23 and 290 s, respectively. The slow response could be accounted for because of the thick LC layer employed. Consequently, it is not suitable for applications that require fast modulation. However, the device is appropriate for instrumentation or apparatuses that require, for example, a fixed beam splitting ratio with occasional fine tuning. The use of highly-birefringent NLCs or dual-frequency LCs could alleviate the problem somewhat. Besides, we demonstrated a grating-structured phase shifter array that can be used as the THz shifter and THz beam steerer. A phase shift as large as 11.24 rad was achieved. Using a designed voltage gradient biasing on the grating structure, broadband THz signal below 0.5 THz can be steered by as much as 8.5°. The experimental results are in good agreement with theoretical predictions.

Acknowledgements

This work was partly supported by the National Science Council of Taiwan (104–2221-E-007-093-MY3), the Academic Top University Program of the Taiwan Ministry of Education, and the U.S. Air Force Office of Scientific Research (FA2386–13–1-4086). Chia-Jen Lin is now with Taiwan Semiconductor Manufacturing Company. Contributions by Mr. Chuan-Hsien Lin are gratefully acknowledged.

Author details

Ci-Ling Pan[1]*, Chia-Jen Lin[2], Chan-Shan Yang[1], Wei-Ta Wu[2] and Ru-Pin Pan[2]

*Address all correspondence to: clpan@phys.nthu.edu.tw

1 Department of Physics, National Tsing Hua University, Hsinchu, Taiwan

2 Department of Electrophysics, National Chiao Tung University, Hsinchu, Taiwan

References

[1] Ferguson B, Zhang X-C. Materials for terahertz science and technology. Nature Materials. 2002;**1**:26-33

[2] Tonouchi M. Cutting-edge terahertz technology. Nature Photonics. 2007;**1**:97-105

[3] Zhang X-C, Xu J. Introduction to THz Wave Photonics. New York: Springer; 2010

[4] Nagatsuma T, Ducournau G, Renaud CC. Advances in terahertz communications accelerated by photonics. Nature Photonics. 2016;**10**:371-379

[5] Alonso-González P, Nikitin AY, Gao Y, Woessner A, Lundeberg MB, Principi A, Forcellini N, Yan W, Vélez S, Huber AJ, Watanabe K, Taniguchi T, Casanova F, Hueso LE, Polini M, Hone J, Koppens FHL, Hillenbrand R. Acoustic terahertz graphene plasmons revealed by photocurrent nanoscopy. Nature Nanotechnology. 2017;**12**:31-35

[6] Yang C-S, Tang T-T, Chen P-H, Pan R-P, Yu P, Pan C-L. Voltage-controlled liquid-crystal terahertz phase shifter with indium-tin-oxide nanowhiskers as transparent electrodes. Optics Letters. 2014;**39**(8):2511-2513

[7] Yang C-S, Tang T-T, Pan R-P, Yu P, Pan C-L. Liquid crystal terahertz phase shifter s with functional indium-tin-oxide nanostructures for biasing and alignment. Applied Physics Letters. 2014;**104**:141106

[8] Yang C-S, Kuo C, Tang C-C, Chen JC, Pan R-P, Pan C-L. Liquid-crystal terahertz quarter-wave plate using chemical-vapor-deposited graphene electrodes. IEEE Photonics Journal. 2015;**7**(6):2200808

[9] Pan CL, Pan RP, Yang CS, Yu P, Voltage-controlled 2 liquid-crystal terahertz phase shifter with indium-tin-oxide (ITO) nanowhiskers as transparent electrodes. In: The XXXI General Assembly of the International Union of Radio Science, paper DFC01.6, Beijing, China. Aug 16–13, 2014

[10] Lin C-J, Li Y-T, Hsieh C-F, Pan R-P, Pan C-L. Manipulating terahertz wave by a magnetically tunable liquid crystal phase grating. Optics Express. 2008;**16**:2995-3001

[11] Lin C-J, Lin C-H, Li Y-T, Pan R-P, Pan C-L. Electrically controlled liquid crystal phase grating for terahertz waves. IEEE Photonics Technology Letters. 2009;**21**:730-732

[12] Mirzaei B, Silva JRG, Luo YC, Liu XX, Wei L, Hayton DJ, Gao JR, Groppi C. Efficiency of multi-beam Fourier phase gratings at 1.4 THz. Optics Express. 2017;**25**:6581-6588

[13] Chen H-T, Padilla WJ, Cich MJ, Azad AK, Averitt RD, Taylor AJ. A metamaterial solid-state terahertz phase modulator. Nature Photonics. 2009;**3**:148-151

[14] Nouman MT, Kim H-W, Woo JM, Hwang JH, Kim D, Jang J-H. Terahertz modulator based on metamaterials integrated with metal-semiconductor-metal varactors. Scientific Reports. 2016;**6**:26452

[15] Wang M, Vajtai R, Ajayan PM, Kono J. Electrically tunable hot-silicon terahertz attenuator. Applied Physics Letters. 2014;**105**:141110

[16] Hsieh C-F, Lai Y-C, Pan R-P, Pan C-L. Polarizing terahertz waves with nematic liquid crystals. Optics Letters. 2008;**33**:1174-1176

[17] Farid A, Laurita NJ, Tehrani B, Hester JG, Tentzeris MM, Armitage NP. Inkjet printed wire-grid polarizers for the THz frequency range. Journal of Infrared, Millimeter and Terahertz Waves. 2017;**38**:276-282

[18] Park J-H, Yu C-J, Kim J, Chung S-Y, Lee S-D. Concept of a liquid-crystal polarization beamsplitter based on binary phase gratings. Applied Physics Letters. 2003;**83**: 1918-1920

[19] Berry CW, Jarrahi M. Broadband terahertz polarizing beam splitter on a polymer substrate. Journal of Infrared, Millimeter and Terahertz Waves. 2012;**33**:127-130

[20] Ung BS-Y, Fumeaux C, Lin H, Fischer BM, Ng BW-H, Abbott D. Low-cost ultra-thin broadband terahertz beam-splitter. Optics Express. 2012;**20**(5):4968-4978

[21] Mo G-Q, Li J-S. Compact terahertz wave polarization beam splitter using photonic crystal. Applied Optics. 2016;**55**:7093-7097

[22] Filin A, Stowe M, Kersting R. Time-domain differentiation of terahertz pulses. Optics Letters. 2001;**26**:2008-2010

[23] Garet F, Coutaz J–L, Narzarov M, Bonnet E, Parriaux O, Racine G, THz time-domain spectroscopy study of grating couplers and segmented grating filters. In: IEEE 2004 Joint 29th Int. Conf. on Infrared and Millimeter Waves and 12th Int. Conf. on Terahertz Electronics. Oct 2004. pp. 181-182

[24] Kersting R, Strasser G, Unterrainer K. Terahertz phase modulator. Electronics Letters. 2000;**36**:1156-1158

[25] Kleine-Ostmann T, Dawson P, Pierz K, Hein G, Koch M. Room-temperature operation of an electrically driven terahertz modulator. Applied Physics Letters. 2004;**84**:3555-3557

[26] Libon IH, Baumgärtner S, Hempel M, Hecker NE, Feldmann J, Koch M, Dawson P. An optically controllable terahertz filter. Applied Physics Letters. 2000;**76**:2821-2823

[27] Chen J, Bos PJ, Vithana H, Johnson DL. An electro-optically controlled liquid crystal diffraction grating. Applied Physics Letters. 1995;**67**:2588-2590

[28] Chen CY, Hsieh CF, Lin YF, Pan RP, Pan CL. Magnetically tunable room-temperature 2π liquid crystal terahertz phase shifter. Optics Express. 2004;12:2625-2630

[29] Wu HY, Hsieh CF, Tang TT, Pan RP, Pan CL. Electrically tunable room-temperature 2π liquid crystal terahertz phase shifter. IEEE Photonics Technology Letters. 2006;**18**:1488-1490

[30] Chen C-Y, Hsieh C-F, Lin Y-F, Pan C-L, Pan R-P. Liquid- crystal-based terahertz tunable Lyot filter. Applied Physics Letters. Mar. 2006;**88**:101107

[31] Ho I-C, Pan C-L, Hsieh C-F, Pan R-P. Liquid- crystal- based terahertz tunable Solc filter. Optics Letters. 2008;**33**:1401-1403

[32] Ghattan Z, Hasek T, Wilk R, Shahabadi M, Koch M. Sub- terahertz on–off switch based on a two-dimensional photonic crystal infiltrated by liquid crystals. Optics Communication. 2008;**281**:4623-4625

[33] Jewell1 SA, Hendry E, Isaac TH, Sambles JR. Tuneable Fabry–Perot etalon for terahertz radiation. New Journal of Physics. 2008;**10**:033012

[34] Pan R-P, Hsieh C-F, Chen C-Y, Pan C-L. Temperature- dependent optical constants and birefringence of nematic liquid crystal 5CB in the terahertz frequency range. Journal of Applied Physiology. 2008;**103**:093523

[35] Neu J, Beiggang R, Rahm M. Metamaterial-based gradient index beam steerers for terahertz radiation. Applied Physics Letters. 2013;**103**:041109

[36] Xu B, Hu H, Liu J, Wei X, Wang Q, Song G, Xu Y. Terahertz light deflection in doped semiconductor slit arrays. Optics Communication. 2013;**308**:74-77

[37] Xu Z, Mazumder P. Terahertz beam steering with doped GaAs phase modulator and a design of spatial-resolved high-speed terahertz analog-to-digital converter. IEEE Transactions on Electron Devices. 2014;**61**:2195-2202

[38] Bin Shams MI, Jiang Z, Qayyum J, Rahman S, Fay P, Liu L. A terahertz reconfigurable photo-induced fresnel-zone-plate antenna for dynamic two-dimensional beam steering and forming. In: Microwave Symposium (IMS), 2015 IEEE MTT-S International, Phonenix, AZ. May 2015. pp. 1-4

[39] Kamoda H, Kuki T, Fujikake H, Nomoto T. Millimeter-wave beam former using liquid crystal. In: Microwave Conference, 2004. 34th European, Amsterdam, The Netherlands. Oct 2004. pp. 1141-1144

[40] Brosseau C. Fundamentals of Polarized Light. New York: Wiley; 1998

[41] Goodman JW. Introduction to Fourier Optics. 2nd ed. McGraw-Hill; 1996

[42] Fowles JR. Introduction to Modern Optics. 2nd ed. Holt, Rinehart and Winston; 1975

[43] Yang C-S, Lin C-J, Pan R-P, Que CT, Yamamoto K, Tani M, Pan C-L. The complex refractive indices of the liquid crystal mixture E7 in the terahertz frequency range. Journal of the Optical Society of America B: Optical Physics. 2010;**27**:1866-1873

[44] Yang C-S, Lin M-H, Chang C-H, Yu P, Shieh J-M, Shen C-H, Wada O, Pan C-L. Non-Drude behavior in indium-tin-oxide nanowhiskers and thin films by transmission and reflection THz time-domain spectroscopy. IEEE Journal of Quantum Electronics. 2013;**49**:677-690

[45] Pan C-L, Hsieh C-F, Pan R-P, Tanaka M, Miyamaru F, Tani M, Hangyo M. Control of enhanced THz transmission through metallic hole arrays using nematic liquid crystal. Optics Express. 2005;**13**:3921-3930

[46] Chen C-W, Lin Y-C, Chang C-H, Yu P, Shieh J-M, Pan C-L. Frequency-dependent complex conductivities and dielectric responses of indium tin oxide thin films from the visible to the far-infrared. IEEE Journal of Quantum Electronics. 2010;**46**:1746-1754

[47] Kamoda H, Kuki T, Fujikake H, Nomoto T. Millimeter-wave beam former using liquid crystal. In: Proc. 34th European Microwave Conf., Amsterdam, Netherlands. 2004. pp. 1141-1144

[48] Cheng DK. Field and Wave Electromagnetics. 1st ed. Addison-Wesley; 1983. p. 465

[49] Ku C-P, Shih C-C, Lin C-J, Pan R-P, Pan C-L. THz optical constants of the liquid crystal MDA-00-3461. Molecular Crystals and Liquid Crystals. 2011;**541**:303-308

[50] Kamoda H, Kuki T, Nomoto T. Conductor loss reduction for liquid crystal millimeter-wave beam former. IEICE Electronics Express. 2005;**2**:471-476

[51] Jewell SA, Taphouse TS, Sambles JR. Rapid switching in a dual-frequency hybrid aligned nematic liquid crystal cell. Applied Physics Letters. 2005;**87**:021106

[52] Golovin AB, Shiyanovskii SV, Lavrentovich OD. Fast switching dual-frequency liquid crystal optical retarder, driven by an amplitude and frequency modulated voltage. SID 03 Digest. 2003;**2**(1472):55

[53] Hsieh CT, Huang CY, Lin CH. In-plane switching dual-frequency liquid crystal cell. Optics Express. 2007;**15**:11685-11690

[54] Chen C-C, Chiang W-F, Tsai M-C, Jiang S-A, Chang T-H, Wang S-H, Huang C-Y. Continuously tunable and fast-response terahertz metamaterials using in-plane-switching dual-frequency liquid crystal cells. Optics Letters. 2015;**40**:2021-2024

Chapter 11

Design and Fabrication of Ultra-Short Throw Ratio Projector Based on Liquid Crystal on Silicon

Jiun-Woei Huang

Additional information is available at the end of the chapter

http://dx.doi.org/10.5772/intechopen.72670

Abstract

One of applications for liquid crystal on silicon (LCoS) could be an emitted light panel for display and projection. Among optical projectors, the most challenging work is to design ultra-short throw projection systems for LCoS projector for home cinema, virtual reality (VR), head-up display (HUD) in automobile. The chapter discloses the design and fabrication of such kind of projector. In fact, such design is not only to design wide angle projection optics but also to optimize illumination for LCoS in order to have high-quality image. The projector optical system is simply with a telecentric field lens and inlet optics of symmetric double gauss or a large angle eyepiece, with a conic aspheric mirror, thus the full projection angle large than 155°. Applying Koehler illumination, the resolution of image is increased; thus, the modulation transfer function of image in high spatial frequency is increased to form the high-quality illuminated image. Based on telecentric lens type of projection systems and Koehler illumination, optical parameters are provided. The partial coherence analysis has verified that the design is reached to 2.5 lps/mm within 2 × 1.5 m. The best performance of systems has been achieved. The throw ratio is less than 0.25 with HD format.

Keywords: LCoS, Koehler illumination, telecentric, ultra-short throw ratio

1. Introduction

The mass requirement for display as viewing accessory has been applied toward the smart phone, HUD, and computers, home games, and home cinema. Projector has been one of the display tools for classrooms, family rooms or theaters, while the LCD display cannot be fully replaced due to its unique characteristics of adjustable view angle and size, and environmental protection issue.

Since two decades ago, rear projection TV has been used the projection in the back to image in the back of screen then formed in the front screen, to the forward projection LCoS refractive projection systems, the throw ratio is 1.4 to 1.6, recently, Sony short throw projection LSPX-W1 has announced the throw ratio to less than 0.5 [2].

In the past decade, several companies [3] had announced reflective-type projectors could provide the short ratio projection under 0.5, and Sony [4] has announced the throw ratio to less than 0.2, even it is posted in **Figure 1**, yet it is still no unique solution for such kind design disclosure. Thales [5, 7] reports that several reflective design are suggested by using optical deviated and tilted method, yet the engineering is enable to be carried.

1.1. The advantage of LCoS applicable in show throw ratio projector

Digital light processing (DLP) has been a display device based on optical micro-electro-mechanical structures. In DLP projectors, the image is created by microscopically small mirrors laid out in a matrix on a semiconductor chip, known as a digital micro-mirror. In addition, it has been popularly used in the projectors, and due to each element in DLP mechanical driving, it has limited to drive in high-speed image display. However, LCoS display modulates the emitted liquid crystal, which is much faster than mechanical driving panel. The other evaluating factor is color contrast and duration. Based on reflective coating in silicon, the efficiency of true color and duration is superior to DLP. The typical LCoS is shown in **Figure 2**.

1.2. Short throw ratio projection optics

For those forward-projecting diffractive lens are quite popular, yet the throw ratios almost is above 1, thus, the project is especially to design and fabricate a lens for throw ratio under 0.5 or less and is hard to achieve. To design short throw ratio of projection lens which is different from other projection lens with refractive lens, the performance of system is short

Figure 1. Sony short throw ratio projector for home theater.

Figure 2. Typical LCoS panel and output circuit.

throw ratio and compact. Especially, for LCoS projector, the emitted panel, such as liquid crystal on silicon chip (LCoS), has to be emitted by light. The three panels of LCoS projector are shown in **Figure 3**. Emitted RGB panels, carrying modulated LCD video information, pass through di-chloic filters and polarized beam splitters into lens system to form image on screen. Obviously, the lens for delivering image is crucial part in the projector. Because reflective can reduce the size, enlarge the projection angle, and make system packed, it is the best for short throw projector. The image has to be delivered out from LCoS, the optical system, thus, it is proper to design projector by a reflective mirror to project a wide screen with very short distance and wide angle system.

1.3. The HD format and the short throw ratio

To classify the projectors' format, the standard video format for ultra-high definition television is shown in **Table 1**. In this case, the image format DCI 4K: 4096 × 2160 is applied.

1.4. The definition of throw ratio

Usually, the definition of throw ratio is D/W, but for the short throw projector, the throw ratio is defined W/D'. D' is the distance between the bottom side of screen to the last lens or mirror.

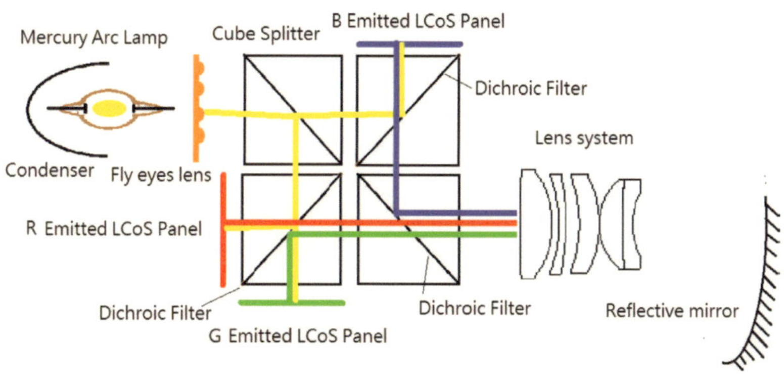

Figure 3. Three-panel LCoS projection diagrams.

Format	Resolution	Display aspect ratio	Pixels
Ultra high definition television	3840 × 2160	1.78:1 (16:9)	8,294,400
Ultra wide television	5120 × 2160	2.37:1 (21:9)	11,059,200
WHXGA	5120 × 3200	1.60:1 (16:10)	16,384,000
DCI 4K (native resolution)	4096 × 2160	1.90:1 (19:10)	8,847,360
DCI 4K (CinemaScope cropped)	4096 × 1716	2.39:1	7,028,736
DCI 4K (flat cropped)	3996 × 2160	1.85:1	8,631,360

Table 1. The format for ultra-high-definition television.

In **Figure 4**, the throw ratio is the ratio of the distance from the lens to the screen (throw) to the screen width. A larger throw ratio corresponds to a more tightly focused optical system.

This projection lens is to reimage of each pixel of LCoS to the projected screen truly, the performance is emphasized the image of LCoS elements, and the size of the object less one LCoS element becomes less important. In three pieces LCoS Panels system, the pixel is 4 μm, and the minimum size is for 2048 × 1080 mm DCI 4K2K projected screen.

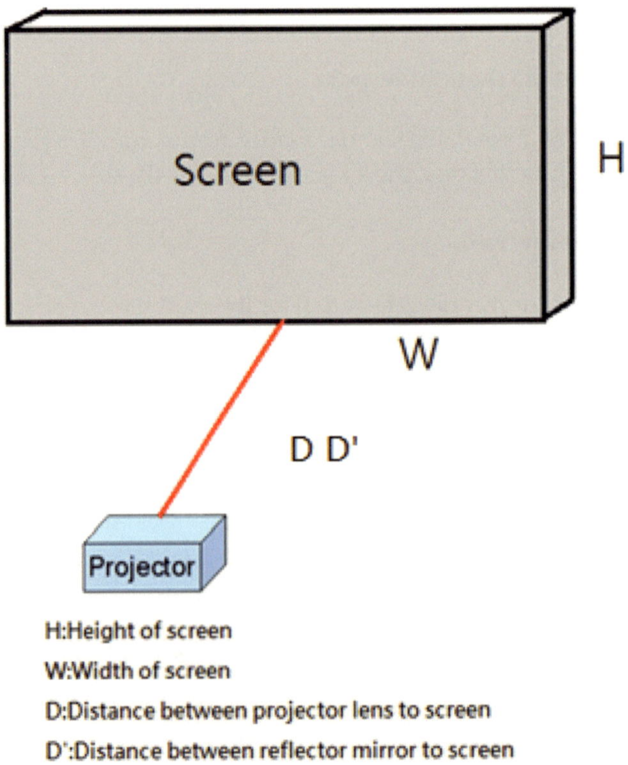

Figure 4. The view to show throw ratio.

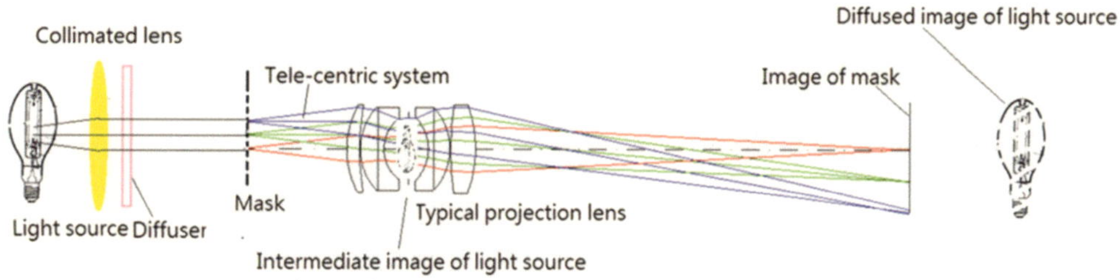

Figure 5. Projection system with illumination and optics.

1.5. The projection system

To design and fabrication of projection system has included two parts: one is imaging, and the other is illumination, [1] and it is desirable that the deterministic optical parameters to optimize the contrast of image requirement must be considered with full consideration of illumination and imaging. **Figure 5** is an illustration of projection system with illumination and optics.

2. Theory

To design projection system has considered two aspects: one is imaging, and the other is illumination, [1] and the theories are explained in the following [6].

2.1. Koehler illumination and telecentric

The liquid crystal on silicon (LCoS) projector includes a cube formed RGB panel with collimated source and projection lens. The illumination requires a cube formed RGB collimated source to introduce the light into pattern by the LCoS module. By a projection lens, the meaningful video information in LCoS is projected into a wall or screen with large field of view, up to throw ratio less than 0.4. In the design, illumination and image require the optimized parameters to form highly demanded images. In order to optimize these systems, the theory of Koehler illumination and telecentric optical system are chosen.

2.1.1. Telecentric system

The telecentric is best fitted for short throw projection system because telecentric in the object side can separate the central object rays and margin rays with different optical path and increase the contrast of image plane [10, 12, 14]. Moreover, the telecentric system can provide the non-distorted image or object along optical axis, and the projection system is easy to be optimized.

2.1.2. Koehler illumination

Koehler illumination [1] is the light source imaged in aperture, while the rays are collimated in LCoS panel. Due to the panels requiring uniform intensity in optical axis in order to keep polarization and coherence, the high performance platform often selects Koehler illumination.

Being designed by telecentric and Koehler illumination, the best performance of systems, having the optimized optical parameters, could be achieved.

2.2. Model of components build up

First, the emitted panel is stimulated by using a paralleled light bundle of mercury light source to hit LCoS panel. Rays, spreading as an emitted object, are shown in **Figure 4**. In the system, three LCoS panels are combined paths as one object, which is pseudo Lambertian distribution in ray field. It acts as telecentric source, shown in **Figure 6**.

2.3. Optical design

The optical system requirement is listed in **Table 2**. In order to enlarge the projection angle, the optical system of the short throw projection includes two parts, one is refractive lens system, and the other is reflective mirror. In **Figure 7**, a telecentric system is defined, and the principal rays from three fields as parallel rays, from the object, will be focused in one position as common aperture, while the other rays form image. Thus, the image will be formed with reduction value greater than 1. The intermediate image is relayed to the final screen with a

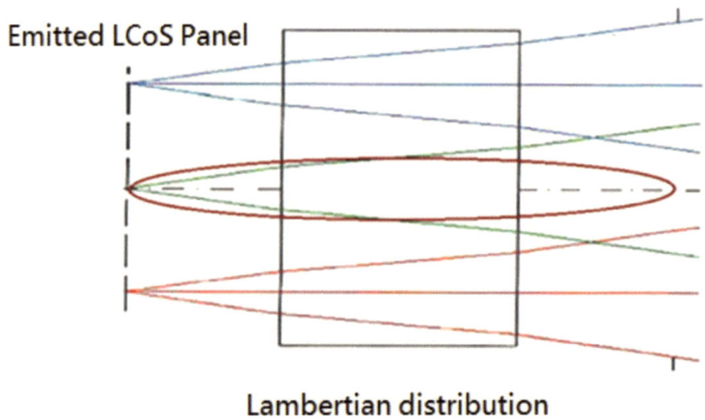

Figure 6. The common aperture and intermediate image.

Specification for short throw factor lens	
Items	**Specification**
Projected screen	2500 mm in diagonal
LCoS size	1 inch
Pixels size	4 μ^2
Video format	4K2K
Lens type	Refractive and reflective
Short distance to project between last mirror and screen	666 mm

Table 2. The specification for theater short throw factor lens.

similar method by adjusting the conic curvature of reflective mirror. The image on screen can be expanded and optimized by adjusting the conic value and radius of reflective mirror.

In a typical case, the design has passed through three stages, which are given as follows.

2.3.1. Initial stage

The initial stage is setup delivering optics. The lens, such as double gauss or eyepiece types in **Figure 8**, will pull instant image of emitted panel out to pass through common aperture and spread out to form an intermediate image and to broad the field of view, as shown in **Figure 9**.

2.3.2. Middle stage

In **Figure 10**, each principal rays formed common aperture due to telecentric effect. Light passing through the aperture and convergent, the projection angle can be enlarged. Because

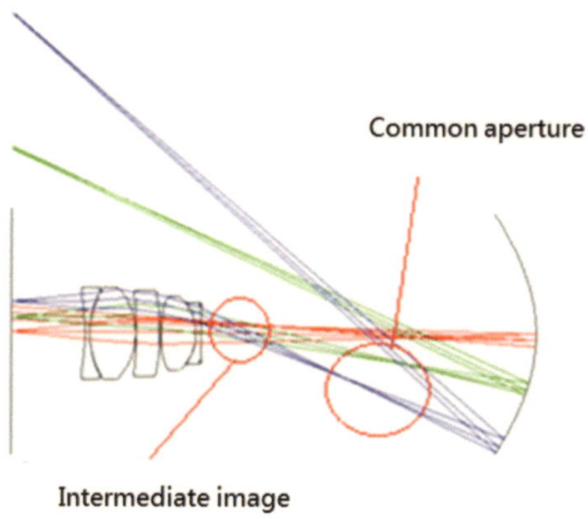

Figure 7. The common aperture and intermediate image.

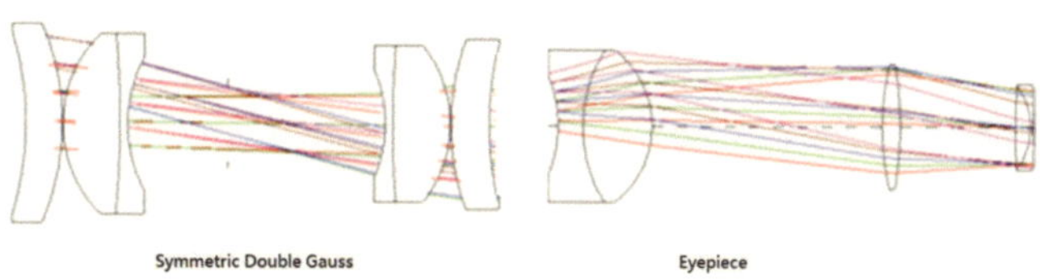

Figure 8. Symmetric double gauss and wide eye-piece types.

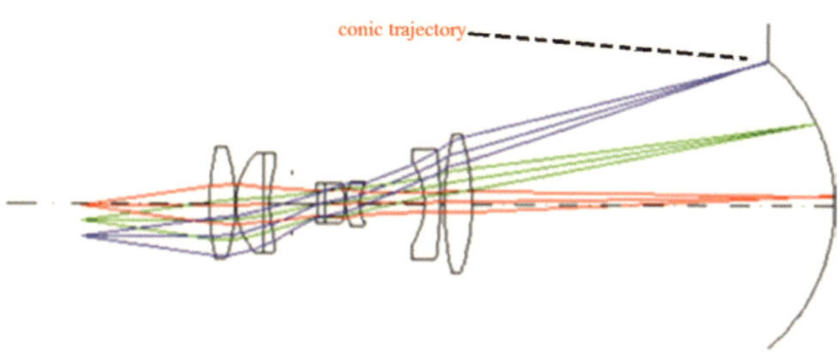

Figure 9. The wide-angle eyepiece to broad the field of view.

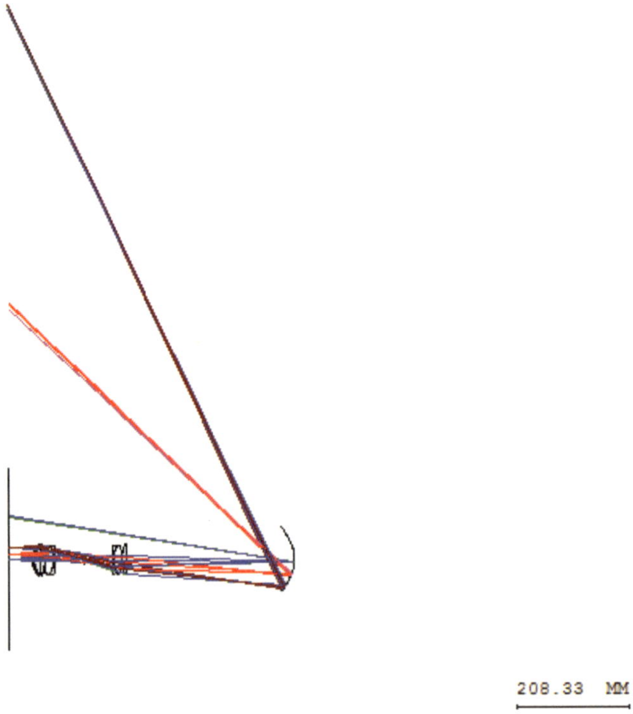

Figure 10. Design in middle stage.

the final image has to be formed in plane, the previous intermediated image has to be a conic trajectory. By mirror theory, the object forms a conic trajectory to broaden the field angle, as in **Figure 10**. Thus, the final image will be placed through the reflective mirror.

2.3.3. Final stage

In the final stage, the intermediate image, coupling with a mirror, forms a broad view image on the screen. The intermediate image relays on the final screen with a similar method by a reflective mirror, shown in **Figure 11**. The whole system has to be optimized in the constraint optical parameters. The optimized condition can be reached [8].

2.4. Image evaluation

Image evaluation has been carried out by evaluation of spot diagrams, image distortion diagram, to instant adjust the lens size, space and material in order to reach the small and less aberration spot. The third aberration for each lens provides the instant information to adjust lens shape and other parameters to the final stage.

2.5. Partial coherence

Illumination could be calculated by illumination program. The condenser and collimated lens could be designed by optimizing the uniformity of illumination of LCoS. For detailed calculation

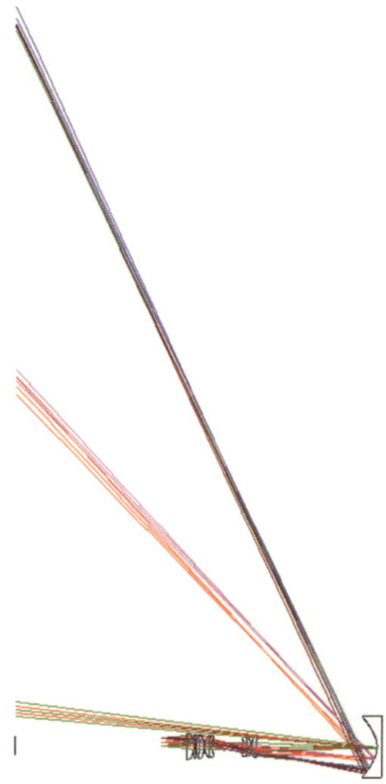

Figure 11. Design in the final stage.

of image formation, each emitted element in LCoS, two vicinities will not be correlated according to the calculation. If the polarization is considered, two coherence sources may just be passing through, and two bundles of coherent photo will never be interfered. In practical condition, this always happens, that is, once the two bundle groups of photo are partially coherent, the partial coherence should be calculated.

2.6. Fabrication

The system has been verified in a projector, without lens, by a 3-panel LCoS system. The optics system totally replaces the previous lens to demo the function.

3. Simulation

Simulation is mentioned below [3, 5, 6].

3.1. LCoS projection structure

LCoS reflective-type projection system is shown in **Figure 12**. The illumination requires a cube formed RGB collimated source to introduce the light into pattern by LCoS module.

The emitted LCoS panel as modulate image reflector, while the collimated rays propagate through the open-state liquid crystal and reflecting back to lens system, thus it can be modified type Koehler illumination. By a projection lens, the meaningful video information in LCoS is projected on a wall or screen with large field of view, up to throw ratio less than 0.3 [9, 11, 13].

3.2. Numerical calculation

codeV (an optical design software) is used in the study [5]. codeV provides MTF, wave-front, coupling efficiency and longitudinal aberration methods to reach the optimized solution. The program-constrained conditions are defined, and at least 100 runs of optimization are performed. The result for requirement of specification is reached. The tolerance of tilted optics has also performed by tilting different elements. The results are expressed in MTF graphs. codeV almost can trace each point in the space through the lens system to target.

3.3. Image quality

The design is shown in **Figure 13**. The refractive lens is to apply double gauss delivering the emitted to spread out due to the symmetric structure and less aberration induced. **Figure 14** shows the comparison between input object and output image, and the output image could truly describe the object with slight pincushion distortion. **Figure 15** shows

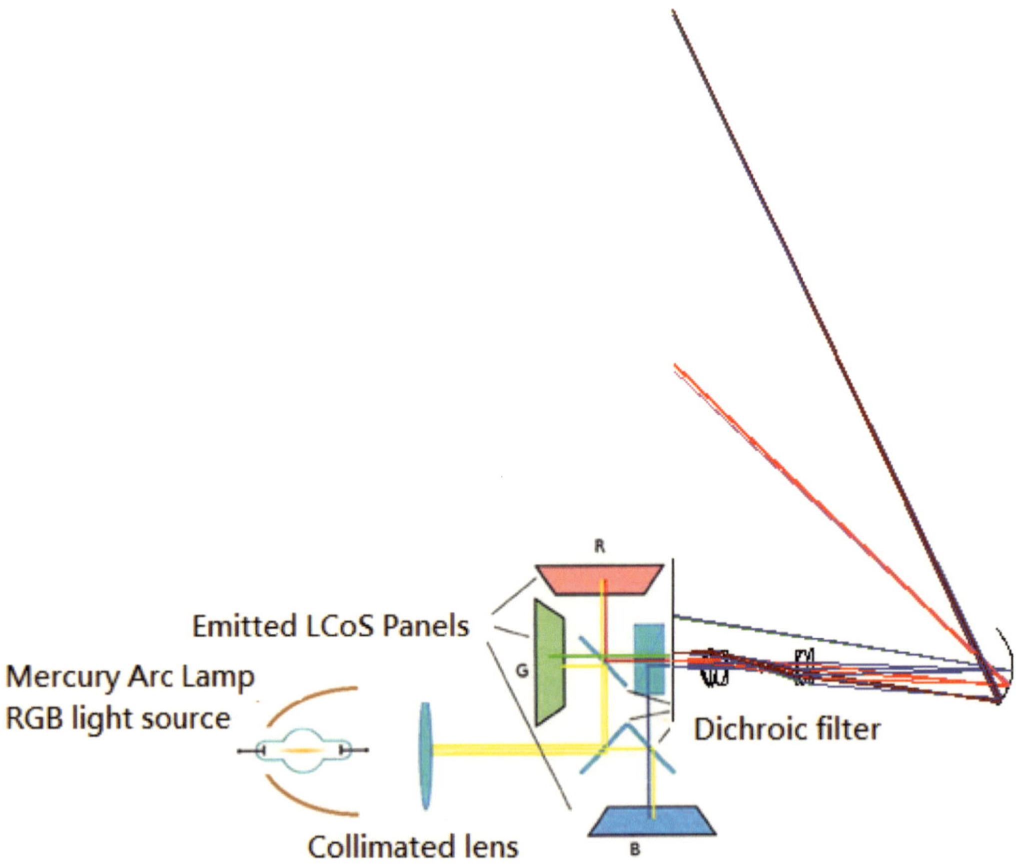

Figure 12. The short throw ratio LCoS projection system with Koehler illumination and telecentric optics.

http://dx.doi.org/10.5772/intechopen.72670

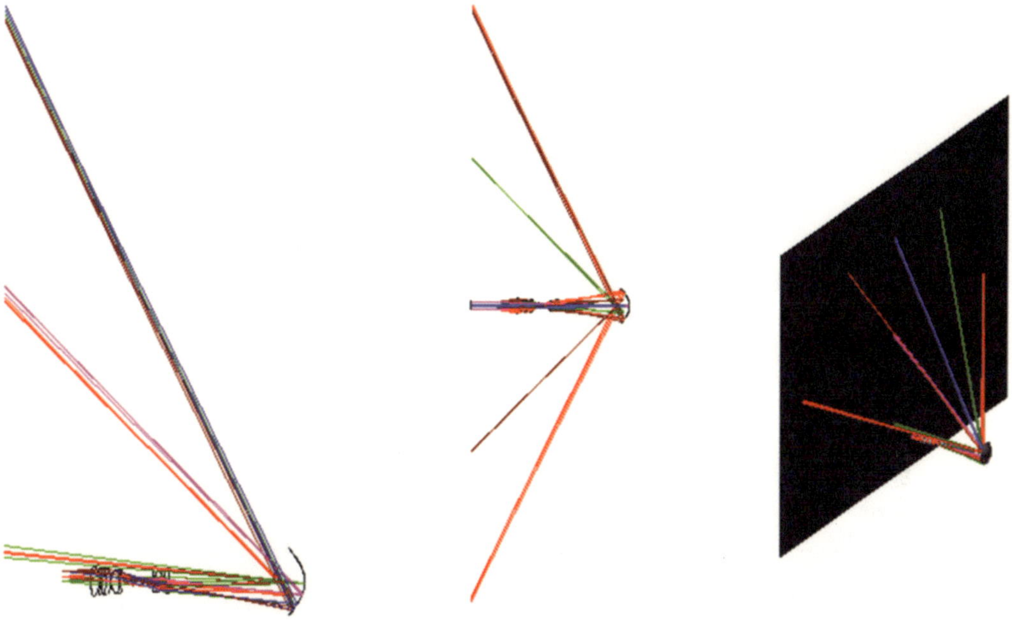

Figure 13. The optical design for short throw ratio projector.

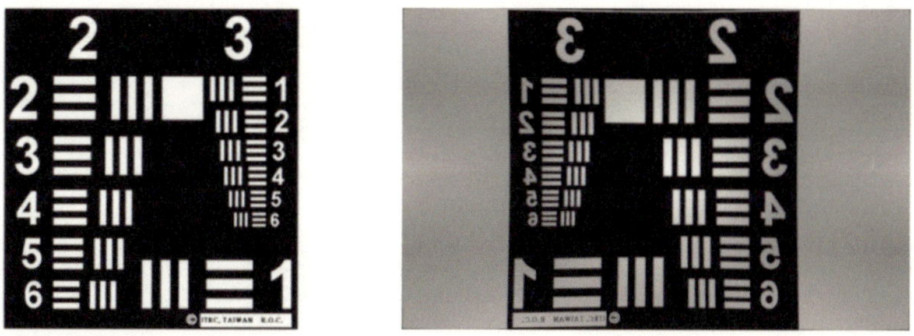

Figure 14. The simulated object and image.

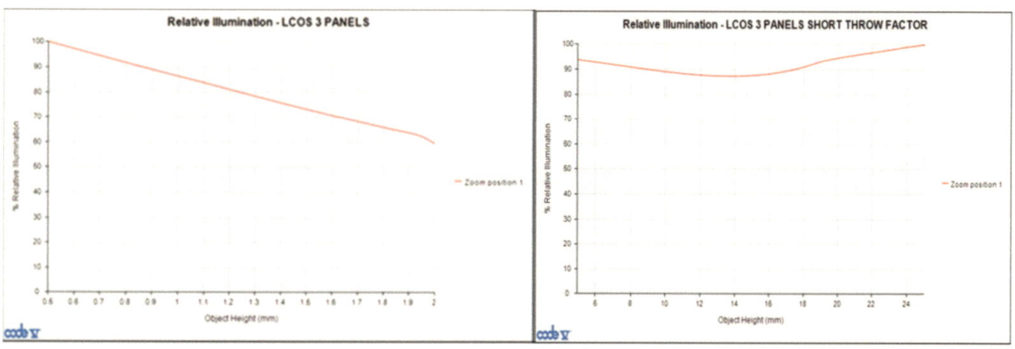

Figure 15. The relative illumination between before and after tuning the lens shapes.

the relative illumination between before and after optimization. Optimize illumination shows that the full screen is above 70%, for the the full field of view is fitted for human eyes.

3.4. Aberration

Figure 16 shows the field distortion between before and after optimization. Initially, the distortion was high, and after adjusting the conic constant of reflective mirror, the distortion in the margin becomes straight and less distorted. **Figure 17** shows the third-order aberration of surfaces. The image aberration is almost reduced to the least in the image

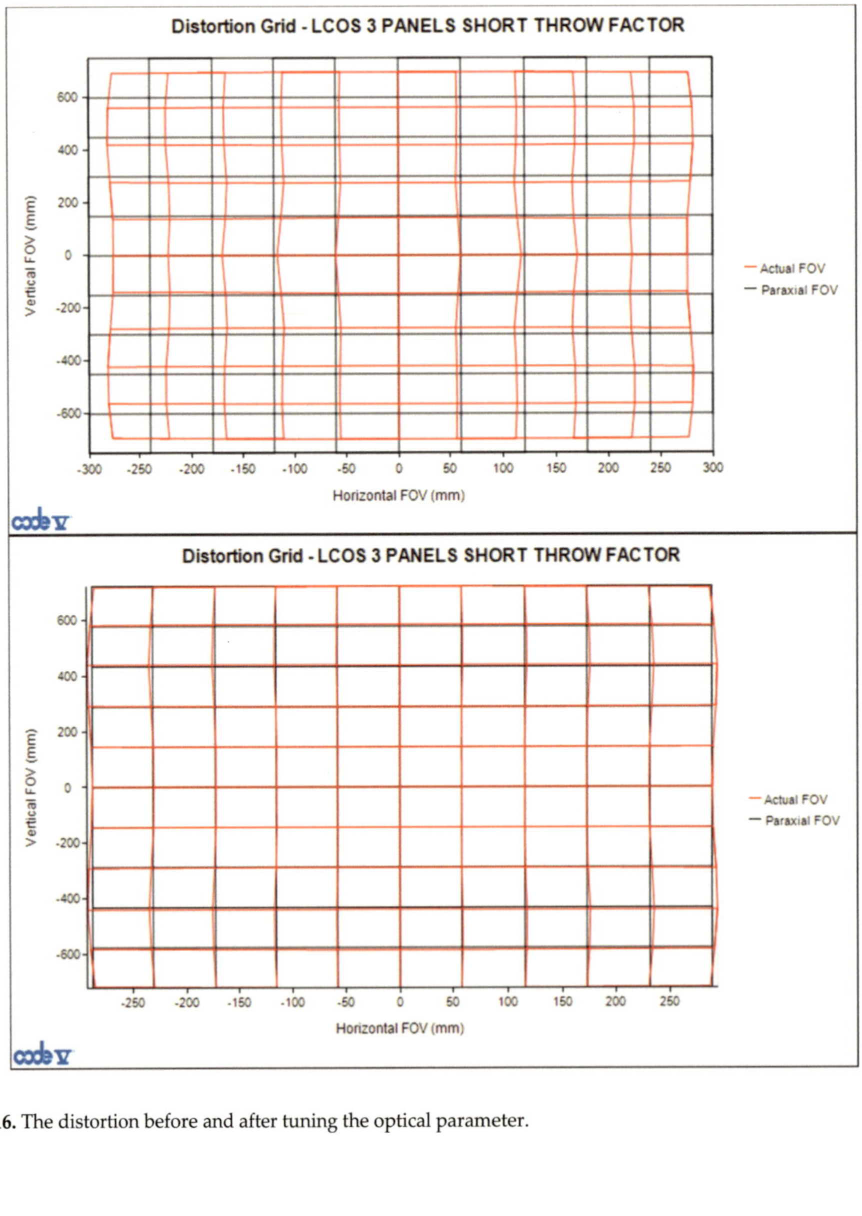

Figure 16. The distortion before and after tuning the optical parameter.

http://dx.doi.org/10.5772/intechopen.72670

plane. The MTF is shown in **Figure 18**. It shows the corresponded spatial frequency for the LCoS is still above 0.25.

3.5. Optical evaluation

The conic aspheric mirror is used to relay the intermediate image on screen. The conic constant keeps in 1.55–1.8 to form image and distortion is less than 1%. The optical system for 2048 mm × 1080 m DCI 4K2K projected screen with projecting distance 670 mm with throw ratio 0.33 is designed, and the minimum size 500 μm for the corresponded 4 μm pixel of LCoS is reached, except for the bottom of the screen.

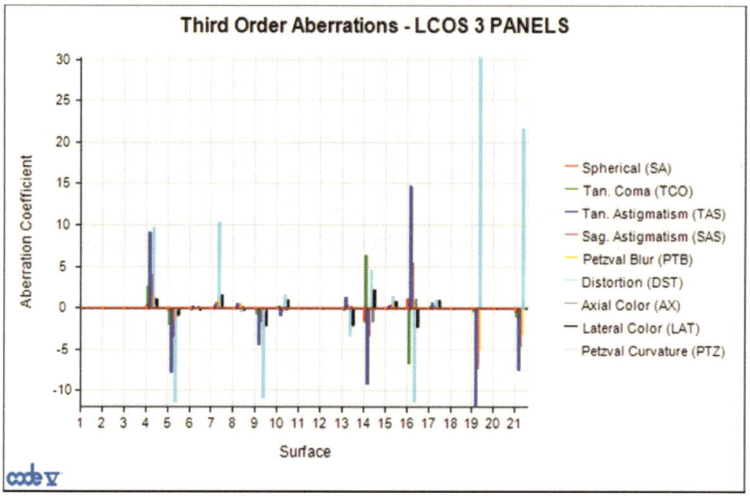

Figure 17. The third-order aberration for each surface.

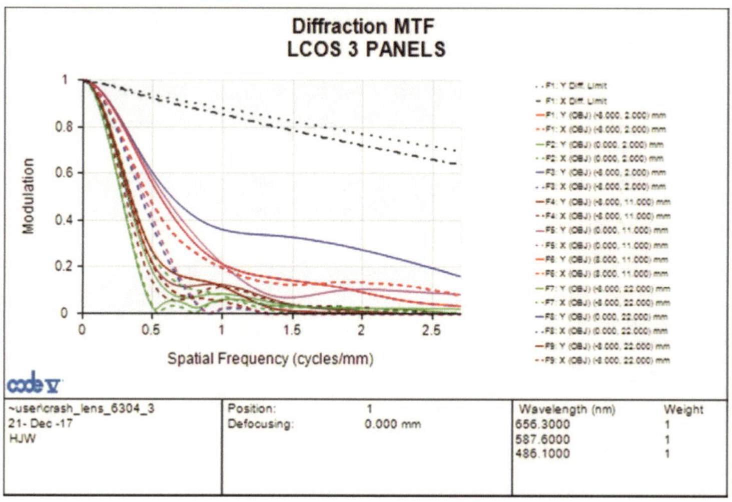

Figure 18. The MTF of short throw ratio projector.

4. Prototype and analysis

4.1. Fabrication for demonstration

In **Figure 19**, the utmost Co LCoS projector with the lens is adapted as emitted panel or effective light source. The opto-mechatronic mount between LCoS engine and short throw ratio projection are built. The performance is ready to be verified at room light on and at room light off, as shown in **Figures 20** and **21**, respectively.

In this case, 2048 mm × 1080 m DCI 4K2K projected screen with projecting distance 670 mm with throw ratio 0.3 is designed, and the minimum size 500 μm for the corresponded 4 μm pixel of LCoS is reached, except for the bottom of the screen.

In **Table 3**, the analysis of the system is explained as follows:

1. The throw ratio of projection is less than 0.3 and reaches the requirement.
2. The contrast of projected image can be enhanced by using telecentric system.
3. By modifying Koehler illumination in reflective LCoS panel, the clear image with high contrast will be viewed.

4.2. The throw factor ratio

The short throw and wide-angle projection lens is different from other lens systems. According Eq. (1), D′ is the distance between the bottom side of screen to the last lens or mirror.

$$\text{Throw ratio} = W/D' \tag{1}$$

In this case, the throw ratio 666/2000 mm = 0.333.

4.3. Illumination and partial coherence analysis

The partial coherence analysis has been applied to this case. Because MTF cannot provide the full view to verify the requirement of 4K2K, the partial coherence distribution can fully

Figure 19. Utmost Co. LCoS projection engine with the self-designed optics and opto-mechatronic mount.

Figure 20. At room light on (day).

present pixels of LCoS in each field to fulfill the requirement of 4K2K. A 4-bar pattern with the pitch of 0.004 μ corresponded to one pixel of LCoS is projected in each field respect to each field. The width of pitch projected on screen is 0.5 mm, and with the full screen, it is 2000 × 1000 mm^2. For each pair, the resolution has reached 2.5 lps/mm, and has 5K2.5K fulfilled 4K2K, beside the field 0°. By setting relative numerical aperture (RNA) 0.6, the partial coherence analysis is shown in **Figure 22**. All the fields are resolvable except blur bottom of the screen.

4.4. The number aperture and F/#

The F/# for the short throw and wide-angle projection, different from other lens systems, has considered two f/#; one is optical system and the other is illumination. Here, the LCoS as light source projects onto the screen to form image, as the source of the Koehler model. The F/# of projection lens is 3 and F/# of each pixel of as effective light source is 40, thus the relative numerical aperture is applied.

Figure 21. At room light on (night).

No.	Specification for short throw factor lens		
	Item	*Design specification*	*Actual*
1	Pixels size in 1 m	1 mm	<0.5 mm
2	Lens type	Refractive and reflective	Refractive and reflective
3	Short throw ratio	<0.4	<0.3

Table 3. The comparison between design specification and actual value.

The result and analysis are mentioned as follows: The design has fulfilled the requirement as shown in **Table 2**. Two configurations of short throw ratio projection lens are fabricated and under test.

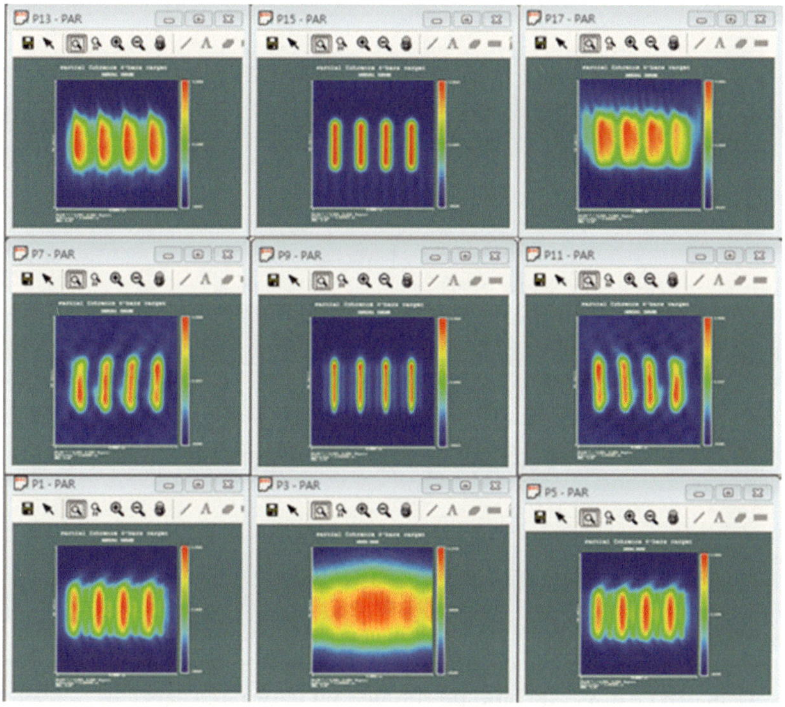

Figure 22. 4-bar pattern with the pitch of 0.004 μ is projected in each field.

5. Conclusion

Being designed by Koehler illumination and telecentric optics, the systems, having the optimized optical parameters, could be achieved. The results for two system requirement are promising, better than other structures. As the systems are built, the optimized performance should be expected. The partial coherence analysis has been applied to verify the system reaching 4K2K.

The chapter presents the design of a short throw projector optical system, and the throw ratio is 0.3, with a single aspheric mirror and the least number of spherical lens. It provides a wide angle projection lens system with partial coherence source parameter provided. The throw ratio is less than 0.3. The procedures of the optical design and illumination are presented.

5.1. Very short projecting distance optical system

With regard to throw ratio that can be less than 0.2, the design has also been carried out. While tilting the mirror at 8°, the throw ratio is 0.186663, as shown in **Figure 23**. However, the image may be keystone, and it can be corrected by the Scheimpflug effect. The image may be corrected by slightly tilting the object, and the corrected nondistorted image may be formed.

Figure 23. The tilt-mirror projector.

5.2. Other issues

The zoom system can be done. It could be zoomed by various distance of first lens, and the 1 to 1.33, by replacing with zoom optics of relay lens.

In conclusion, the optical system has been fabricated and installed in a three-panel forward looking projector system. It has been demonstrated and reached to 4K2K format. The application can be used for other projections such as automobile HUD, VR, glass-type display, smart phone external display, and others can be easily applied.

Acknowledgements

The study is supported by the grants from Ministry of Science and Technology, Taiwan, Republic of China (MOST 104-2221-E-492-040-)and(MOST-2622-E-492_029-CC3). We thank the cooperation of Professor C.H. Chen, and Mr. Wei-Cheng Lin. We deeply thank Dr. Shih-Feng Tseng for his sincerely instruction and encouragement.

Author details

Jiun-Woei Huang[1,2*]

*Address all correspondence to: jwhuang@narlabs.org.tw

1 National Applied Laboratories, Instrument Technology Research Center, Hsin Chu, Taiwan, R.O.C

2 Institute of Applied Mechanics, National Taiwan University, Taipei, Taiwan, R.O.C

References

[1] Chung F-C, Ho F-C, Wu Y-L. High-resolution 60-in liquid crystal rear-projection TV. In: Projection Displays 2000: Sixth in a Series", Proc. SPIE 3954. 2000

[2] http://www.sony.net/Products/4k-ultra-short-throw/

[3] Matsumoto S, Amano R, Okuda M, Adachi T, Okuno S. Ultra-short Throw Distance Front Projector with Mirror-Lens Hybrid Projection Optical System. 9.4-2 Consumer Electronics, 2008. ICCE 2008. Digest of Technical Papers. In: International Conference on Date 9-13 January 2008

[4] Abe T, Mashitani K, Kanayama H. Floor-Projected 3D system by 3D ready Ultra Short Throw Distance Projector. In: IEEE International Conference on Consumer Electronics (ICCE); 2011. pp. 757-758

[5] DLP, https://en.wikipedia.org/wiki/Digital_Light_Processing

[6] Sidney F. Ray, Applied Photograpic Optics, Ch.64. London: Focal Press; 1988. pp. 453-468

[7] Charbonneau M. Short throw projector system. Thales Co; 2012

[8] Code V 11. Optimization: Synopsys Co; 2017

[9] Huang J-W. The optical design of ultrashort throw system for panel emitted theater video system. In: International Conference on Optical and Photonic Engineering (icO-PEN 2015), Proc. SPIE 9524-134; 2015

[10] Huang J-W. The design and fabrication of common optical components lithography lens. In: International Conference on Optical and Photonic Engineering (icOPEN 2015) Proc. SPIE 9524-69; 2015

[11] Huang J-W. Optical design of ultra-short throw LCoS projection system. In: ODF2016 International Conference on Optics-Photonics Design & Fabrication 2S3-05. Weigarten Germany; 2016

[12] Huang J-W. Optical design of a 1-to-1lithography projection. Optical Review. 2016;**23**(5): 870-877

[13] Huang J-W. Optical design of ultrashort throw liquid crystal on silicon projection system. Optical Engineering. 2017;**56**(5):051408

[14] Huang J-W. Chap.15.7.2 and Chap. 38. Design and fabrication of optro-mechatronics system. Taipei, Taiwan, R.O.C.: Wunan Bookstore Co. http://www.wunan.com.tw/bookdetail.asp?no=11702

Chapter 12

Recent Dispersion Technology Using Liquid Crystal

Yuji Yamashita

Additional information is available at the end of the chapter

http://dx.doi.org/10.5772/intechopen.74156

Abstract

Lyotropic liquid crystals have prospective potentials for several industrial applications and also being a key technology in terms of the quality assurance of a product, drug carrier, as well as interpretation of biological phenomena. This chapter will provide the recent topics on several applications of liquid crystals in the cosmetic and pharmaceutical fields and review how to generate the lyotropic liquid crystals in the amphiphilic material system on the basis of the phase behavior and why the liquid crystal structure can impact the respective application.

Keywords: liquid crystal-based emulsification, nanoemulsion, vesicle, cubosome, hexosome, drug delivery vehicle

1. Introduction

In a few decades, formulation technology in the fields of cosmetics and pharmaceutics has evolved owing to the advanced nanotechnologies involving theory, computational simulation, and analytical devices, and nowadays, various forms such as a capsule, tablet, poultice, and liquid emulsion can be designed in consideration of usability, quality assurance, as well as efficacy of an active ingredient. Colloid science is a very strong tool to understand and control these points and eventually most of formulations regardless of soft and hard matters. In addition, the stuff we are made of, blood, organ, and bone, contains colloidal particles. Since the industrial era, new kinds of colloid-containing products, including paint, foam, pastes, and so on, have been developed.

The colloidal system is referred to be a system in which one phase is homogeneously dispersed in another phase. It seems to be the similar relation of solute and solvent, while this dispersion system should be little soluble mutually. Both the dispersed phase and continuous phase are in gas, liquid, solid states, and may be in liquid crystal (LC) state, and generally many industrial products can be categorized by these phase states (**Figure 1**) [1]. Typically, the emulsified products such as milk consist of oil (oil phase) and water (aqueous phase), namely being a liquid-liquid dispersion system. When one liquid is dispersed in another liquid, the dispersion system would be called "lotion" for a transparent solution or "emulsion" for a turbid solution. Thus, the colloid dispersion system can be classified into two types, in which one is "molecular colloid" or "association colloid," and the other "dispersed colloid." The molecular colloids are known to be formed in polymer solutions such as starch and protein, and the association colloids are micellar solution consisting of surfactant molecules. These two colloid systems are thermodynamically stable and spontaneously formed in a solvent, generally called "solubilizing system." This system has been utilized for cleansing, the targeting drug delivery of a poorly soluble compound encapsulated in micelle, and so on. On the other hand, the dispersed colloid is unstable and separated into two phases sooner or later, and many of formulations, such as liquid-liquid emulsion and liquid-solid suspension, are concerned.

This chapter will introduce unstable colloid dispersion systems using LC. One may have doubt on the relation between the colloid dispersion and LC. However, this intermediate state has a potential to generate new value and some liquid crystals have been already contributed to the formulation technology. Here, the following two topics will be separately mentioned because LCs are applied in different manners.

(1) Emulsification technology using self-assemblies

(2) LC dispersions

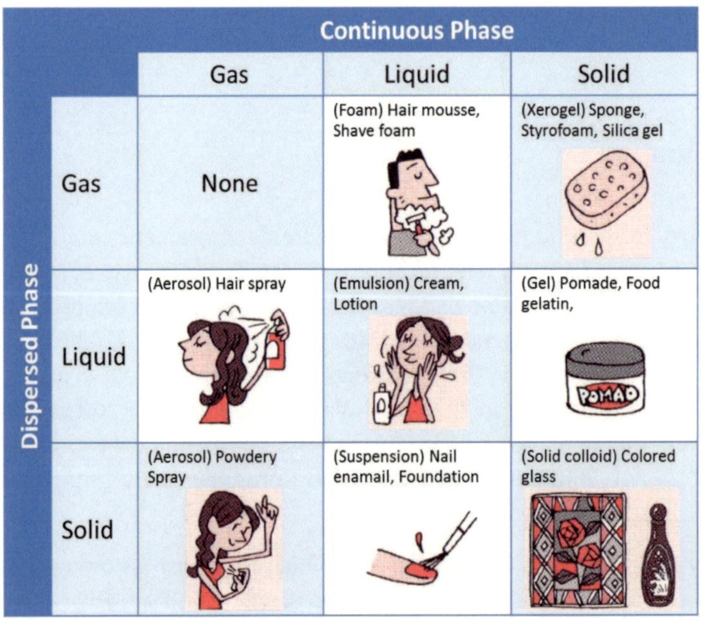

Figure 1. Various colloidal dispersion systems [1].

The first topic will explain emulsion systems stabilized by LCs and the unique properties, and note that LC can be used as a stabilizer for emulsion. The second topic may be more common for the recent researchers of LCs and will review various LC dispersions that are prospective vehicles for the drug delivery system.

2. Lyotropic liquid crystal formed in surfactant system

First, LC used for cosmetics and pharmaceutics is explained in brief. As known well, the LC can be classified into "lyotropic" and "thermotropic" LCs, which may be defined by their dependent parameters, concentration and temperature, respectively. In some cases, they cannot be definitively distinguished by their features, for example, the nematic phase is often observed in the thermotropic LC, but a peculiar surfactant solution system forms it at certain temperature [2]. In addition, the identical LC structure may be termed independently, for example, hexagonal LC for lyotropic system and columnar LC for thermotropic one. The principal difference between two LCs is constituent; the representative compound to form the lyotropic LC is surfactant, and the thermotropic LC is formed by anisotropic molecules with a mesogen group. Some surfactants have mesogen groups in the molecule as well, whereas the important interactions in the lyotropic LC system should be solvation and hydrophobic interaction rather than molecular interactions via the mesogen group that can provide the translational order and optical anisotropy. Therefore, most of the surfactant cannot work in the absence of solvent and rarely forms LC by itself.

The surfactant is paraphrased by amphiphiles which have the dual character, hydrophilicity and lipophilicity, derived from hydrophilic and lipophilic groups. The thermodynamic properties of amphiphiles in aqueous solution are controlled by the hydrophilic group to avoid contact with water, referring to "hydrophobic effect" [3]. This leads to spontaneous formation of micelle at lower concentration of surfactant (above critical micellar concentration, CMC) and generally liquid crystals at higher concentration. The formation of self-assembled bodies is predominantly determined by an entropic contribution which arises from the local structuring of water, known as iceberg structure.

At high concentrations, surfactants can self-assemble into lyotropic LCs and their structures depend on the concentration. **Figure 2** shows schematic structures of the series of typical lyotropic LCs formed in a surfactant system. Cubic LC is very stiff and optically isotropic, basically divided into two types: discontinuous (I_1) and bicontinuous cubic LC (V_1). These cubic LCs are furthermore classified into 230 kinds of the crystal lattice with symmetries called space group. The space group can be assigned by the characteristic reflection plane relevant to Miller indices. Hexagonal LC (H_1) has the two-dimensional structure that the infinitely elongated rod-like micelles are packed in the hexagonal array and shows optical anisotropy. Lamellar LC ($L\alpha$) consists of one-dimensionally stacked bilayers and also shows optical anisotropy. The reverse-type micelle and LCs except for $L\alpha$ are formed in the surfactant solution; reverse micelle (L_2), reverse discontinuous (I_2) and bicontinuous cubic LC (V_2), and reverse hexagonal LC (H_2).

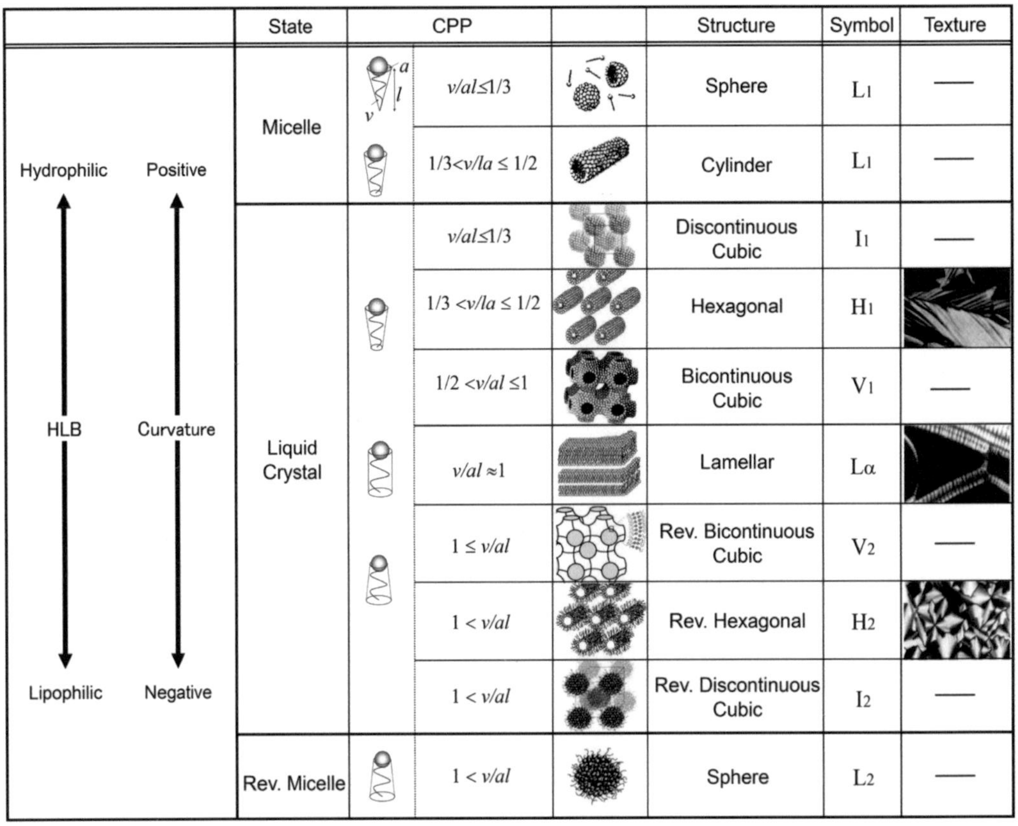

Figure 2. Summary of self-assembly structures formed in surfactant systems, and relationship between the structure and three parameters, critical packing parameter (CPP), hydrophilic-lipophilic balance (HLB), and interfacial curvature.

A schematic phase diagram in a binary surfactant/water system is demonstrated in **Figure 3**, indicating that all LCs not always appear over the concentration range. The type of LC formed in the system depends on the kind of surfactant, added oil, and additive as well as surfactant concentration. Temperature is also a factor to determine the micelle and LC structure. The temperature-dependent phase transitions can be observed in **Figure 3**, for example, micellar solution (L_1) → two phase (II), and hexagonal LC (H_1) → L_1. Any phase transitions in a surfactant system are always relevant to interaction between surfactant and solvent, and three important parameters, interfacial curvature, critical packing parameter (CPP), hydrophile-lipophile balance (HLB) number, prevailing in the academic and industrial fields are applied to understanding and controlling the self-assembly structures and the phase transition phenomena. The concentration-dependent LC structures can also be interpreted by these parameters.

2.1. Interfacial curvature

The LC structure is characterized by the interfacial curvature (main curvature). In principle, overall area on the interface can be defined as the mean curvature (H) and Gaussian curvature (G) using the radii of the main curvatures, R_1 and R_2

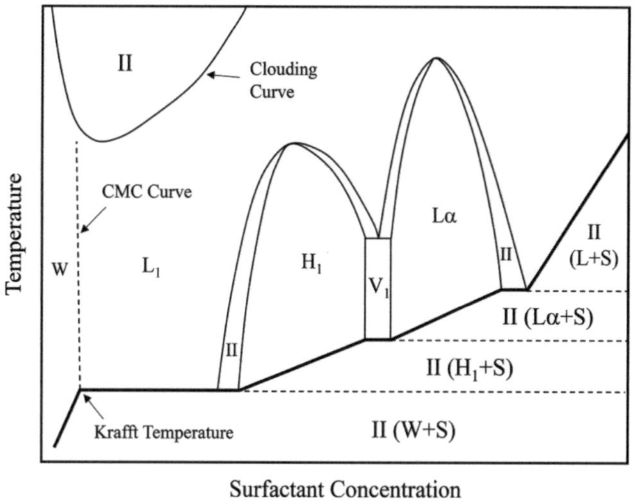

Figure 3. Schematic phase diagram of a binary surfactant/water system. W: monodispersed solution, L_1: micellar solution, H_1: hexagonal LC, V_1: bicontinuous LC, La: lamellar LC, S: surfactant solid, and II: two phase.

$$\text{Mean curvature:} \quad H = \frac{1}{2}\left(\frac{1}{R_1} + \frac{1}{R_2}\right) \tag{1}$$

$$\text{Gaussian curvature:} \quad G = \frac{1}{R_1} \times \frac{1}{R_2} \tag{2}$$

In the case of a spherical micelle, which is formed at the low surfactant concentration, the mean curvature corresponds to $H = 1/R$ ($R = R_1 = R_2$) and the Gaussian curvature is $G = 1/R$ [2] because of its isotropic structure. On the other hand, anisotropic structures, such as cylindrical micelle and bilayer structure, give different curvatures; $H = 1/(2R_1)$ and $G \sim 0$ for the cylindrical structure, $H \sim 0$ and $G \sim 0$ for the bilayer structure. In general, the positive curvature indicates convex toward the water phase, and contrarily the negative one is concave. Thus, the curvature continuously changes from positive to negative or from large to small in the order corresponding to L_1, I_1, H_1, V_1, $L\alpha$, V_2, H_2, I_2, L_2 as shown in **Figure 2**.

2.2. Critical packing parameter (CPP)

The LC structures are governed geometrically by the volume fraction of the self-assembly occupied in space of the solution and the molecular structure of surfactant composed in the system. The surfactant molecules can be arranged in a self-assembly under a given condition so that the interfacial area per molecule will be minimized in order to avoid the contact of the alkyl chain and water. The morphology of the self-assembly is determined by the balance of two opposing forces, hydrophobic attraction at the alkyl chain-water interface, and repulsive force between the head groups of surfactants (ionic repulsion, hydration force, steric hindrance, etc.). The interfacial free energy per surfactant molecule (μ_N^0) can be written as follows [4]:

$$\mu_N^0 = \gamma a + \frac{K}{a} \tag{3}$$

where K is the constant, γ is the interfacial tension, and a is the cross-sectional area of the surfactant head group at the interface. The first and second terms in the equation represent attraction and repulsion, respectively. Assuming that these interactions would operate within the same interfacial area, the optimized effective cross-sectional area per molecule (a_S) is estimated from the minimum μ_N^0.

Israelachivili proposed "critical packing parameter (CPP)," which allows one to predict the morphology of the self-assembly [4]. CPP has the non-dimensional unit and can be calculated using the volume of alkyl chain (V_L), the length of the extended alkyl chain (l), and a_S

$$\mathrm{CPP} = \frac{V_L}{a_S l} \tag{4}$$

CPP gives a geometric characterization of a surfactant molecule and will be seen to be very useful when discussing the type of self-organized structure formed by a given amphiphile. Considering what surfactants fall into the different categories of the self-assembly structures shown in **Figure 2**, we note that CPP characterizes the self-assembly structure, for example, the CPP < 1/3 for the spherical micelles (L_1, I_1), 1/3 ~ 1/2 for the cylindrical micelles (H_1), ~1 for the bilayer structure ($L\alpha$). For the nonionic surfactant, CPP becomes smaller with increasing the polymerization degree of the hydrophilic group [5–7], indicating that curvature changes toward positive.

2.3. Hydrophile-lipophile balance (HLB) number

The HLB number has been utilized as a parameter which characterizes the surfactant and would be widely spreading in the industrial field because of the chain length distribution of the commercial surfactants.

HLB denotes the nature of surfactant in terms of hydrophilicity and lipophilicity. Griffin [8, 9] codified the HLB numbers for nonionic surfactants. Till now, several equations have been proposed to calculate the HLB number for different surfactants including ionic surfactants [10–14]. Generally, the HLB number can be calculated from the hydrophilic and lipophilic portions of the molecule. The HLB number is a useful parameter for selection of surfactants suitable for various applications (e.g., emulsifier, solubilizer, wetting agent, and antifoamer).

3. Formulation utilizing self-assembly

3.1. Liquid crystal emulsification

Since an emulsion is a thermodynamically unstable system, the state and stability are greatly influenced by the preparation process. This can be understood from the several emulsification

http://dx.doi.org/10.5772/intechopen.74156

methods such as the phase inversion temperature (PIT) method [15], D-phase emulsification [16], quenching method [17], and liquid crystal (LC) emulsification [18], which are attributed to stability at the oil-water interface accumulated by surfactant molecules or self-assemblies.

The LC emulsification method was discovered by Suzuki et al. and referred to the process that an oil phase was added directly to a lamellar liquid crystal ($L\alpha$) phase, and then dispersed by agitation to produce an emulsion (**Figure 4**) [18]. The key to LC method is to select an appropriate surfactant that preferentially forms $L\alpha$ phase, as well as constituents of aqueous phase. According to CPP, the surfactant with a balanced HLB number, in general, a two tails surfactant tends to form $L\alpha$. This emulsification method is achieved in two steps corresponding to the arrows in **Figure 4**. In the first step, the oil phase is added and dispersed into the $L\alpha$ phase composed of surfactant/glycerol/water. In the second step, water is poured into the oil in $L\alpha$ (O/LC) to form an O/W emulsion. This LC method can form a stable emulsion because the LC phase is present as "third phase" surrounding the dispersed oil phase and physically prohibits coalescence of emulsion droplets. The stabilization mechanism of emulsion can be referred to the other emulsification technologies such as Pickering emulsification [19] and three-phase emulsification [20], collectively named "Active Interfacial Modifier (AIM)" [21, 22].

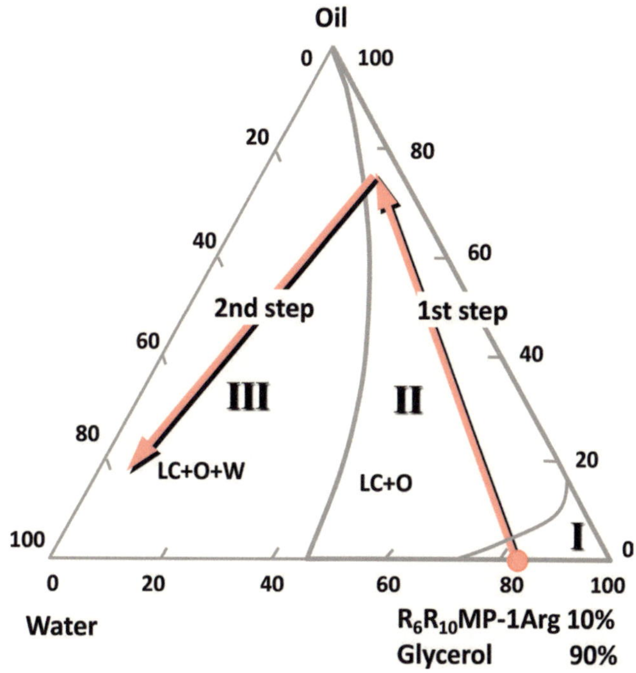

Figure 4. Procedure of LC emulsification in the ternary phase diagram of b-branched L-arginine hexyldecyl phosphate (R_6R_{10}MP-1Arg)/glycerol/oil/water system [18]. Premixture of R_6R_{10}MP-1Arg/glycerol/water forms the lamellar LC, in which oil is added then in order to form the two phase LC+O (O/LC) (first step). Finally, O/W emulsion (LC+O+W) is obtained by adding water to the O/LC solution (second step). W: water phase, O: oil phase, and LC: liquid crystal phase.

3.2. Nanoemulsion prepared by cubic liquid crystal

Nanoemulsions are nanosized emulsions, typically, a size of tens to hundreds nanometer, which can be expected to improve the stability of emulsion and the delivery of active ingredients. The term "nanoemulsion" also refers to a mini-emulsion which is fine oil/water or water/oil dispersion stabilized by an interfacial surfactant film. According to the droplet size, the nanoemulsions are apparently transparent or translucent [23–25]. Contrary to the microemulsions, the nanoemulsions are thermodynamically unstable, yet they may have high kinetic stability. Disruption of the nanoemulsions would be processing within hours, days, or weeks through general flocculation, coalescence, and Ostwald ripening. These characteristic properties have put the nanoemulsions to practical use, such as cosmetics [26–28], pharmaceutics [29–34], reaction media for polymerization [35, 36], and agrochemicals [37].

In industrial fields, it has been paid attention to how to formulate and prepare a stable emulsion. Two major methods for the preparation of fine emulsions are well known: dispersion or high-energy methods, and condensation or low-energy methods [23]. The high-energy method is the most popular procedures to produce a fine emulsion using specific equipment, such as high-shear stirring, high-pressure homogenization, and ultrasonication [24].

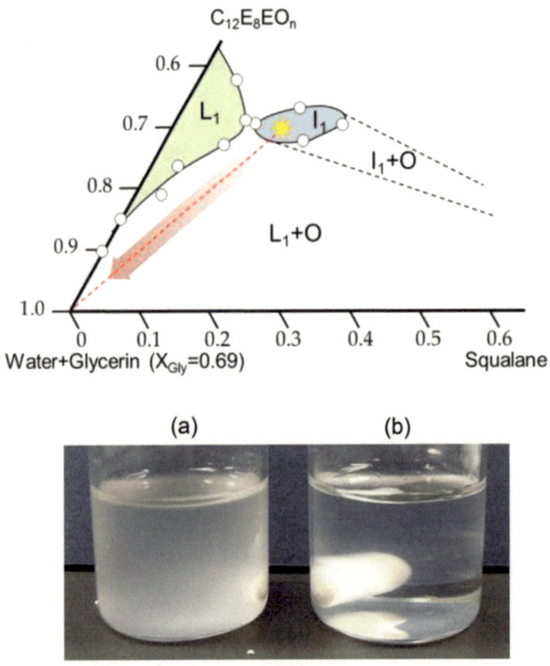

Figure 5. Quasi-ternary equilibrated phase diagram in the polyoxyethylene octyldodecyl ether ($C_{12}C_8EO_n$)/water/glycerol/squalane system at 25°C (top) [46]. The weight ratio of water/glycerol is fixed at 31/69. The arrow in the phase diagram indicates the preparation route of the novel emulsion. Bottom pictures show sample appearances of the solutions prepared by (a) simple mixing and (b) dilution method utilizing the cubic liquid crystal (I_1). L_1: micellar solution, I_1: discontinuous cubic LC, and O: excess oil (O).

This method, however, is not preferable from the point of environmental view because of a large amount of energy loss. On the other hand, the low-energy methods utilize unique properties of surfactant and in particular the phase transitions that take place during the emulsification process as a result of a change in the spontaneous curvature of the surfactant. The phase transition with the drastic curvature change can be driven by the phase inversion temperature (PIT) method [15, 38] and the phase inversion composition (PIC) method [39]. The preparation methods of nanoemulsions have been widely reported in the nonionic and ionic surfactant systems, by using both of the high- and low-energy methods [23, 24]. Solans et al. had thoroughly investigated low-energy input methods using PIT and PIC and successfully produced finely dispersed nanoemulsions [40–45]. It was also demonstrated that a liquid crystal formation would play an essential role in forming a fine nanoemulsion [44].

Yamashita et al. proposed a unique nanoemulsion using a discontinuous cubic LC (I_1) [46]. This nanoemulsion is simply obtained by diluting I_1 without any high-energy input (**Figure 5**). Contrary to the common emulsions, the I_1-based nanoemulsion has an abnormal shear-response: the semi-stable structure of the nanoemulsion is breaking down gradually by applying a mechanical energy (**Figure 6**). On the other hand, H_1 and V_1 do not form such transparent nanoemulsion. Such a new type of emulsion would be applicable for cosmetics and pharmaceutics as an external application. Since the solution transforms from nanoemulsion to emulsion when shearing force is applied, the solubility of active agent loaded in the hydrophobic compartment of the nanoemulsion should be varied. This can also modulate partition between the formulation and the skin surface (stratum corneum), which is a key factor for transdermal drug delivery systems [47, 48].

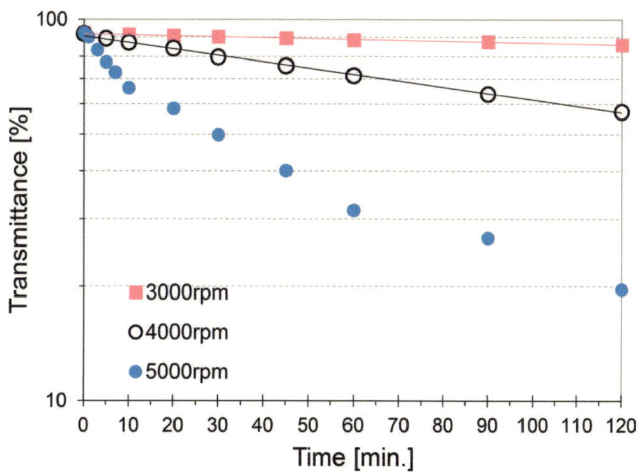

Figure 6. Change in transmittance of the nanoemulsion formed in the polyoxyethylene octyldodecyl ether ($C_{12}C_8EO_n$)/water/glycerol/squalane system as a function of time under different shearing rates; 3000 rpm (■), 4000 rpm (○), and 5000 rpm (•) [46]. The transmittance measurements were carried out using the monochromatic light source (l = 550 nm) at room temperature. The surfactant concentration is 1.4 wt.%.

4. Liquid crystal dispersion

Liquid crystal dispersions are promising drug carriers and typically referred to vesicle (liposome), cubosome, and hexosome that have two domains to accumulate both hydrophilic and lipophilic ingredients, although the micelle or reverse micelle has either compartment.

4.1. Vesicle and liposome

A vesicle is a hollow aggregate with a shell made from one or more amphiphilic bilayers. According to the number of bilayer shell, vesicles can be roughly categorized: a vesicle with a single bilayer is called "unilamellar vesicle" and the one with a shell of several bilayers is "multilamellar vesicle (MLV)." MLV is sometimes called "onion vesicle." **Figure 7** exhibits a unilamellar vesicle. Vesicles formed by lipids are termed "liposomes," which are of great interest and have been widely studied because they are simple membrane models for cell. Vesicles or liposomes have no biological functionality, while vesicle formation and fusion should be important in many physiological processes. Liposomes are also important technology in cosmetics and for drug delivery. In both cases, the liposome acts as a delivery vehicle for active material contained inside. The aims of encapsulating the active materials (or drugs) in the liposome are mainly targeting and release control, whereby not only effective delivery but reduction of side-effect can be attained. However, this targeting technology has not been established yet, although gradually developed by recent studies such as protein recognition and stealth vehicle.

Vesicles (or liposomes) are usually not in thermodynamic equilibrium, while they can be kinetically stable for quite long period. As seen in **Figure 8** [49], vesicles are formed in a two-phase region, $L\alpha$ + W, where excess water is separated from the $L\alpha$ phase. In such systems, the constituent molecules cannot transform to another LC when diluted with water because of their packing restriction of lipophilic chain, and instead vesicles are formed to minimize the energy loss of lamellar membrane edge (E_{edge}) [50, 51].

$$E_{edge} = 2\pi R\,\gamma_L \tag{5}$$

R is the radius of lamellar sheet (disk) and γ_L is the line tension. On the other hand, the bending energy (E_{bend}) should be required to form the vesicles, expressed by the following equation [52]:

$$E_{bend} = 8\pi\kappa \tag{6}$$

where k is the bending modulus. When E_{bend} is smaller than E_{edge}, vesicles are preferentially formed. The unit structure of vesicle is same with $L\alpha$ and CPPs of both morphologies are assigned to be nearly unity. According to the morphological similarities, the concentric $L\alpha$phase can be reversibly transformed to a multi-lamellar vesicle (MLV) by applying a certain shearing force (**Figure 9**) [39].

Practically, unilamellar vesicles with different sizes are used for the drug delivery carrier and cell model, while the bilayers of these vesicles may have different physicochemical properties

http://dx.doi.org/10.5772/intechopen.74156

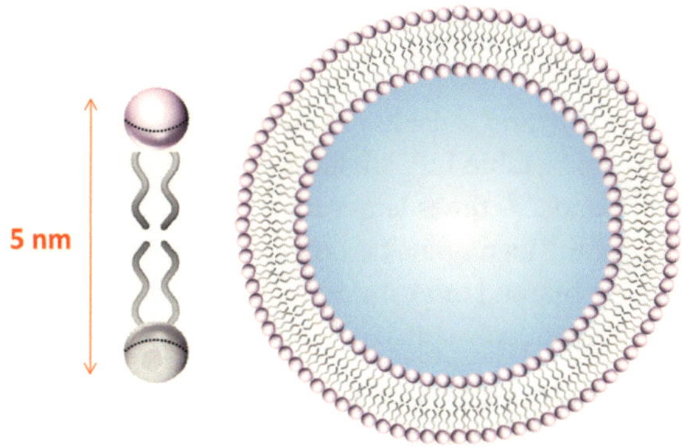

Figure 7. Schematic representation of unilamellar vesicle.

depending on the size. General unilamellar vesicles are listed in **Table 1**, where one can compare to various cell sizes [53, 54]. Regarding the topological effect of the vesicle, the surface energy depends on the curvature as expressed by Laplace equation

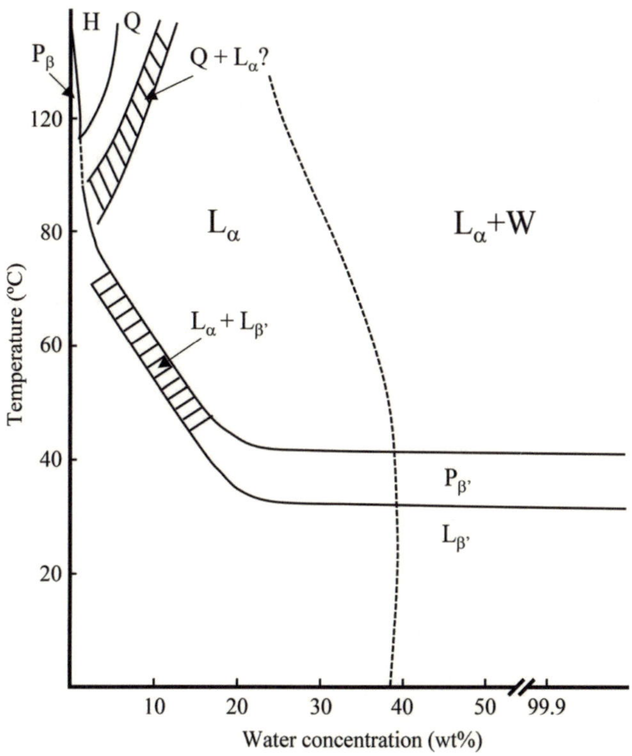

Figure 8. Phase diagram of the binary DPPC/water system [49]. La: lamellar LC, H: reverse-type hexagonal LC, Q: reverse-type bicontinuous cubic LC, P_b: flat ripple phase, P_b: non-flat ripple phase, L_b: gel phase, and W: excess water.

$$P_{in} = P_{out} + \frac{2\gamma}{r} \tag{7}$$

where P_{in} and P_{out} are the inside and outside pressure, γ is the interfacial tension, and r is the radius of curvature. As shown in **Table 1**, P_{in} for large unilamellar vesicle (LUV) and small unilamellar vesicle (SUV) are 25 and 250 times larger than giant unilamellar vesicle (GUV), respectively. In other words, E_{bend} of the membrane becomes larger with decreasing the vesicle size and then the molecules are less mobile and more ordered. Sakamoto suggested that the bilayer curvature had a significant effect on not only stiffness, but also function of the bilayer membrane [55].

Many methods can be applied to prepare various vesicles, which result in different types of vesicles and size distributions [56, 57]. First of all, it should be noted that vesicles are formed in a specific composition range depending on the kind of surfactant and phospholipid used in the system, and generally in the diluted lamellar phase which refers to the region coexisting the lamellar LC (Lα) and excess water (W) in the phase diagram. In this region, vesicles can be easily prepared by simple shaking, but many of them are MLV. Sonication is typical treatment to form vesicles with single bilayer; the high-frequency sound waves can break up the inhomogeneous stacked bilayers, inducing reassembly of bilayer. Such rough preparation produces SUV with a broad size distribution since the mechanical action is very uneven. Instead, an alternative procedure can be taken to form in particular LUV and GUV, referring to the thin film method: (1) the amphiphile is dispersed in an organic solvent, (2) the organic solvent is distilled away under vacuum to form a thin film of the amphiphile, and then (3) an excess of water is added to the thin film. In addition, dialysis and filtration (extrusion) are often utilized to fractionate the different sizes of vesicles. However, these methods deliver only formation of vesicles with a desirable size and membrane structure, and further technical methods are required to attain the prospective functions of uniform vesicles such as targeting and a large encapsulating ratio.

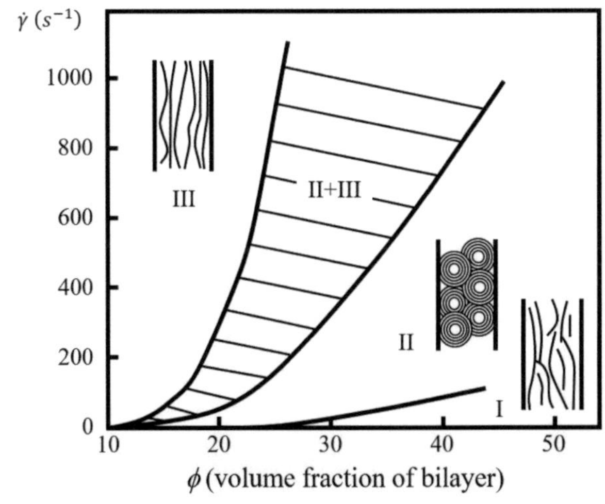

Figure 9. Dynamic phase diagram of SDS/pentanol/water/dodecane system as functions of the volume fraction of bilayer (f) and shear rate (γ') [77]. I region: defected lamellar LC, II region: multilamellar vesicle (MLV), and III region: non-defected lamellar LC.

http://dx.doi.org/10.5772/intechopen.74156

	Size	Relative pressure difference $P_{in} - P_{out} = 2\gamma/r$
Thickness of cell membrane	*ca.* 5 nm	
Small virus	30 nm	
Small unilamellar vesicle (SUV)	~ 40 nm	1
Lysosomes	200–500 nm	
Large unilamellar vesicle (LUV)	~200 nm	50
E. coli—a bacterium	2 mm	
Human red blood cell	9 mm	
Giant unilamellar vesicle (GUV)	10 mm	250
Human egg	100 mm	

Table 1. Classification of vesicles by size and relative pressure difference by Laplace equation [53, 54].

4.2. Cubosome and hexosome

Cubosome and hexosome are aqueous dispersions of inverted-type bicontinuous cubic [58–62] and hexagonal LCs [63, 64], respectively. Such nanostructured aqueous dispersions with internal hierarchical self-assemblies have received much attention because of their potential applications such as functional food and drug carriers [65–68]. **Figure 10** shows one example of phase diagram in the monoolein/water system [69], where bicontinuous cubic LC (Ia3d, Pn3m) are observed in the composition and temperature ranges. In addition, likely vesicles, two phase, Pn3m + water, is present in the water-rich region where cubosome can be formed. The fully hydrated inverted-type LCs with distinctive nanostructures are internally confined in the

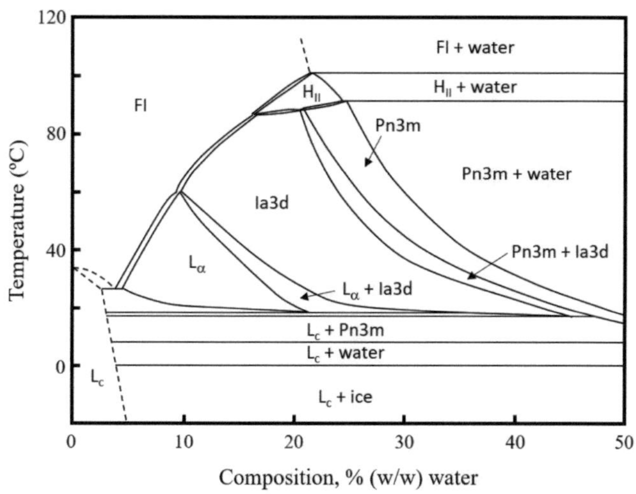

Figure 10. Phase diagram of monoolein/water system [69]. FI: fluid isotropic phase, La: lamellar LC, H_{II}: reverse-type hexagonal LC, Ia3d and Pn3m (space group): reverse-type bicontinuous cubic LC.

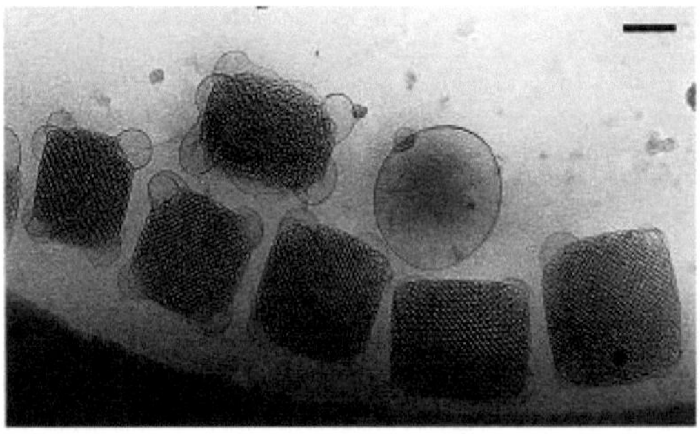

Figure 11. Cryo-TEM micrograph of cubosome [70]. The bar corresponds to 100 nm.

kinetically dispersed particles upon application of high-energy input in the presence of a suitable stabilizer like surfactant [58, 59, 68]. The internal nanostructures are controlled by CPP of amphiphilic molecule and have specific curvatures *H* and *G*. These aqueous dispersions, cubosome and hexosome, are often characterized by small-angle X-ray scattering (SAXS) and cryo-TEM. As seen in **Figure 11**, the cryo-TEM micrograph clearly demonstrates the internal nanostructures in the dispersions [70].

The feasibility of the nanostructured aqueous dispersions as drug carrier has been investigated since the 2000s, and the advantages of utilizing these dispersions have been reported, for example, solubilization of drug, bioavailability, efficient delivery, reduction of side effects, percutaneous penetration, protection of drug degradation, and release control [71–76]. However, the number of studies on drug delivery system utilizing these dispersions is still limited regardless of the unique properties so far, and further investigations will be required to understand their potentials for drug carries and also to reveal the interaction of bioactive materials and LC carries while taking the phase behavior of LC into consideration.

5. Conclusion

Beyond expectation, lyotropic liquid crystals are the soft matter familiar to our life, even managing biological functions such as homeostasis in the living system. Recently, we intend to learn or mimic many things from nature to construct artificial products with some function; on the other hand, the scientific technologies that we have ever accumulated would be applicable to reveal a new mechanism of biofunction by integrating several academic fields.

The formulations utilizing the liquid crystals have been contributed to the development of industry and supported our life. This may be a reason why the liquid crystals are constructed by self-assembling of numerous molecules and possess the properties of both liquid and solid. Still, there are many questions on the several applications utilizing the liquid crystals, and thus further investigations of the liquid crystals will clarify and find out unknown phenomena leading to novel functions of the liquid crystals.

http://dx.doi.org/10.5772/intechopen.74156

Author details

Yuji Yamashita

Address all correspondence to: yyamashita@cis.ac.jp

Faculty of Pharmacy, Chiba Institute of Science, Chōshi, Chiba, Japan

References

[1] Fukui H. Cosmetics Made Absolutely Simple (in Japanese). Tokyo: Nikkan Kogyo Shinbunsya; 2009. p. 59

[2] Acharya DP et al. Phase behavior and effect of enantiomerism on potassium N - dodecanoyl alaninate/water/decanol systems. Journal of Oleo Science. 2003;**52**(8):407

[3] Tanford C. The Hydrophobic Effect. New York: Wiley; 1980

[4] Israelachivili JN et al. Theory of self-assembly of hydrocarbon amphiphiles into micelles and bilayers. Journal of the Chemical Society, Faraday Transactions II. 1976;**72**:1525

[5] Huang K-L et al. Progress in Colloid and Polymer Science. 1998;**110**:171

[6] Kunieda H et al. The Journal of Physical Chemistry. B. 1997;**101**:7952

[7] Kunieda H et al. The Journal of Physical Chemistry. B. 2001;**105**:5419

[8] Griffin WC. Journal of the Society of Cosmetic Chemists. 1949;**1**:311

[9] Griffin WC. Journal of the Society of Cosmetic Chemists. 1954;**5**:249

[10] Davies JT et al. Interfacial Phenomena. Academic Press; 1961. p. 371

[11] Nihon Emulsion Co. Ltd. 2017. https://www.nihon-emulsion.co.jp/en/tech/organic.html [Accessed: December 4, 2017]

[12] Kawakami Y. Surfactant. Kagaku. 1953;**23**:546

[13] Yamashita Y et al. Novel indexation to express interfacial properties of surface active materials. Bulletin of Chiba Institute of Science. 2013;**6**:89

[14] Ontiveros JF et al. Structure–interfacial properties relationship and quantification of the amphiphilicity of well-defined ionic and non-ionic surfactants using the PIT-slope method. Journal of Colloid and Interface Science. 2015;**448**:222

[15] Shinoda K et al. The effect of temperature on the phase equilibria and the types of dispersions of the ternary system composed of water, cyclohexane, and nonionic surfactant. Journal of Colloid and Interface Science. 1968;**26**:70

[16] Sagitani H. Formation of O/W emulsions by surfactant phase emulsification and the solution behavior of nonionic surfactant system in the emulsification process. Journal of Dispersion Science and Technology. 1988;**9**:115

[17] Nakajima H et al. Ultrafine emulsions and application to cosmetics. Journal of Society of Cosmetic Chemists of Japan. 1988;**23**:288

[18] Suzuki T et al. Formation of fine three-phase emulsions by the liquid crystal emulsification method with arginine β-branched monoalkyl phosphate. Journal of Colloid and Interface Science. 1989;**129**:491

[19] Pickering SU. Emulsions. Journal of the Chemical Society, Transactions. 1907;**91**:2001

[20] Tajima K et al. Three-phase emulsification of hexadecane with dimyristoylphosphatidylglycerol sodium salt in water: An interpretation by new phase transition in bilayer assembly. Journal of Oleo Science. 2001;**50**(6):475

[21] Sakai K et al. Active interfacial modifier: Stabilization mechanism of water in silicone oil emulsions by peptide–silicone hybrid polymers. Langmuir. 2010;**26**(8):5349

[22] Sakai K et al. Water-in-oil emulsions prepared by peptide-silicone hybrid polymers as active interfacial modifier: Effects of silicone oil species on dispersion stability of emulsions. Journal of Oleo Science. 2013;**62**(7):505

[23] Solans C et al. Surfactant in solution: Fundamentals and applications. In: Mittal KL, Shah DO, editors. Surfactant Science Series. New York: Marcel Dekker; 2002

[24] Sudol ED et al. In: Lovell PA, El-Aasser MS, editors. Emulsion Polymerization and Emulsion Polymers. Chichester, U.K.: John Wiley & Sons Ltd.; 1997. pp. 700-722

[25] Sagitani H. In: Friberg SE, Lindman B, editors. Organized Solutions. New York: Marcel Dekker; 1992. pp. 259-271

[26] Nakajima H. In: Solans C, Kunieda H, editors. Industrial Applications of Microemulsions. New York: Marcel Dekker; 1997. pp. 175-197

[27] Friedman DI et al. Submicron emulsion vehicle for enhanced transdermal delivery of steroidal and nonsteroidal antiinflammatory drugs. Journal of Pharmaceutical Sciences. 1995;**84**:324

[28] Sagitani H. Making homogeneous and fine droplet O/W emulsions using nonionic surfactants. Journal of the American Oil Chemists' Society. 1981;**58**:738

[29] Jumaa M et al. The effect of oil components and homogenization conditions on the physicochemical properties and stability of parenteral fat emulsions. International Journal of Pharmaceutics. 1998;**163**:81

[30] Lundberg BB. A Submicron lipid emulsion coated with amphipathic polyethylene glycol for parenteral administration of paclitaxel (Taxol). The Journal of Pharmacy and Pharmacology. 1997;**49**:16

[31] Calvo P et al. Development of positively charged colloidal drug carriers: Chitosan-coated polyester nanocapsules and submicron-emulsions. Colloid and Polymer Science. 1997;**275**:46

[32] Baluom M et al. Absorption enhancement of calcitonin in the rat intestine by carbopol-containing submicron emulsions. International Journal of Pharmaceutics. 1997;**154**:235

[33] Sznitowska M et al. Increased partitioning of pilocarpine to the oily phase of submicron emulsion does not result in improved ocular bioavailability. International Journal of Pharmaceutics. 2000;**202**:161

[34] Müller-Goymann CC. Physicochemical characterization of colloidal drug delivery systems such as reverse micelles, vesicles, liquid crystals and nanoparticles for topical administration. European Journal of Pharmaceutics and Biopharmaceutics. 2004;**58**:343

[35] Blyte PI et al. Miniemulsion polymerization of styrene using the oil-soluble initiator AMBN. Journal of Polymer Science, Part A: Polymer Chemistry. 1999;**37**:4449

[36] Wang ST et al. Emulsion and miniemulsion copolymerization of acrylic monomers in the presence of alkyd resins. Journal of Applied Polymer Science. 1996;**60**:2069

[37] Jon DI et al. 1999. International Patent Application No. WO9919256 A2

[38] Shinoda K et al. The Stability of O/W type emulsions as functions of temperature and the HLB of emulsifiers: The emulsification by PIT-method. Journal of Colloid and Interface Science. 1969;**30**:258

[39] Marszall L. Nonionic surfactants. In: Shick MJ, editor. Surfactant Series. Vol. 23. New York: Marcel Dekker; 1987. pp. 493-547

[40] Esquena J et al. Progress in Colloid and Polymer Science. 1998;**110**:235

[41] Izquierdo P et al. Formation and stability of nano-emulsions prepared using the phase inversion temperature method. Langmuir. 2002;**18**:26

[42] Usón N et al. Formation of water-in-oil (W/O) nano-emulsions in a water/mixed non-ionic surfactant/oil systems prepared by a low-energy emulsification method. Colloids and Surfaces, A: Physicochemical and Engineering Aspects. 2004;**250**:415

[43] Solè I et al. Optimization of nano-emulsion preparation by low-energy methods in an ionic surfactant system. Langmuir. 2006;**22**:8326

[44] Maestro A et al. Influence of the phase behavior on the properties of ionic nanoemulsions prepared by the phase inversion composition method. Journal of Colloid and Interface Science. 2008;**327**:433

[45] Solè I et al. Nano-emulsions prepared by the phase inversion composition method: Preparation variables and scale up. Journal of Colloid and Interface Science. 2010;**344**:417

[46] Yamashita Y et al. Shear-response emulsion prepared through discontinuous cubic liquid crystal. Chemistry Letters. 2013;**42**(4):433

[47] Higuchi T. Physical chemical analysis of percutaneous absorption process from creams and ointments. Journal of the Society of Cosmetic Chemists. 1960;**11**:85

[48] Hyakutake M. Japanese: Journal of Society of Cosmetic Chemists of Japan; 1985;**19**:30

[49] Stillwell W. An Introduction to Biological Membranes. London: Elsevier; 2013. p. 191

[50] Jung HT et al. Proceedings of the National Academy of Sciences of the United States of America. 2002;**99**:15318

[51] Shioi A et al. Model for formation and growth of vesicles in mixed anionic/cationic (SOS/CTAB) surfactant systems. Langmuir. 2002;**18**:7341

[52] Helfrich W. Elastic properties of lipid bilayers: Theory and possible experiments. Zeitschrift für Naturforschung. 1973;**28c**:693

[53] Wikibooks. 2017. https://en.wikibooks.org/wiki/Cell_Biology/Introduction/Cell_size [Accessed: November 11, 2017]

[54] Takaichi Y et al. Netsu Sokutei. 1992;**19**:103

[55] Sakamoto K. The importance of planarity for lipid bilayers as biomembranes. Advances in Biomembranes and Lipid Self-Assembly. 2016;**23**:1

[56] Akashi K et al. Preparation of giant liposomes in physiological conditions and their characterization under an optical microscope. Biophysical Journal. 1996;**71**:3242

[57] Mueller P et al. Formation and properties of cell-size lipid bilayer vesicles. Biophysical Journal. 1983;**44**:375

[58] Gustafsson J et al. Submicron particles of reversed lipid phases in water stabilized by a nonionic amphiphilic polymer. Langmuir. 1997;**13**:6964

[59] Larsson K. Aqueous dispersions of cubic lipid–water phases. Current Opinion in Colloid and Interface Science. 2000;**5**:64

[60] Spicer PT et al. Novel process for producing cubic liquid crystalline nanoparticles (cubosomes). Langmuir. 2001;**17**:5748

[61] de Compo L et al. Reversible phase transitions in emulsified nanostructured lipid systems. Langmuir. 2004;**20**:5254

[62] Barauskas J et al. Self-assembled lipid superstructures: Beyond vesicles and liposomes. Nano Letters. 2005;**5**:1615

[63] Yaghmur A et al. Emulsified microemulsions and oil-containing liquid crystalline phases. Langmuir. 2005;**21**:569

[64] Dong YD et al. Bulk and dispersed aqueous phase behavior of phytantriol: Effect of vitamin E acetate and F127 polymer on liquid crystal nanostructure. Langmuir. 2006;**22**:9512

[65] Drummond CJ et al. Surfactant self-assembly objects as novel drug delivery vehicles. Current Opinion in Colloid and Interface Science. 1999;**4**:449

[66] Spicer PT. Progress in liquid crystalline dispersions: Cubosomes. Current Opinion in Colloid and Interface Science. 2005;**10**:274

[67] Larsson K. Lyotropic liquid crystals and their dispersions relevant in foods. Current Opinion in Colloid and Interface Science. 2008;**14**:16

[68] Yaghmur A et al. Characterization and potential applications of nanostructured aqueous dispersions. Advances in Colloid and Interface Science. 2008;**147-148**:333

[69] Qiu H et al. The phase diagram of the monoolein/water system: Metastability and equilibrium aspects. Biomaterials. 2000;**21**:223

[70] Gustafsson J et al. Cubic lipid–water phase dispersed into submicron particles. Langmuir. 1996;**12**:4611

[71] Chung H et al. Self-assembled "nanocubicle" as a carrier for peroral insulin delivery. Diabetologia. 2002;**45**:448

[72] Esposito E et al. Cubosome dispersions as delivery systems for percutaneous administration of indomethacin. Pharmaceutical Research. 2005;**22**:2163

[73] Boyd BJ et al. Hexosomes formed from glycerate surfactants—formulation as a colloidal carrier for irinotecan. International Journal of Pharmaceutics. 2006;**318**:154

[74] Lopes LB et al. Reverse hexagonal phase nanodispersion of monoolein and oleic acid for topical delivery of peptides: *In vitro* and *in vitro* skin penetration of cyclosporin A. Pharmaceutical Research. 2006;**23**:1332

[75] Swarnakar NK et al. Enhanced oromucosal delivery of progesterone via hexosomes. Pharmaceutical Research. 2007;**24**:2223

[76] Kuntsche J et al. Interaction of lipid nanoparticles with human epidermis and an organotypic cell culture model. International Journal of Pharmaceutics. 2008;**354**:180

[77] Diat O et al. Effect of shear on a lyotropic lamellar phase. Journal de Physique II (France). 1993;**3**:1427